About IFPRI

The International Food Policy Research Institute (IFPRI), a CGIAR Research Center established in 1975, provides research-based policy solutions to sustainably reduce poverty and end hunger and malnutrition. IFPRI's strategic research aims to foster a climate-resilient and sustainable food supply; promote healthy diets and nutrition for all; build inclusive and efficient markets, trade systems, and food industries; transform agricultural and rural economies; and strengthen institutions and governance. Gender is integrated in all the Institute's work. Partnerships, communications, capacity strengthening, and data and knowledge management are essential components to translate IFPRI's research from action to impact. The Institute's regional and country programs play a critical role in responding to demand for food policy research and in delivering holistic support for country-led development. IFPRI collaborates with partners around the world.

Advancing Gender Equality through Agricultural and Environmental Research: Past, Present, and Future

Edited by Rhiannon Pyburn and Anouka van Eerdewijk

A Peer Reviewed Publication

International Food Policy Research Institute
Washington, DC

A catalogue record for this book is available from the Library of Congress.

DOI: https://doi.org/10.2499/9780896293915

ISBN: 978-0-89629-391-5

Recommended citation: Pyburn, Rhiannon, and Anouka Van Eerdewijk (eds). *Advancing Gender Equality through Agricultural and Environmental Research: Past, Present, and Future.* Washington, DC: International Food Policy Research Institute, 2021.

International Food Policy Research Institute, 1201 Eye Street NW, Washington, DC 20005, USA Telephone: 202-862-5600

Cover Design: Lee Dixon/IFPRI
Book Layout: Joan Stephens/JKS Design
Chapter Photo Credits: Chapter 2/H. Rutherford/CIP, Chapter 3/S.Malyon/CIAT, Chapter 4/ Mokhamad Edliadi/CIFOR, Chapter 5/CGIAR, Chapter 6/ Wade Fairley/WorldFish, Chapter 7/ Prashanth Vishwanathan/CCAFS, Chapter 8/CIMMYT, Chapter 9/ Rod Waddington/flickr, Chapter 10/ Peter Lowe/CIMMYT

Contents

Figures, Tables, and Boxes

Figures

Tables

Boxes

Acknowledgments

The work on this book was undertaken as part of the CGIAR Research Program on Policies, Institutions, and Markets (PIM) led by the International Food Policy Research Institute (IFPRI). The authors and editors are grateful for the funding support provided by the CGIAR Fund contributors to PIM (https://pim.cgiar.org/about/funders/) and by Canada's International Development Research Centre. We thank all CGIAR Centers and Research Programs that supported the authors in contributing to this book—it was a systemwide effort reflecting insights from a wide array of gender researchers with diverse backgrounds and perspectives.

A very warm thank-you to all contributors to the book for your time, commitment, enthusiasm, and patience, and especially to Marlène Elias and Agnes Quisumbing—the "book committee"—who guided the process from conception to fruition. A special thanks to Ewen LeBorgne and Andrea Vos for behind-the-scenes support throughout.

This book has been peer reviewed and benefited tremendously from thoughtful and detailed comments from many known and anonymous reviewers. Special thanks to Pamela Stedman-Edwards at IFPRI for managing the editing and production process.

The opinions expressed here belong to the authors, and do not necessarily reflect those of PIM, IFPRI, or CGIAR.

Foreword

Women make up 43 percent, on average, of the agricultural labor force in low- and middle-income countries. Understanding their roles, knowledge and needs, and the barriers they face, is essential to transforming food, land, and water systems to meet the demands of the 21st century, including the demands of women themselves. Gender research is thus a crucial component of inclusive and impactful agricultural research and innovation that contribute to advancing agricultural productivity, incomes, nutrition, environmental sustainability, and gender equality.

Over the past few years, I have witnessed a growing cadre of dedicated CGIAR gender researchers develop and apply innovative research methods to solve important development problems. Their achievements and perseverance have helped generate new funding, new partnerships, and new areas for gender research. And their efforts to strengthen collective action on gender in CGIAR have helped shift our focus, bringing gender equality to the fore in the world's largest agricultural research network and establishing CGIAR as a globally important source of research on issues around gender in food systems.

This shift is reflected in some important recent developments. At the start of 2020, we launched the Generating Evidence and New Directions for Equitable Results (GENDER) Platform to take CGIAR's gender research forward. And in CGIAR's new 10-year research and innovation strategy for food, land, and water system transformation, launched at the start of 2021, gender equality is one of five priority impact areas. As such, CGIAR's significant body of gender research—elegantly synthesized in this book—will be a critical input as we shape our research agenda to carry out this strategy.

This book stands to become the new go-to resource on gender in agriculture. Bringing together contributions from almost 60 authors who expertly straddle gender research and agricultural science, it offers important insights for the wider agricultural research and development communities. A comprehensive synthesis of CGIAR gender research to date, it not only illuminates what we know—and what we don't know—about the contributions of gender research to development outcomes, but also, and especially, investigates the contribution of agricultural development to gender equality outcomes. The lessons emerging from this synthesis have important implications for work that supports countries to achieve their national development objectives, as well as for our collective approach to meeting global targets such as the Sustainable Development Goals.

This book will be of value to many, informing the direction of research on gender and food, land, and water systems for years to come. I congratulate all those who have contributed to this significant publication, for the quality of their work and their continued commitment to advancing research in this important area.

Elwyn Grainger-Jones
Managing Director, Institutional Strategy and Systems
CGIAR

Preface

Frank Place

I have seen over the years a growing number of talented gender researchers within CGIAR developing and applying innovative research methods toward important development problems. These include the Women's Empowerment in Agriculture Index suite of tools, the GENNOVATE project, and collaborative work on gender and seed systems, gender and value chains, and the feminization of agriculture. Unfortunately, apart from these and other specific initiatives, gender research has more often than not been appended to other projects, and has been allocated marginal financial resources. This has undermined the visibility and impact of gender research, despite attempts to strengthen coordination and collaboration through a network (2012–2016) and then a platform (2017–2019).

The first scientific conference on CGIAR gender research in December 2017 at KIT Royal Tropical Institute in Amsterdam was eye-opening for me, and allowed me to more clearly appreciate the richness of research being undertaken and the many ways in which it connected up. The impetus for this book emerged at the second annual scientific conference of the CGIAR Collaborative Platform for Gender Research, held in September 2018 at the campus of the International Livestock Research Institute in Addis Ababa, Ethiopia. Many conversations emphasized the need to re-examine research to date on key themes and to develop a forward-looking gender research agenda that cuts across CGIAR research institutes and programs.

In this context, a committee was put in place in early 2019 to guide the development of this book process, and a survey was shared with CGIAR gender researchers to start defining themes and identifying potential co-authors. Nine themes were selected that cut across gender research of multiple institutes and research programs. The committee aimed for cross-cutting themes, rather than those tied to a specific research program or center, so as to stimulate a high-level discussion and to make it possible to pull out larger trends and insights. Some of the nine themes have been studied within CGIAR for over 40 years; others are much more recent. In the end, over 55 co-authors contributed to these chapters, mostly, though not entirely, gender researchers from CGIAR.

Of particular interest is the perspective that has guided the literature reviews of all chapters. Reviewers not only looked at the literature in terms of what gender research has contributed to the goals of productivity, incomes, nutrition, and environment but also, and especially, drew out what research on agriculture has contributed to gender equality outcomes. This intentionally chosen perspective is significant because gender equality is an important Sustainable Development Goal, and of paramount importance for many governments and funding agencies. The new CGIAR 2030 Research and Innovation Strategy elevates gender equality to one of its five priority impact areas.

The lessons emerging from the synthesis have implications for development practitioners, but even more so for the research community at large. In the coming years, CGIAR will be a leading research body for issues around gender and agri-food systems. The 2012–2021 phase of the CGIAR Research Programs has allowed the generation of new funding, the forging of new partnerships and the development of new research areas for gender researchers. The new GENDER platform is moving forward and building on this momentum. It is important that gender be featured as a significant research component throughout the new portfolio of One CGIAR to be launched in 2022, and that gender researchers are active in shaping our future research agenda.

Acronyms

5DE	women's empowerment across five domains in agriculture
AAS	CGIAR Research Program on Aquatic Agricultural Systems
ACIS	Agro-Climate Information Services project
ANGEL	Agriculture, Nutrition, and Gender Linkages
AR4D	agricultural research for development
A-WEAI	Abbreviated-WEAI
BCC	behavior change communication
BMI	body mass index
CAPRI	CGIAR System-wide Program on Collective Action and Property Rights
CCAFS	CGIAR Research Program on Climate Change, Agriculture and Food Security
CHANGE	Creating Homestead Agriculture for Nutrition and Gender Equity
CHCS	Custom-Hiring Centers
CIFOR	Center for International Forestry Research
CIMMYT	International Maize and Wheat Improvement Center
CSA	climate-smart agriculture
CSISA-BD	Cereal Systems Initiative for South Asia-Bangladesh
CSVS	climate-smart villages
E-HFP	Enhanced Homestead Food Production Program
FGD	focus group discussion
GAAP2	Gender Assets and Agriculture Project Phase 2

GAD	Gender and Development
GALS	Gender Action Learning System
GBI	CGIAR Gender and Breeding Initiative
GEI-CSV	Gender Empowerment Index for Climate-Smart Villages
GIMT	Gender Indicator Monitoring Tool
GREAT	Gender-Responsive Researchers Equipped for Agricultural Transformation
GTA	gender transformative approaches
HFP	homestead food production
HKI	Helen Keller International
ICRISAT	International Crops Research Institute for the Semi-Arid Tropics
ICT	information and communication technology
IFPRI	International Food Policy Research Institute
IMMANA	Innovative Methods and Metrics for Agriculture and Nutrition Actions
IP	Innovation Platform
IPCC	Intergovernmental Panel on Climate Change
IPV	intimate partner violence
IRRI	International Rice Research Institute
ISSD	Integrated Seed Sector Development
IT	information technology
IWMI	International Water Management Institute
LED	low emissions development
LLL	laser land-leveling
MEL	monitoring, evaluation, and learning
MNFY	micronutrient-fortified yogurt
MUS	Multiple Use Water Services
NC	Nurturing Connections
NGOS	non-governmental organizations
NIRDPR	National Institute of Rural Development and Panchayati Raj

NRM	natural resource management
NSAPS	nutrition-sensitive agriculture programs
OSP	orange-fleshed sweet potato
PCA	Principal Component Analysis
PIM	CGIAR Research Program on Policies, Institutions, and Markets
PPB	participatory plant breeding
PRO-WEAI	project-level Women's Empowerment in Agriculture Index
R4D	research for development
RCT	randomized controlled trial
REDD+	Reducing Emissions from Deforestation and Forest Degradation
REU	HarvestPlus Reaching End User project
SDGS	Sustainable Development Goals
SHGS	self-help groups
SNAP	Social Norms Analysis Plot
USAID	United States Agency for International Development
VCD	value chain development
VFL	Village Farmer Leader
VMFS	Village Model Farms
VSLAS	village savings and loans associations
WDI-GAI	Women's Decision-Making Index and Gender Attitudes Index
WEAI	Women's Empowerment in Agriculture Index
WEFI	Women's Empowerment in Fisheries Index
WEI	Women's Empowerment Index
WELI	Women's Empowerment in Livestock Index
WEMEASR	Women's Empowerment—Multidimensional Evaluation of Agency, Social Capital and Relations
WENI	Women's Empowerment in Nutrition Index
WID	Women in Development

Chapter 1

CGIAR RESEARCH THROUGH AN EQUALITY AND EMPOWERMENT LENS

Rhiannon Pyburn and Anouka van Eerdewijk

Over the past decade or so, there has been a renewed, and more concerted and comprehensive, interest in gender equality and women's empowerment in the agricultural development sector. Renowned development organizations have put gender dynamics back in the spotlight by means of a series of publications focused on gender equality.[1] This momentum has generated a unique opportunity to advance gender equality within and through agricultural research and development, and to institutionalize gender research within agricultural research for development (AR4D) organizations. This has also given us a chance to ask whether enough traction has been gained to mark a true turning point. As one key thinker in this field has stated that, "with agriculture now firmly back on the development agenda, it is time to re-socialize the ways that agricultural research, policy and practice deal with women and men, and analyze them both in relation to one another and their wider context" (Okali 2012, 2). This book is both an expression and a consequence of this overall momentum.[2]

Ester Boserup's landmark book Woman's Role in Economic Development (1970) was groundbreaking in its examination of women in African farming systems, addressing women's invisibility in the agricultural economy (Okali 2012) and opening the door for the next 50 years of efforts to embed gender

1 See, for example, the *Gender in Agriculture Sourcebook* (World Bank, FAO, and IFAD 2009); *Gender and Governance in Rural Services—Insights from India, Ghana and Ethiopia* (World Bank, and IFPRI 2010); *Gender Dimensions of Agricultural and Rural Employment: Differentiated Pathways out of Poverty: Status, Trends and Gaps* (FAO, IFAD, and ILO 2010); *The State of Food and Agriculture 2010–2011: Women in Agriculture—Closing the Gender Gap for Development* (FAO 2011); and the *World Development Report 2012: Gender Equality and Development* (World Bank 2012).

2 Note: this book uses a broad definition of agriculture, to include livestock, aquaculture, forestry, and fisheries.

analysis in development. Since then, the evolution from "women in development" (WID) initiatives to "women and development" (WAD) and eventually "gender and development"(GAD)[3] and gender mainstreaming[4] has not only echoed in, but also been firmly rooted in, agricultural and natural resource domains. This book reflects on selected[5] past gender and agricultural and environmental research by taking stock of progress. What makes it distinctive is that it reexamines this past research using an explicit gender equality and women's empowerment lens. This implies a shift away from more instrumentalist frames that consider how gender analysis contributes to agricultural and environmental research objectives such as improved productivity or tailoring technologies to ensure better user uptake. Instead, contributors to this book intentionally flip the question to ask: How does agricultural and environmental research and development **contribute to gender equality and women's empowerment?** The book revisits gender research and development interventions to date across nine themes, reframing the analysis in a way that puts gender equality and women's empowerment at the center. Each of the thematic chapters uses an adapted version of the "flipped" guiding question as a compass for reassessing the evidence.

This guiding question responds to recent developments within CGIAR—an international partnership of agricultural and environmental research institutes. As an international development research partnership, CGIAR is committed to the Sustainable Development Goals (SDGs) including SDG 5, to "achieve gender equality and empower all women and girls." In 2020, CGIAR prioritized gender equality as one of the five impact areas—specifically "gender equality, youth and social inclusion"[6]—for the more streamlined reorganization

---

3 WID espoused an economic argument making women's work visible and positioning women as productive members of society (Miller and Razavi 1995). Driven by the "efficiency approach," the idea was that allocating development funding to women, as well as men, made economic sense. WAD brought in neo-Marxist feminist thinking, shifting the focus from women's relationship to development to the relationship between patriarchy and capitalism (Rathgeber 1990). GAD conceptualizes empowerment—in particular women's individual and collective agency—as a way of addressing women's subordination: it addresses the social constitution of gender relations and their context-specific and dynamic nature (Okali 2012).

4 The 1995 Fourth World Conference on Women (FWCW), of the United Nations, in Beijing marked a turning point for gender and development: women's rights became central and gender mainstreaming was taken on board as a bold new strategy (Mukhopadhyay 2016) with transformative potential. Gender mainstreaming recognized that policymaking and institutions reproduced gender inequalities and needed to be a part of any strategy to address them (van Eerdewijk 2016).

5 Note: book chapters include a selective rather than *exhaustive* review of the gender research on each theme being tackled.

6 See CGIAR 2030 Research and Innovation Strategy: https://cgspace.cgiar.org/bitstream/handle/10568/110918/OneCGIAR-Strategy.pdf

of the 15 research institutes into "One CGIAR." This provides a clear direction and mandate for future gender research. CGIAR both reflects and is a part of evolutions in gender research in/for agricultural and environmental development (see Annex 1: CGIAR background).

The nine thematic chapters are authored by over 55 gender researchers[7] and reflect cross-cutting areas of CGIAR gender research. The literature and studies reviewed come from both CGIAR institutes and other authors. Six of the nine thematic chapters examine gender equality and women's empowerment in key agricultural and environmental domains—namely, animal and plant breeding; seed systems; value chains; nutrition-sensitive agriculture; natural resources (specifically water, land, and forests research); and climate adaptation and mitigation. The remaining three thematic chapters examine gender-specific research related to the feminization of agriculture; assessing women's empowerment; and gender transformative approaches (GTAs). The thematic chapters engage with the "flipped question" in coherent yet diverse ways, in terms of both their focus on research or approaches and interventions and their attention to gender equality and/or empowerment. Chapters 2, 3, 6, 8, and 9 look more at research; Chapters 4, 5, 7, and 10 pay more attention to analyzing approaches and interventions.

Important to note is that this book is directed toward a research audience—both gender researchers and researchers within the broader field of AR4D, within and outside of CGIAR. Co-authors seek to present sharply articulated insights, reflections, and questions on gender equality and women's empowerment in agricultural and environmental research in an accessible and clear manner for policymakers and researchers who may not be gender specialists. The nine thematic chapters speak to this diverse research audience; by putting gender equality and women's empowerment at the center, the book allows gaps as well as strengths in the AR4D knowledge base to emerge. It also exposes institutional obstacles to prioritizing gender equality and women's empowerment in AR4D. The chapters recognize, celebrate, and contextualize significant contributions that have advanced gender equality in and through agricultural and environmental research, and craft a next set of research questions and approaches to support researchers in navigating critical shortcomings. We[8] hope that this focus on CGIAR—its gender research history and the future-oriented

7 Most contributors are CGIAR (gender) researchers; others are external gender consultants who work regularly with CGIAR partners. See the "Contributors" section of this book for a complete list.

8 In this chapter, "we" refers to the editors of the book.

recommendations—will not only respond to the interests of gender and other researchers within CGIAR but also hold significance and relevance more broadly within AR4D.

This introductory chapter has three parts. Part 1 positions the insights of the book within the evolution of CGIAR gender research, tracing developments since the 1970s and highlighting key programs, evaluations, and contributions. We look at how the institutionalization of this gender research has shuffled forward in fits and starts over time, outlining achievements as well as remaining challenges. Part 2 offers a guide to reading the book. We position the concepts of gender equality and women's empowerment used across the thematic chapters, introduce the chapters by highlighting their key contributions and outcomes, and look at the evidence on which the chapters base their findings and discussions. Part 3 analyzes and reflects on the substance and implications of these thematic chapters. We begin by synthesizing the conceptual threads, looking at the different levels and aspects of women's empowerment and gender equality that the chapters address. We then take steps toward a forward-looking research agenda by crafting a number of meta-level gender research questions and articulating imperatives for the advancement of a gender research agenda. Finally, we discuss the institutional implications of this research agenda before concluding with some final reflections.

Part 1: The evolution of gender research within CGIAR

The past 50 years of efforts to integrate gender analysis into CGIAR's agricultural and environmental research and interventions have unfolded alongside and in iteration with the progression in thinking on gender and development more broadly. To advance toward gender equality, robust gender research and knowledge are essential: reliable qualitative and quantitative data—generated through gender analysis—can expose the nuance, variation, and reality on the ground vis-à-vis gender inequalities and how they affect rural women and men. Solid evidence on gender relations and women and men's resources and constraints is key, as is contextual embedding (Doss et al. 2018). Mandated with purveying agricultural research as an international public good, CGIAR Centers are prominent players in responding to this need for research and knowledge, which encompasses theory, analytical frameworks, methodologies, and valid data. In fact, the capacity to work collaboratively at different scales on improving the quality of data has been recognized as a point of comparative advantage for CGIAR gender research, as has the role of myth-busting

(Baden et al. 2017, 27): challenging "zombie" statistics and ensuring robust evidence to guide practice. CGIAR research clearly has a role to play in advancing gender equality.

As we write this book, CGIAR is at a crossroads vis-à-vis gender research. The past nine years have seen two phases of CGIAR Research Programs (CRPs). The period saw the establishment of an active Gender and Agriculture Research Network,[9] which established and provided a foundation for the CGIAR Collaborative Platform for Gender Research,[10] operational for three years (2017–2019). In January 2020, a new phase began with the CGIAR GENDER—**G**enerating **E**vidence and **N**ew **D**irections for **E**quitable **R**esults—Platform,[11] which was repositioned at par with other CGIAR Platforms and Research Programs. This better profiling of the Platform is an important signal that gender research is a priority for the network and its partners. In addition, CGIAR is at a critical moment as it moves toward One CGIAR, with the aim of streamlining the research themes of its institutes. Gender equality and social inclusion feature prominently as one of the five impact areas[12] and thus, in moving forward, gender research has a strategic and visible position within the CGIAR research agenda. Gender researchers are now tasked with crafting a gender research agenda that responds to the aspirations and vision of this next phase.

These two developments—the new GENDER Platform and One CGIAR—mean this is an opportune moment to take stock of the history and wealth of gender knowledge that has been generated on key agriculture and environmental themes to date. In this book, authors look at these bodies of work with a fresh eye and the ambition to build an engaging, forward-looking research agenda. Such a future-oriented gender research agenda can and will draw on past CGIAR gender research but, to be credible and impactful, it also needs to speak and respond to broader developments in gender research, beyond

9 The Gender and Agriculture Research Network was a CGIAR System Management Office (SMO) initiative that was coordinated by the International Center for Tropical Agriculture (CIAT) between 2012 and 2016.

10 The CGIAR Collaborative Platform for Gender Research was housed in the CGIAR Research Program on Policies, Institutions, and Markets (PIM) and was coordinated by KIT Royal Tropical Institute in Amsterdam. This book is a key output of that phase of CGIAR-wide collaboration on gender research.

11 The CGIAR GENDER Platform is hosted by the International Livestock Research Institute (ILRI) in Nairobi, Kenya. See https://gender.cgiar.org/

12 The other four One CGIAR impact areas are nutrition, health, and food security; poverty reduction, livelihoods, and jobs; climate adaptation and mitigation; and environmental health and biodiversity.

CGIAR and beyond agricultural and environmental domains: we return to this later in the chapter.

The focus of this book is gender research; in this chapter, we also address institutional dimensions within CGIAR that interfere with, block, advance, or facilitate such research, so we can understand the context in which that research is undertaken. Institutions can be either formal or informal and can be understood as "*the rules of the game in a society or, more formally... the humanly devised constraints that shape human interaction*" (North 1990, 3). For our purposes, formal institutional dimensions include policies, strategies, budget allocations and resourcing, staff expertise (hiring), accountability mechanisms and performance monitoring, coordination, and leadership on gender at higher levels within the governance system. Examples of informal institutions that we address include norms and practices, as well as prevailing beliefs as to what constitutes good science and how research is prioritized, designed, and organized (see Figure 1.1).[13] This part 1 of the chapter looks at gender research themes and trends as well as the institutionalization of gender research in CGIAR. We refer to notable gender events, CGIAR gender research initiatives, gender reviews, evaluations and assessments, and institutional developments, all in relation to broader milestones in gender and development since the 1970s (see Annex 2: Timeline of key CGIAR gender developments). The gender "struggle" within CGIAR has had two, often conflated, yet distinct, fronts—namely, gender analysis within CGIAR research; and gender equality in CGIAR workplaces. What we deliberately do not address in detail is gender in the workplace and the related human resource issues.[14]

---

13 In Figure 1.1, the dark purple box indicates the focus of this book—namely, research on gender equality and women's empowerment. Specifically, this is explored through the nine themes listed in the light purple box. The figure shows that research on gender equality and women's empowerment is embedded within an institutional context (green). As such, we look at both the evolution of CGIAR gender research and the related institutional developments. Figure 1.6 later in the chapter elaborates how this piece—this book—fits into the broader gender theory and empirical context.

14 We heed the call of the *Stripe Review of Social Sciences in the CGIAR*, to challenge an implicit assumption that hiring women scientists leads to more and/or better gender research. The review makes clear the importance of a sharp distinction between gender research and gender equality in the workplace and states, "*Many Centers have troubled histories of mistakenly assuming that female scientists should be the ones to do the work on gender issues. [...] should be everyone's business, not just that of women with PhDs*" (CGIAR Science Council 2009, 60). For more on gender in the workplace in agricultural research institutes, refer to the *Gender and Diversity Program* of CGIAR, which ran from 1999 to 2011 and focused in particular on career development and staffing. The contributions of AWARD—African Women in Agricultural Research for Development—since its inception in 2007 are also notable, in particular the fellowships, mentoring, and training to support women agricultural scientists.

FIGURE 1.1 The focus of this book

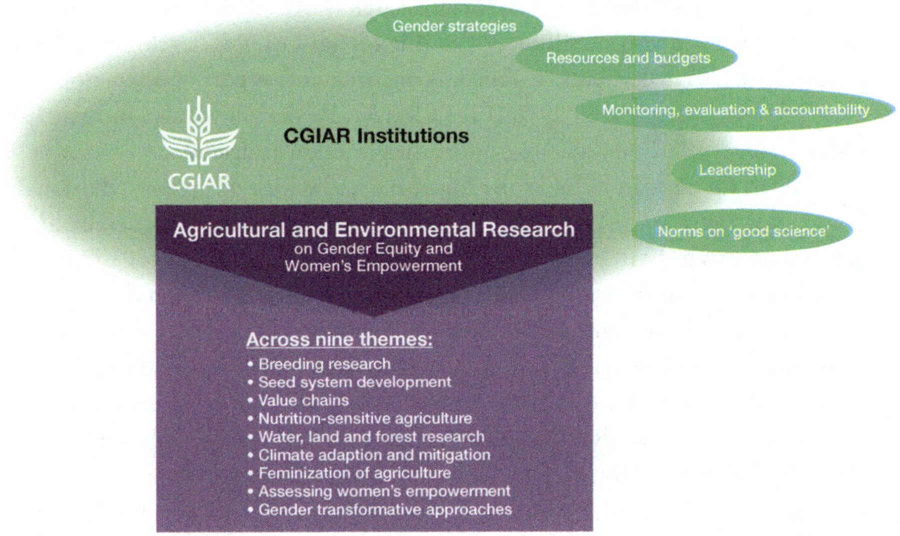

Gender research themes and trends

Gender research across the CGIAR has followed a non-linear path, generating significant and innovative programs, initiatives, and knowledge at different points in time. The 1970s saw the first mentions of women (not yet gender) in CGIAR research—for example, acknowledging the "family factor" in agricultural systems, recognizing female-headed households,[15] and mentioning women in relation to family planning and nutrition (van der Burg 2018). However, these early mentions tended to be observational rather than analytical and did not lead to specific women-related policy or research (van der Burg 2019). Since then, CGIAR gender research has both reflected and contributed to the wider debates and trends on gender and agricultural and environmental development. In this section, we do not attempt to provide a complete or comprehensive overview of all contributions but rather highlight key milestones;[16] detail related to specific themes can be found in the thematic chapters.

---

15 CGIAR gender researchers have challenged concepts that were important at particular moment in time (e.g., recognizing female-headed households) but that later research has proven wrong or invalid. In the case of female-headed households, later research demonstrated that using headship as a gender category was imperfect (see Doss and Kieran 2014).

16 For comprehensive overviews of CGIAR gender research over specific time periods, see, for example, Feldstein (1995, 1998), Okali (2017), and van der Burg (2019). See also Annex 2: Timeline of key CGIAR gender developments.

One of the early research trends offering entry points for gender analysis within CGIAR was Farming Systems Research (FSR), which reflected the increasing relevance of smallholder farming for agricultural development in the 1960s and 1970s (Okali 2012). This was documented as part of discussion at a CGIAR-wide meeting in 1978 (1981 CGIAR Review in van der Burg 2019). FSR was particularly amenable to addressing women and gender relations as it looked at the human dimension of farming and wider "farm constellations" (ibid.). Notably, the work of Hilary Sims Feldstein and Susan Poats embedded gender analysis in FSR; their two-volume publication captured this.[17] FSR played a key role in the integration of gender analysis in agricultural research, and some regard it as central to today's recognition of gender as a significant analytical category in agricultural research (Feldstein and Jiggins 1994; Okali 2012). FSR continued until the mid-1990s, by which time it was critiqued as being expensive, overly time-consuming for results (Okali 2012), and generally too messy. However, the concept had paved the way for participatory approaches to research—Rapid Rural Appraisal, Participatory Rural Appraisal, and Participatory Learning and Action—which further opened up space for women's participation and gender analysis (van der Burg 2019). Within CGIAR, this was in part through the *Participatory Research and Gender Analysis* (PRGA) program, discussed below.

In the 1980s, the International Rice Research Institute (IRRI), with the *Women in Rice Farming Systems initiative* (WIRFS),[18] was a leader on women's research (Chater and Carangal 1996, 9). In 1983, IRRI convened the first international gender conference within CGIAR on the theme of women in rice farming systems in Africa and Asia (Chater and Carangal 1996; van der Burg 2019). The related publication stressed the complexity involved in looking at social norms, family structures, and core social dimensions (e.g., caste, class, economic status, religion, ethnicity, and so on). It covered topics such as the exchange of rice seed through women's communication networks and the involvement of female respondents in the enumeration of edible plants in Kalimantan (Dey 1985 and Warson 1985 in Jiggins 1986, 17). WIRFS, active from 1986 to 1996 (Kauck, Paruzzolo, Schulte 2010; van der Burg 2018) had two objectives: to develop and test technologies that would benefit

17 *Working Together—Gender Analysis in Agriculture* (Feldstein and Poats 1989a, 1989b). Chapter 8 comprises a case study of women in a rice farming systems project (Paris 1989, 209–239), as part of the work discussed in the next paragraph.

18 WIRFS was set up to address the omission of any conscious effort to target research and extension efforts for women's benefit. It was a part of the Asian Rice Farming Systems Network (ARFSN) (Chater and Carangal 1996).

rural women in Asia and to instill an awareness of women's needs in national agricultural research and extension systems (Chater and Carangal 1996, 41). It classified three kinds of technology[19] to benefit women: labor-saving, knowledge-based, and livelihood-oriented (ibid.). A key contribution of WIRFS, captured in the words of then IIRI Director General Swaminathan, was that it "fostered collaboration between social and biophysical scientists and translated insights from gender analysis into targeted actions to reduce women's work and time burdens" (Kauck, Paruzzolo, and Schulte 2010, 5).

In 1985, two years after the conference on women in rice systems, a follow-up at the "senior-CGIAR level" was organized by the International Service for National Agricultural Research (ISNAR)—then a member of the CGIAR consortium—and co-sponsored by The Rockefeller Foundation (RF and ISNAR 1985).[20] The conference—*Women and Agricultural Technology: Relevance for Research*—took place in Bellagio, Italy, and marked the beginning of a system-wide[21] dialogue on gender. It focused on integrating women into the "modernization" of agriculture and user perspectives in technology development—those of women users in particular. Women were framed as farmers or beneficiaries of agricultural research. The concluding statement of the seminar affirmed gender as an important variable and confirmed the heterogeneity among women as users of technologies—as producers of crops and livestock; as participants on family farms in processing, marketing, storing, and preparing food; and as wage laborers and urban consumers. It articulated the need for a "do no harm" approach (van der Burg 2019). The longer-term strategy was to consider women in all phases of research and development, including feedback from female farmers (RF and ISNAR 1985; van der Burg 2018).

It is important to recognize CGIAR gender research in the 1970s and 1980s as these researchers were pioneers who often had to navigate institutional hurdles to carve out space for their work. These initiatives got "women" and "gender" on the CGIAR research agenda and produced a diversity of studies. Key contributions of these early efforts to the evolution of CGIAR gender

19 Examples of the technologies they developed included ultra-lite and "extra-ultra-lite" rice transplanters, a micro rice mill, rice husk stoves, and training schemes on seed handling, as well as income-generating projects such as poultry-keeping in Thailand and glutinous rice in the Philippines (Chater and Carangal 1996: 41–46).

20 Participants included directors-general or representatives from the international Centers, the CGIAR Secretariat, and the Technical Advisory Committee; the president and representatives of The Rockefeller Foundation; and experts with research experience on the role of women in agricultural development in less developed countries (RF and ISNAR 1985).

21 Then referred to by CGIAR partners as "intercenter."

research include (a) impact studies looking at the effect of innovations on women and men, establishing *gender as a variable*; (b) user perspectives that enabled the development of technologies with *women as a target group*; and (c) farming systems approaches that viewed farmers as agents within their livelihood strategies, which led to *the recognition of women and gender relations as part of different farming constellations* (van der Burg 2019). Attempts to integrate gender starting from the mid-1980s started to question gender-biased assumptions and employ gender as a *category of analysis* within social science research (Kauck, Paruzzolo, Schulte 2010). They built a foundation of *gender analysis as part of scientific capacities and systems* and included *more women farmers* in agricultural research and development (ibid.). However, the depth and types of the studies from the 1970s into the 1990s varied considerably. For example, an International Livestock Research Institute (ILRI) review of gender-related research in livestock in the 1980s and 1990s observes that this did not use conceptual or theoretical tenets (e.g., in analyzing gender roles in small ruminants research) and that reference to women or gender relations tended to be "more rhetorical than empirical" (Tangka et al. 2000, 46).

By the mid-1990s, two overview reports of CGIAR gender research had been produced that covered all gender research published in their respective periods (Feldstein 1995, 1998). Feldstein's inventory for 1990 to 1995 found that, when it came to gender-related research and gender training, "*all Centers are doing something*" (Feldstein 1995, 4). For that period, the inventory documented 140 gender-related studies from across CGIAR, with the International Food Policy Research institute (IFPRI) and the Center for Tropical Agriculture (CIAT) being the most prolific, followed by the International Institute of Tropical Agriculture (IITA) (ibid.). The follow-up inventory (1996–1998) observed progress in using gender analysis as a research tool, with the number of studies using gender analysis increasing to 207—a 48 percent increase since the first review (Feldstein 1998). The International Water Management Institute (IWMI) had become a leader in publishing gender research, and there was increasing recognition throughout the system of the value of women's contributions to agriculture and of the usefulness of gender analysis (ibid.). The review noted progress toward more proactive consideration of gender relations in research, training, and dissemination activities, and that Centers were still developing approaches to gender analysis to fit their mandates (ibid.). The two reviews found the vast majority of CGIAR gender studies to be either characterization and diagnostic studies or on women-specific technologies. The second review saw an increasing number of methodology development papers, adoption studies, and impact assessments as well as a sharp increase in literature reviews, workshops, and seminars (Feldstein 1998). It was not until the later 1990s that more concerted programs began to emerge.

Almost 20 years later, a 2016 evaluation recognized the significant bodies of work within CGIAR on women's empowerment; collective action, nutrition, health, and food security; and gender norms and innovation (Baden et al. 2017). This kind of strategic gender research—that which moves beyond the commodity and natural resource domains of CGIAR Centers—has often been undertaken through multi-year cross-Center projects or programs. Examples of thematic research programs include *Strengthening Development Policy through Gender and Intra-Household Research* (IFPRI); the *CGIAR System-Wide Program on Participatory Research and Gender Analysis* (PRGA); *Collective Action and Property Rights* (CAPRi); the *Gender, Agriculture, and Assets Project* (GAAP), including the *Women's Empowerment in Agriculture Index* (WEAI) and its adaptations; *GENNOVATE*—on gender norms and innovation; and the *Gender and Breeding Initiative* (GBI).[22] We elaborate on some of these here as examples; others—CAPRi (Chapter 7), WEAI (Chapter 9), GENNOVATE (Chapter 10 and Annex 3), and GBI (Chapter 2)—are discussed in more detail elsewhere.

IFPRI's Strengthening Development Policy through Gender and Intra-Household Research, which ran from 1992 to 2003, is a notable example of high-quality gender research (Kauck, Paruzzolo, Schulte 2010; van der Burg 2018, 2019). The program was an "example of the transformative use of sex-disaggregated quantitative data to assess and identify ways to reach gender equitable policy outcomes" (Kauck, Paruzzolo, Schulte 2010, 5). It began by testing economic models of household behavior, spurred by the development of the collective model of the household. The finding that individuals within households did not necessarily share the same preferences or completely pool resources—suggesting that it mattered who received cash transfers—had important implications for development policy. This led to an effort to test household models using sex-disaggregated primary data collected in four high-concentration countries and several other countries. A

22 GBI, led by the CGIAR Program on Roots, Tubers, and Bananas and the International Potato Center (CIP) (2016–2018) is notable in its tackling of the long-standing challenge of bringing a gender perspective and gender analysis to breeding research. GBI questioned and problematized the way in which breeding programs work, and this resulted in new approaches to product profiling and engagement with the populations that new varieties and breeds aim to serve. This is an example of how gender analysis can trigger institutional change in the sense of changing practices—in this case how breeding work is undertaken within CGIAR. For more on GBI, see Chapter 2 and http://www.rtb.cgiar.org/gender-breeding-initiative/

2005 multi-country study measured the food policy response impacts of this program and found it was effective; its results were central to policy formation (ibid.).

Participatory research offered an exceptional entry point for gender analysis (see Sachs 1996; Okali 2012). A spearheading program on gender research began in 1997 with the CGIAR System-Wide Program on Participatory Research and Gender Analysis (PRGA).[23] PRGA ran across three phases—1997–2002 (phase 1), 2003–2007 (phase 2), and 2008–2012 (phase 3). By the time of the 2000 internally commissioned review (Prain et al. 2000), there was still a need to consolidate and more consistently integrate gender analysis into the participatory research component of the program. The second phase included a "deeper theoretical exploration of processes of transformational change… investigating lessons about gender-sensitive and pro-poor development, strategies for change, multi-stakeholder development of food chains, and livelihood diversification" (CGIAR Science Council 2007). However, the second phase was critiqued heavily for focusing more on advocating for the mainstreaming of gender analysis rather than building a constituency of gender researchers within CGIAR (ibid.).[24] The third phase strategic platform had three overarching themes: climate change, food security (building on its legacy of research on plant breeding and natural resources), and, despite the critique, gender mainstreaming. Perhaps the most significant contribution of PRGA was in relation to participatory plant breeding (PPB), outside of CGIAR. PRGA passed on small grants to partners and, while PPB did not take hold within CGIAR, it did externally: it was commended for success in mainstreaming gender analysis within the African National Agricultural Research Systems (ibid.).

The Gender, Agriculture, and Assets Project (GAAP)[25] is a research initiative making important contributions toward understanding the dynamics of gender assets in agricultural development programs. It looks at both tangible assets, including productive assets like land, livestock, labor, and finance, and intangible assets like social capital, social networks, and education. The first phase (2009–2013) was co-led by IFPRI and ILRI and worked with eight

23 For more on PRGA, see Gomez (2009).

24 This critique is contradictory to that of the internally commissioned review seven years earlier, which noted the need for advocacy and awareness-building to address entrenched gender blindness vis-à-vis research methods, including in research considered "participatory" (Prain et al. 2000, 23).

25 For more information see https://gaap.ifpri.info/

development projects. Among other findings,[26] the project supported the development of the asset, social capital, and time use components of the Women's Empowerment in Agriculture Index (WEAI).[27] A 2014 evaluation articulates three main contributions of GAAP1 as shedding light on (a) the complexity and cultural specificity of both asset ownership (particularly joint ownership by women and men) and its contribution to women's empowerment; (b) the unintended consequences produced by agricultural projects, which non-gender-sensitive outcome measures can mask; and (c) the importance of measurement of sex-disaggregated assets in agriculture projects to show how men and women make decisions and how to engage them more effectively (Firetail 2014). The second phase—GAAP2[28]—began in 2016 and is ongoing: it builds on the phase 1 work with nutrition and income projects from South Asia and Africa south of the Sahara.

It is possible to link some of the more recent gender research to three CGIAR-wide ambitions[29] to improve gender equality and inclusion within the portfolio: (a) gender-equitable control over productive assets and resources; (b) development and dissemination of technologies that reduce women's labor and energy expenditure; and (c) women's (and youth's) equitable participation in decision-making. A comprehensive 2011–2015 review found that there had historically been more research on the first and third points than on the second: "complex social change processes arising from technological change are not yet widely documented" (Baden et al. 2017, 26).

To sum up, the history of gender research across the CGIAR system can be characterized as uneven, unstable, and enduring. Uneven, because it has evolved in fits and starts, with exciting moments of great potential often followed by fallow periods; Centers have also progressed at very different rates. Unstable, because progress has often been tied to specific funding or projects, research trends that are overshadowed by a subsequent trend, or thought leaders who eventually move on (see the detailed research reviews of the 1990s: Feldstein 1995, 1998). Finally, enduring, as gender research clearly has a long history within CGIAR and continues to grow in prominence: it has stood the test of time and progressed with the support of patient and persistent advocates.

---

26 For more on GAAP1 findings, see Meinzen-Dick et al. (2014a), Njuki et al. (2014), Quisumbing et al. (2015) and Johnson et al. (2016), among others. See also this technical guide produced by the project: http://ebrary.ifpri.org/utils/getfile/collection/p15738coll2/id/128594/filename/128805.pdf

27 See Chapter 9 for more on WEAI.

28 Led by IFPRI.

29 Within CGIAR, these are referred to as sub-Intermediate Development Outcomes, or sub-IDOs.

Enduring also because it is possible to trace the CGIAR gender story, owing to the public nature of CGIAR research and the transparency and availability of evaluations, assessments, and reviews. The thematic chapters ahead will further elaborate substantive progress on gender research.

Institutionalizing gender research within CGIAR

A number of assessments and reviews conducted since the 1980s have fueled waves of effort to institutionalize gender research within CGIAR. These underscore the vital nature of institutionalizing gender research and make recommendations as to how to proceed. The first review, in 1980, assessed the responsibility and value of the CGIAR system to the "most disadvantaged groups, including resource-poor farmers and women" (Consultative Group Meeting, October 1980, in van der Burg 2019, 41). The second review, a year later, was blunt in stating that women must be considered in agricultural production otherwise CGIAR risked reduced impact or the total failure of its programs (van der Burg 2019). In 1984, Janice Jiggins was commissioned to study gender impacts in CGIAR for a 1982–1988 CGIAR-wide assessment (ibid.). She concurred with the 1981 review: that CGIAR would not reach its goals if it did not take women's roles in production and food systems seriously (Jiggins 1986; van der Burg 2018). What stands out in her revealing report are the institutional constraints to doing gender research within CGIAR. She found "in general a conservative record of hesitation and cautious exploration fueled mainly by donor-financed initiatives outside of core budgets and implemented by temporarily attached staff. Moreover, there is much in the way the research process is organized and research criteria are derived which stands in the path of more substantive efforts" (Jiggins 1986, 9). Tellingly, the overall 1982–1988 review reduced the detailed analysis in her paper to a single box (Anderson, Herdt, Scobie 1988).

Points raised in that study from over 35 years ago continue to be relevant today. For example, Jiggins lamented that, "it seems uncommonly hard to convince researchers that women possess skills, knowledge and experience which is not replicated in the knowledge and skills held by men and yet which is of use to breeders in the task of setting the research agenda (Fortmann 1981; Fresco and Jiggins 1985)" (Jiggins 1986, 16–17). Other critiques resurfaced in an internal PRGA review in 2000: "advocacy and awareness building are needed because of the entrenched nature of gender-blind research methods, even among researchers who consider their work to be participatory" (Prain et al. 2000, 23). An input paper for the 2008 review also found that some of the commissioned external program and management reviews claimed that "good science was blind to gender" and that guidelines at the time were "silent on the need for attention to gender in the reviews' assessments of the Centers' science" (Gibbs 2008 in Kauck, Paruzzolo, Schulte 2010, 27). Clearly, epistemological biases remain strong and persistent.

As early as 1995, reviews commended individual Centers for stand-alone efforts to institutionalize gender research: "ICRISAT [the International Crops Research Institute for the Semi-Arid Tropics] and IRRI are incorporating a gender dimension in their priority setting... CIP [the International Potato Center] and ICRISAT now include a question concerning the gender implications of research and training in their project proposal or reporting forms" (Feldstein 1995, 7). However, system-wide this was not the case. It was not until 1990 that the first CGIAR Gender Program was put in place: this ran until 1996 and addressed both gender equality in the workplace and gender in research (Prain et al. 2000). After that, the two prongs of the program were separated: research through PRGA (see above) running 1997–2011 and the Gender and Diversity Program focusing on human resource issues running 1999–2011. All these programs were voluntary and worked with interested Centers.

Despite these efforts and some achievements, CGIAR has struggled to position and mainstream gender analysis, as can be seen in a Science Council comment: "there is need to accelerate gender analysis into the wider CGIAR system (across all research themes)... a real need for more focused research on gender analysis leading to mainstreaming gender analysis into all CGIAR research. This is not being achieved in the current PRGA program (nor was it achieved when the gender analysis was part of the Gender and Diversity program)" (CGIAR Science Council 2007, vii). An important independent review in 2008 concurred, noting that a pattern of "misplaced reliance by CGIAR leaders on sub-managerial staff functions has shunted responsibility for gender away from operations management and professional staff" (CGIAR Independent Review Panel 2008, 4, 46). The *Stripe Review of the Social Sciences in CGIAR* (CGIAR Science Council 2009) that followed also found that gender was not yet well integrated into CGIAR research. It remarked that there was little evidence of consistent attention to gender issues by senior-level Center or System management, either in research prioritization, research design, or performance monitoring. It also provided a gender-specific recommendation that stated that, to "mainstream gender equity as a basic axiom of CGIAR research.... operationally, this requires managers take explicit responsibility for gender equity in research..." (ibid., 60). This strong statement to leadership, along with an explicit recommendation in the 2008 review to adopt a gender strategy and system-wide strategic objectives for gender integration with instruments to ensure accountability (CGIAR Independent Review Panel 2008, 12), paved the way for the advancement of gender research in the decade that followed.

Optimistically, the 2008 review further observed a readiness to adopt both empirical (research-oriented) and institutional approaches—including a system-wide policy, strategy, and results framework—to gender mainstreaming within CGIAR (CGIAR Independent Review Panel 2008, 4, 48). Acknowledging champions and good work to date, it called for a shift from an advocacy-based approach reliant on personal persuasion to an accountability approach addressing institutional standards (ibid.). It framed an accountability approach as the "professional responsibility" of CGIAR leadership, linking it to development effectiveness, and insisted on the need to move beyond ad hoc efforts dependent on individual initiative toward embedded system-wide instruments (e.g., impact assessments) (ibid., 46–47). In 2010, the International Center for Research on Women (ICRW) was commissioned centrally to undertake the *CGIAR Gender Scoping Study* (Kauck, Paruzzolo, Schulte 2010) to further inform the mainstreaming of gender across the soon-to-be-developed CGIAR Research Programs (CRPs). Like earlier reviews, the scoping study acknowledged a number of strategic gender research initiatives but also observed that such initiatives were spotty and inconsistent. It concluded that there had been no attempts to embed gender analysis across the CGIAR system with proper resourcing and supported effort (ibid., 7). The scoping study confirmed that a CGIAR-wide gender policy—including gender strategies and action plans—was needed, and echoed that its absence was part of the reason for the mixed success (Kauck, Paruzzolo, Schulte 2010).

At the same time as the scoping study, the decision to integrate gender system-wide was finally taken and the major task of gender mainstreaming began (Okali 2017). Since that decision, considerable strides have been made toward institutionalizing gender research, including (a) the development of CGIAR-wide and CRP gender strategies, accompanied by budget allocations and operational plans; (b) the integration of gender equality in the overall CGIAR-wide Strategy and Results Framework; (c) the cultivation of a system-wide Gender Network and later Platform; and (d) the hiring of gender researchers and the appointment of gender research coordinators. We look at some of these elements in the paragraphs to come. However, it is worth noting that this system-wide gender mainstreaming effort within CGIAR came very late relative to gender initiatives in development more broadly—over 16 years after gender mainstreaming[30] was introduced in Beijing.[31]

A key milestone in institutionalizing gender equality was the first CGIAR-wide Gender Strategy,[32] put in place in 2011 for all CGIAR Centers, CRPs, the Consortium Office, and Center Boards. It included objectives, deliverables, and accountability mechanisms with related timelines. The overall objective in relation to research was "to improve the relevance of CGIAR research to poor women as well as men (reduced poverty and hunger, improved health and environmental resilience) in all geographical regions where the work is implemented" (CGIAR Consortium Board 2011, 5). It acted as a systematic roadmap for all Centers and CRPs to develop their own strategies. As part of the Gender Strategy, a system-wide gender advisor was appointed on a part-time basis, tasked with initiating a Gender Network and with supporting gender researchers and CRPs in strengthening their gender strategies (CGIAR Consortium Board 2011). As such, gender staffing and coordination were also starting to be strengthened. The 2017 gender evaluation referred to the CGIAR-wide Gender Strategy as having "played a catalytic role" in getting gender on the CGIAR research program agendas (Baden et al. 2017, xiv).

---

30 The aim of gender mainstreaming was ambitious: to fully integrate gender equality concerns into "*analyses and formulation of all policies, programmes and projects; initiatives to enable women as well as men to formulate and express their views and participate in decision making across all issues*" (Mukhopadhyay 2016, 78). The most often-used definition of gender mainstreaming in international development stems from the Economic and Social Council 1997 and reads as follows: "*Mainstreaming a gender perspective is the process of assessing the implications for women and men of any planned action, including legislation, policies or programmes, in all areas and at all levels. It is a strategy for making women's as well as men's concerns and experiences an integral dimension of the design, implementation, monitoring and evaluation of policies and programmes in all political, economic and societal spheres so that women and men benefit equally and inequality is not perpetuated. The ultimate goal is to achieve gender equality*" (United Nations 1997, 1, in e.g. Moser and Moser 2005, 2).

31 The 1985 United Nations Conference on Women in Beijing, marking the end of the United Nations Decade for Women (1976–1985), marked a turning point for gender and development, introducing gender mainstreaming and framing the next decade or more of gender and development work. Women's rights became central and gender mainstreaming was taken on board as a bold new strategy (Mukhopadhyay 2016) with transformative potential. This was based on recognition that policymaking and institutions reproduced gender inequalities and needed to be a part of addressing them (van Eerdewijk 2016).

32 The CGIAR Consortium Level Gender Strategy (CGIAR Consortium Board 2011) addressed both gender mainstreaming in research, and gender and diversity in the workplace (at the Centers), as mutually reinforcing branches of an integral plan (Baden et al. 2017). We focus on the gender mainstreaming in research component, as justified earlier in the chapter.

To adhere to the CGIAR-wide Gender Strategy, all CRPs developed gender strategies over the period 2011–2013.[33] These included (CGIAR Consortium Board 2011):

- An evidence-based statement providing justification and rationale as to why gender should be addressed within the domain of the CRP;
- Gender-responsive goals and objectives of the program;
- A description of impact pathways with explicit attention to gender;
- A description of gender activities;
- A description of the monitoring and evaluation system in place to track progress;
- A budget;
- A management system; and
- A review of capacity within the CRP for gender analysis and gender research.

The gender strategies the Centers and CRPs developed distinguished between gender-integrated and gender-specific (or strategic) research. In addition, many sought to differentiate gender research along a gender awareness continuum from gender-blind to gender-aware, and from accommodative approaches that work within existing gender relations to those that are transformative—that is, actively working to transform existing gender relations.[34]

In 2016, these gender strategies were referred to and further articulated as part of the proposals for a second phase of the CRPs, including a gender narrative and a gender annex, which operationalized the plans. The gender narratives in these proposals covered issues such as gender in the program's theory of change; summaries of key publications and knowledge generated on the theme to date (e.g., throughout the first phase of the related research program); an explanation of how gender is integrated into the CRP; and staff and partner gender research capacities. The annexes were concise overviews that brought to the fore aspects such as how gender would be integrated substantively into the Research Programs' thematic areas (flagships), as well as necessary institutional

---

33 More precisely, this concerned all phase 1 CRPs, which ran from 2011 to 2016. Phase 2 CRPs ran from 2017 to 2021. (See Annex 1: CGIAR background.)

34 Otherwise stated, gender accommodative approaches treat gender as a variable, whereas gender transformative approaches are systemic in nature (van der Burg 2019).

dimensions such as meeting staffing needs, gender budgeting, gender advisor representation on management committees, and so on.

In 2016, six years after the decision to mainstream gender, CGIAR commissioned a thorough assessment of its gender research and gender equality in human resources. This evaluation, the most recent to date, found that the system-wide Gender Strategy[35] had "largely achieved its purpose of catalyzing gender mainstreaming in research" across CGIAR (Baden et al. 2017, 22). However, the CRP gender strategies were described as "overly ambitious and thus difficult to fulfil in relation to the resources that were available, and which fell significantly short of original expectations" (ibid., 22). The report concedes that, despite their "piecemeal nature," system-level decisions created incentives for gender integration and laid the basis for accountability mechanisms to ensure that CRPs delivered on their gender commitments. It further recommended that CRPs update their strategies, bearing in mind resources available and adjusting ambitions accordingly. Allocating appropriate, realistic, and sufficient budgets for gender research and integrating gender into other agricultural and environmental research remain a challenge: budget allocations for both strategic gender research and gender-integrated research are critical.

Some CRP gender strategies were indeed updated and further refined and developed during CRP phase 2 implementation. Another important milestone in this phase was the embedding of gender (and youth[36]) inclusion as a cross-cutting theme in the *2016–2030 CGIAR Strategy and Results Framework*, which set the objectives for that phase and beyond. Gender and youth inclusion were identified as a key ambition[37] for CGIAR research. This explicit articulation of gender research ambitions in strategic planning provided institutional anchoring. Most recently, the One CGIAR[38] framework has elevated institutional commitment, with gender equality, youth, and social

---

35 The report refers to this as the "Consortium Level Gender Strategy" (CGIAR Consortium Board 2011).

36 The linking of gender to youth has been debated and criticized, with good reason. The 2017 CGIAR gender evaluation report remarks that the conflation of youth and gender in the Strategy and Results Framework (SRF) and CRP gender strategies is *"unhelpful to the conceptual clarity needed to support gender mainstreaming"* (Baden et al. 2017, 22).

37 CGIAR refers to these key ambitions as Intermediate Development Outcomes (IDOs). The Gender and Youth Inclusion IDO has three sub-IDOs: gender-equitable control over productive assets and resources; the development and dissemination of technologies that reduce women's labor and energy expenditure; and women's (and youth's) equitable participation in decision-making.

38 One CGIAR is described as a *"dynamic reformulation of CGIAR partnerships, knowledge, assets and global presence."* It comprises a sharper mission statement, unified governance, institutional integration (regarding policies, services, and regional presence), a new research modality, and more and pooled funding. For details see https://www.cgiar.org/impact/one-cgiar/

inclusion as one of five impact pillars. This important milestone marks how, after decades of effort, gender equality has finally moved to the center of the CGIAR research and impact agenda.

As discussed earlier, coordination among gender researchers grew from a network focused on institutional issues into the CGIAR Collaborative Platform for Gender Research, and then into the GENDER Platform. This responded to one of the recommendations in the 2010 review, to institute a platform for gender knowledge-sharing and collaboration (Kauck, Paruzzolo, Schulte 2010). In around 2014, more concerted efforts began to hire social scientists, gender researchers, and post-doctoral fellows,[39] with the aim of building a critical mass of gender experts and responding to growing demand for gender integration in agricultural and environmental research. Alongside this development, each CRP appointed a gender research coordinator to participate in system-wide discussions among gender researchers. Even though gender expertise was still heterogeneous across the CRPs, with an overreliance on junior staff (Baden et al. 2017), the growing institutional commitment, including the hiring of gender staff, fueled further coordination system-wide on gender research, strategies, and institutional concerns.

To sum up, developments in institutional support and the advancement of gender research and gender equality itself are inextricably entwined. As such, assessments and reviews have articulated the need to address and remove institutional stumbling blocks within CGIAR over and over again, with increasingly compelling arguments, vigor, and specificity. Gender researchers, individual Centers, and other champions have advocated and fought for the recognition of gender research and continue to push the needle. But institutional anchoring is necessary to safeguard progress and avoid the risk of advances remaining ad hoc, ephemeral, and reliant on individual motivations. Looking at the institutionalization of CGIAR gender research, it has clearly been something of a slog since the 1980s. Yet persevere it has. In fact, the speed of institutional progress over the past 10 years is quite remarkable given the late and cumbersome start. Today, gender analysis is being positioned within the established research agendas of these international agricultural and environmental research centers and programs. This strengthened institutional support creates space for gender

---

39 The CGIAR Gender Post-Doctoral Fellowships Program awarded 18 2-year social science fellowships over 2 calls (in 2015 and 2016). The intention was for these young professionals to develop unique expertise "on the job" and apply gender analysis in technical research on agriculture and natural resource management. For more on the Program see https://cgspace.cgiar.org/bitstream/handle/10568/108588/CGIAR%20Gender%20Research%20Action%20Plan-Fellowship%20Program_Brief3.pdf?sequence=1&isAllowed=y

research, as the growing number of gender-related or gender-integrated publications, research projects, and programs shows. Nonetheless, as the 2017 review states, "the process of institutional mainstreaming is ongoing, and incentives, accountability systems, resources and networks are needed to retain the growing momentum" (Baden et al. 2017, xiii). Institutional challenges persist. In particular, norms and practices, which include long-standing biases as to what constitutes knowledge and "good science," remain a challenge. This goes hand in hand with recognizing and revaluing qualitative alongside quantitative research, and the social sciences alongside the technical or life sciences. Overcoming implicit epistemological preferences and biases is a next obstacle.

Part 2: Guide to reading this book

Here we provide guidance on reading this book. We begin by defining key concepts and positioning the work. We then introduce the thematic chapters and their contributions to the book as a whole. Finally, we provide some insights and analysis on the evidence base for the chapters.

Conceptual positioning

Two concepts at the heart of this book are gender equality and women's empowerment—concepts that tend to surface in similar research, policies, or debates. Some chapters in the book focus on gender equality; others take women's empowerment as central. Despite this variation, the chapters share conceptual foundations in their understanding of gender equality and women's empowerment. Both concepts are concerned primarily with gender as a social *marker*, manifesting itself in social hierarchies affecting the lives of women and men. Both are concerned with the power imbalances, and the resulting inequalities and disempowerment—and hence with gender as a social *relation*. Both also express a more desirable state, where gender as a social relation no longer limits people's lives.

Gender equality refers to:

> "the equal rights, responsibilities and opportunities of women and men and girls and boys. Equality does not mean that women and men will become the same but that women's and men's rights, responsibilities and opportunities will not depend on whether they are born male or female. Gender equality implies that the interests, needs and priorities of both women and men are taken into consideration, recognizing the diversity of different groups of women and men. Gender equality is not a women's issue but should concern and fully engage men as well as women. Equality between women and men is seen both as a human

rights issue and as a precondition for, and indicator of, sustainable people-centered development."[40]

The understanding of gender equality and women's empowerment in this book problematizes and challenges unequal power relations between, and among, women and men. Gender inequality and women's disempowerment are understood in terms of unequal social relations and hierarchies, rather than as a characteristic of individual women, or concerning only women. And, while recognizing gender as a primary social marker and conceptualizing it as a social relation, both concepts acknowledge the diversity and heterogeneity among women and among men. We return to this point about intersectionality later.

Gender and development work uses many frameworks, from the *Harvard Analytical Framework*[41] of the 1980s to the *Gender and Development Framework*[42] (Moser and Levy 1986) and the *Social Relations Framework*[43] (Kabeer 1994), among others. In this book, we analyze gender equality and women's empowerment outcomes against the backdrop of the *Reach-Benefit-Empower* framework (Johnson et al. 2018). This framework comes up in several thematic chapters and is useful for categorizing interventions in terms of the types of gender outcomes aimed for or realized (see also Danielsen et al. 2018). It distinguishes between agricultural interventions aimed at reaching, benefiting, or empowering women. Interventions focused on *reaching women* tend to assess women and men's participation in (project) activities, whereas strategies designed to *benefit women* emphasize outcome indicators, such as income, productivity, or assets. The third category, interventions that aim to *empower women*, however, focuses on "outcomes that are inherently

40 https://trainingcentre.unwomen.org/mod/glossary/view.php?id=36&mode=letter&hook=G&sortkey=&sortorder=

41 Developed initially by a team at Harvard University—C. Overholt, M. Anderson, K. Cloud, J. E. Austin—and published in 1985, the *Harvard Analytical Framework* (also referred to as the *Gender Roles Framework*) was developed at the request of the World Bank and in collaboration with the United States Agency for International Development as an assessment on women in development projects. It comprises a set of data collection and analysis tools to analyze gender roles and access to and control over assets.

42 The *Gender and Development Framework* for gender policy and programming focused on women's triple roles—reproductive, productive, community management—making a distinction between women's practical and strategic needs, and women's interests (Moser and Levy 1986). It redirected attention from women to gender relations and recognized the inherent male bias of efficiency approaches (Elson 1991; Moser 1993; Miller and Razavi 1995) that shifted the costs of welfare to the household, where women did a greater share of labor and thus bore the brunt of such initiatives.

43 The *Social Relations Framework* squarely shifted attention to gender relations, rather than men and women's roles in development, and positioned gender relations as relations of power (Okali 2012): "*... do not make the assumption that raising women's productivity is simply a matter of reallocating resources, nor will reallocating resources to women necessarily lead to women's equality or autonomy*" (Kabeer 1994, 97).

FIGURE 1.2 Gender outcomes typology

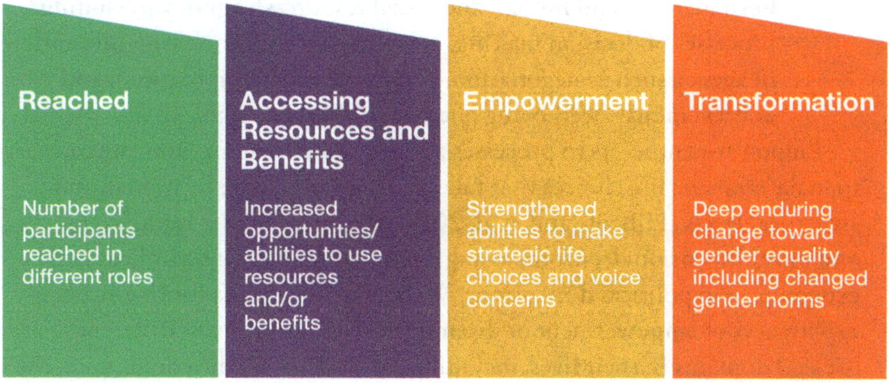

Source: Kleiber et al. (2019) and CGIAR Research Program on Fish Agri-Food Systems (2020) based on Johnson et al. (2018).

empowering (e.g., women's agency), inherently disempowering (e.g., gender-based violence, time burden) or indicators of women's position relative to men (e.g., degree of control over income, participation in joint decision making, gender-asset gap)" (Johnson et al. 2018, 5). The CGIAR Research Program on Fish Agri-Food Systems further developed the original framework to include "transform" as a fourth change category, to emphasize interventions that aim to *transform constraining gender norms, attitudes and behaviors* towards those that support gender equality (CGIAR Research Program on Fish Agri-Food Systems 2017; Kleiber et al. 2019). Figure 1.2 presents the adapted framework, including transformation as a fourth dimension. In our understanding of gender equality and women's empowerment, they are both concerned with shifting and transforming the (formal and informal) structures that underpin inequalities and the marginalization and disempowerment of women.

Our understanding of *empowerment* is informed by Naila Kabeer's conceptualization as "the expansion of the capacity to make strategic and meaningful choices by those who have previously been denied this capacity, but in ways that do not merely reproduce, and may indeed actively challenge, the structures of inequality in their society" (Kabeer 2017, 651). Specifically, we use the following definition of empowerment:

> "... the process by which those who have been denied the ability to make strategic life choices acquire such an ability... The ability to exercise choice incorporates three inter-related dimensions: **resources**

> (defined broadly to include not only access, but also future claims, to both material and human and social resources); **agency** (including processes of decision making, as well as less measurable manifestations of agency such as negotiation, deception and manipulation); and **achievements** (well-being outcomes)" (Kabeer 1999, 435).

Empowerment refers to processes and outcomes shaping women's capacity to make strategic life choices that take place at individual, relational, and structural levels (Hillenbrand et al. 2015). Empowerment is multilevel and multidimensional, encompassing a number of often-interlinked social, economic, and political dimensions. Women's—and men's for that matter—experiences of empowerment or disempowerment vary across different societal domains in their lives, including the household, the community, the market, and the state (Kabeer 1999; van Eerdewijk et al. 2017).

Each thematic chapter focuses on particular aspects of women's empowerment or gender equality—for example choice and power to choose (Chapter 2); participation (Chapter 4); agency, decision-making, work burden, access to and use of productive resources, and collective action (Chapter 7); decision-making and the division of labor and resources (Chapter 5); or ways of measuring or assessing changes in different domains of women's empowerment (Chapter 9). Others argue for a shift in how a theme is approached—for example, embedding gender perspectives more meaningfully within seed system development to move beyond simply "reaching" women as beneficiaries (Chapter 3); iterating between feminist analyses and natural resource discourses to the benefit of both (Chapter 6); reframing studies on the "feminization" of agriculture to instead interrogate gender equality advances or lost ground in rural transformation processes (Chapter 8); or getting to the roots of gender inequality and using approaches that address structural change (Chapter 10).

To come back to diversity and heterogeneity among and between women—as well as men—different women experience inequalities in different ways, depending on their age, race, class, marital status, caste, religion, ability, position within the family, education, or sexual orientation. Empowerment and gender equality are fundamentally entwined with other intersecting axes of social power relations, and women—just like men—are a socially heterogeneous group (Colfer, Basnett, Ihalainen 2018). Whereas feminist thinking has introduced gender as a concept so as to be able to analyze women's disempowerment and the inequalities they experience, this by no means implies that gender is the only basis on which these are experienced. *Intersectionality* is hence a core concept to underline and unpack how such other social markers intersect with gender in hierarchies, inequalities, and marginalization. It draws attention to the intertwining effects of multiple deprivations and overlapping disadvantages, and warns of the risks of reducing all analysis of and approaches to gender inequalities to "gender alone." Such intersecting inequalities mean that different women may face different constraints and have different opportunities. Intersectionality is also a key concept as it allows and calls for a more explicit understanding of how intersecting social markers and inequalities affect the most marginalized groups of women (Yuval-Davis 2006; van Eerdewijk et al. 2017).

In our understanding of gender equality and women's empowerment for this book, the change that is entailed is inherently transformative in nature. Empowerment is directly tied to inequalities and disempowerment, and the ways in which women previously lacked "the capacity to make strategic and meaningful choices"; gender equality and women's empowerment directly challenge the structural inequalities and the way power operates in women's lives (Kabeer 2017; van Eerdewijk et al. 2017). Gender equality and women's empowerment are, however, not always, or often not, approached and used in this way. The depoliticization, technocratization, and instrumentalization of core feminist concepts as empowerment has turned them into "light" versions, which emphasize women's individual self-improvement rather than collective transformative struggle. This is problematic, because it isolates women from the social inequalities that govern their lives; these versions tend to end up trying to "fix women" or, alternatively, leaving the responsibility for change in their hands while institutional dimensions go untouched and unquestioned (Batliwala 2007; Batliwala and Dhanraj 2007; van Eerdewijk 2016; Cornwall 2018).[44]

Navigating the chapters

This book comprises nine thematic chapters. What all of them have in common is a fresh analysis of evidence related to the theme tackled, which is reviewed using a gender equality and women's empowerment lens, responding

44 In fact, after its introduction in mainstream development in the 1990s, the term "empowerment" was widely adopted, and in the process co-opted, diluted, and watered down. "Light" versions of empowerment tend to emphasize economic aspects of gender equality and empowerment over other elements, and often focus on measurable and tangible dimensions. Women's empowerment is not the only transformative concept that, in its introduction to and uptake in research, institutions, and development, has become subject to "discursive ambiguity" (Eyben 2010, 57), and "simplification" and "sloganization" (Cornwall, Harrison, Whitehead 2007a, 4). It is in this "struggle for interpretive power" (ibid.) that the intentional and explicit positioning of our understanding of gender equality and women's empowerment gains significance.

to the guiding question of the book: **How does agricultural and environmental research and development contribute to gender equality and women's empowerment?** We begin with a set of themes under which technical topics have integrated gender (breeding, seed systems, value chains). We then move into themes where gender analysis is well integrated and even pivotal to the theme (nutrition-sensitive agriculture, climate adaptation and mitigation, natural resource governance). Finally, we have a set of strategic gender-specific themes (the feminization of agriculture, assessing women's empowerment, gender transformative approaches). Roughly sketching the progression of chapters, gender analysis becomes a more integrated and significant component of the theme as we move further into the book.

Chapter 2 tackles gender in a long-standing and foundational domain within CGIAR: breeding. Animal and plant varieties developed are key technologies that CGIAR contributes as international public goods to agricultural development internationally. This chapter—*Examining choice to advance gender equality in breeding research*—zeroes in on the work done to integrate gender into breeding programs. It dissects the institutional dimensions of breeding research, investigating in detail the steps needed to advance toward gender equality, starting at the beginning: with setting breeding objectives. The chapter not only points to critical entry points for considering gender along the breeding cycle but also explains the implications of integrating gender needs, preferences, and constraints for institutional practices in breeding.

Chapter 3 focuses on seed, as a vital and first component of food systems. The chapter examines how gender equality and women's empowerment are or could be connected to seed system development and performance, and vice versa. This chapter—*Moving beyond reaching women in seed systems development*—examines work from across CGIAR on this relatively new domain for gender research. The chapter finds that seed system interventions often reach women, and that some may benefit them. However, stronger recognition of women's involvement and leadership in seed production and management would offer further benefits (like access to quality seed) as well as opportunities for empowerment outcomes. That said, the latter is not well documented in most seed systems research. Gender norms and power asymmetries demand attention for seed system interventions.

Chapter 4 draws on 20 years of research on gender and value chain development, bringing critical analysis to a theme that virtually all CGIAR research programs cover. This chapter—*Promise and contradiction: Value chain participation and women's empowerment*—finds evidence on the complexity of the relationship between participation in value chain development and women's empowerment. Participation can be empowering by enhancing positioning, social capital, confidence, skills, and leadership; it can be disempowering through loss of control over production processes, loss of social status, exploitative labor conditions, marginalization, and time poverty. And it is both at the same time. Gains in one aspect of women's empowerment do not necessarily coincide with gains in another, and the broader socio-political and economic context—including the structure of the value chain and gender relations and norms in households, communities, and society at large—plays a role in mediating this. Notably, the chapter finds that most interventions with empowerment outcomes have a deliberate, and often exclusive, focus on women and gender relations.

Chapter 5 re-examines the body of work on gender and nutrition, putting gender equality at the center of analysis, rather than positioning it as an instrumental means to achieve nutrition objectives. This chapter—*Nutrition-sensitive agriculture for gender equality*—digs into a domain that has long been associated with women and the household. It finds that most nutrition-sensitive agricultural programs (NSAPs) are gender accommodative (work within existing gender norms) rather than transformative in nature. Where interventions do consider gender relations, they tend to aim to "reach" or "benefit" women, but rarely to empower them. Where empowerment is an objective, poor implementation—lack of intentionality and consideration of an impoverished range of empowerment domains—often leads to failure. Where NSAPs do empower women, monitoring, evaluation, and learning systems are unlikely to capture this. The chapter outlines features emerging from NSAPs with the most potential for empowerment outcomes.

Chapter 6 goes beyond agriculture per se, entering into a second major research domain within CGIAR, that of natural resources. This chapter—*A gender-natural resources tango: Water, land, and forest research*—reviews significant research from within CGIAR and beyond. It explores and critically analyzes how gender research paves the way for analysis of social and political dimensions of natural resource management. The field's shift from management to governance, and then to plurality of rights, not only opened up space for gender analysis but also proved to be a "game-changer": it enabled better understanding of how inequalities based on gender and other social markers played out around natural resources so they could be addressed. The chapter underscores the need to go further in tackling root causes and systemic barriers to gender equality and women's empowerment, including through stronger synergies between research for development agendas on natural resources and feminist approaches.

Climate change research, including climate-smart agriculture, is a more recent, but critically important, topic for CGIAR. Chapter 7 re-examines and reframes work on gender and climate change in the light of gender equality objectives. Climate change research and interventions tend to be systemic in nature—looking at human and natural conditions and focused on increasing resilience and reducing vulnerability for climate change adaptation and mitigation. This chapter—*From vulnerability to agency in climate adaptation and mitigation*—emphasizes women's agency in relation to climate resilience and the role of collective action in supporting enhanced agency. The chapter argues that the next generation of climate change research will need to do more than diagnose women's vulnerability or understand gender differences in perceptions and impact of and response to climate change. The focus must shift to understanding how to promote gender equality and increase climate resilience.

Chapter 8—*From the "feminization" of agriculture to gender equality*—looks at bigger picture trends and drivers of rural transformation that affect the livelihoods of rural women and men, including migration, war, and conflict; the commercialization of both large-scale and smallholder agriculture (including value chain development); technologies and innovation; and climate change. This chapter unpacks how the term "feminization of agriculture" is used and reviews state-of-the-art research, looking at the impacts of these rural transformations on both women and men in terms of labor patterns and workload, decision-making, and/or management roles and visibility. It sets a clear and comprehensive research agenda along two lines. The first is about reconceptualizing the feminization of agriculture and more systematically and comprehensively measuring the domains in which gender equality is advancing (or not) amid the process of rural transformation. The second is about addressing data gaps with improved measures. Importantly, the authors make a call to broaden the scope to look beyond agriculture and explore what else may be empowering for women, or other domains that may (better) advance gender equality for rural men and women.

Chapter 9—*Assessing women's empowerment in agricultural research*—reviews, compares, interrogates, and reflects on key qualitative and quantitative measures of empowerment developed and used within CGIAR and more broadly. Assessment can advance women's empowerment, especially when done in ways that engage with existing inequalities and challenge obstructive power relations. How empowerment is conceptualized and operationalized can limit or enable the extent to which both institutional dimensions of inequalities and change, as well as immaterial aspects such as norms and sense of self, come into the analytical frame. The chapter challenges practitioners and researchers to shake free of instrumental uses of assessments of "empowerment" and to return to its political and transformative foundations. It reflects on the challenges in assessing women's empowerment, from the difficulties of capturing certain aspects of empowerment over others (e.g., economic versus psychological) to the politics and complexity of measurement over the dynamics of space and time. The ways in which the research process contributes to gender equality are central.

Chapter 10—*Toward structural change: Gender transformative approaches*—pulls together relatively recent pilots, think pieces, and research, arguing for approaching gender analysis and gender and development differently. It calls for CGIAR researchers to go beyond gender integration and start engaging with deeper inequalities—like entrenched gender norms as to how women and men can/should behave—including institutional constraints to this. In addition to a clear articulation of the distinctive features and ambitions of GTAs, the chapter's value lies in the timely collection and review of emerging studies on actual experiences with and outcomes of GTA efforts. A challenge for work on gender norms and for GTAs is that, despite recognizing the need for structural change, they tend to focus on norms and local dynamics, rather than engaging with structural issues and trends, especially beyond local scales. This chapter, particularly in the research agenda laid out, calls for scaling up (and out) of GTAs, including these higher-level issues.

The evidence

All thematic chapters described above are based on extensive literature reviews. At the end of each chapter, just before the reference list, the reader will find four visuals that aim to provide an impression of the coverage of the evidence on which the chapter discussion and conclusions are based. These visuals consist of:

- A map that illustrates the geographies of case studies in the cited publications;
- A pie chart that shows the proportion of publications from the CGIAR versus those with other institutional origins (non-CGIAR);

- A second pie chart showing the methodology used in the publications reviewed—whether the publication is based on qualitative, quantitative, or mixed-methods data, or methodological in nature (citations that do not fit into these categorizes are considered "other"); and
- A timeline illustrating when the papers reviewed were published.

Here, we briefly synthesize the compiled coverage of evidence from across all chapters, offering visuals that represent the book as a whole (see Annex 3 for details).

To begin, Map 1.1 illustrates where the studies cited are located **geographically.**

Unsurprisingly, geographic coverage of the evidence is uneven globally, with more studies across the book coming from Asia and East Africa: India (92 publications), Bangladesh (83), Kenya (69), Uganda (67), and Nepal (57) are the most cited. However, there are significant variations from one chapter to another, with clusters of studies linked to specific themes. In considering the evidence, this variation—the regional specificities—needs to be weighed against more generalizable trends in drawing conclusions. The geographic prevalence of studies cited may correspond loosely with either having CGIAR Centers based there (e.g., Bangladesh, Ethiopia, India, Kenya) or being priority countries for CGIAR research or donor investments. It may also reflect data limitations, with more studies possible in countries where the sex-disaggregated data exist. Next generations of gender research in agricultural and environmental domains may want to consider the geographies studied in order to build up a body of work in specific regions or to break new ground by venturing into less-known terrain. Annex 3 provides more detail on the geographies and variations across the thematic chapters.

A second visual pertains to the **institutional origins** of the reviewed studies: CGIAR versus non-CGIAR. Writing teams reviewed both CGIAR research, from 1 or more of the 15 CGIAR Centers, as well as key studies carried out and/or published elsewhere. A publication was considered to be CGIAR if (a) one of the co-authors was based at a CGIAR Center at the time of publication; (b) the publication was funded by a CGIAR Center or a CRP; or (c) it was published by CGIAR. Figure 1.3 shows that a slightly higher proportion of the studies cited are from CGIAR sources. All chapter teams began with a wide outreach within CGIAR for related publications so the visibility of CGIAR publications is not surprising and was, in fact, purposive.

MAP 1.1 Geography of study sites for publications cited in this Book

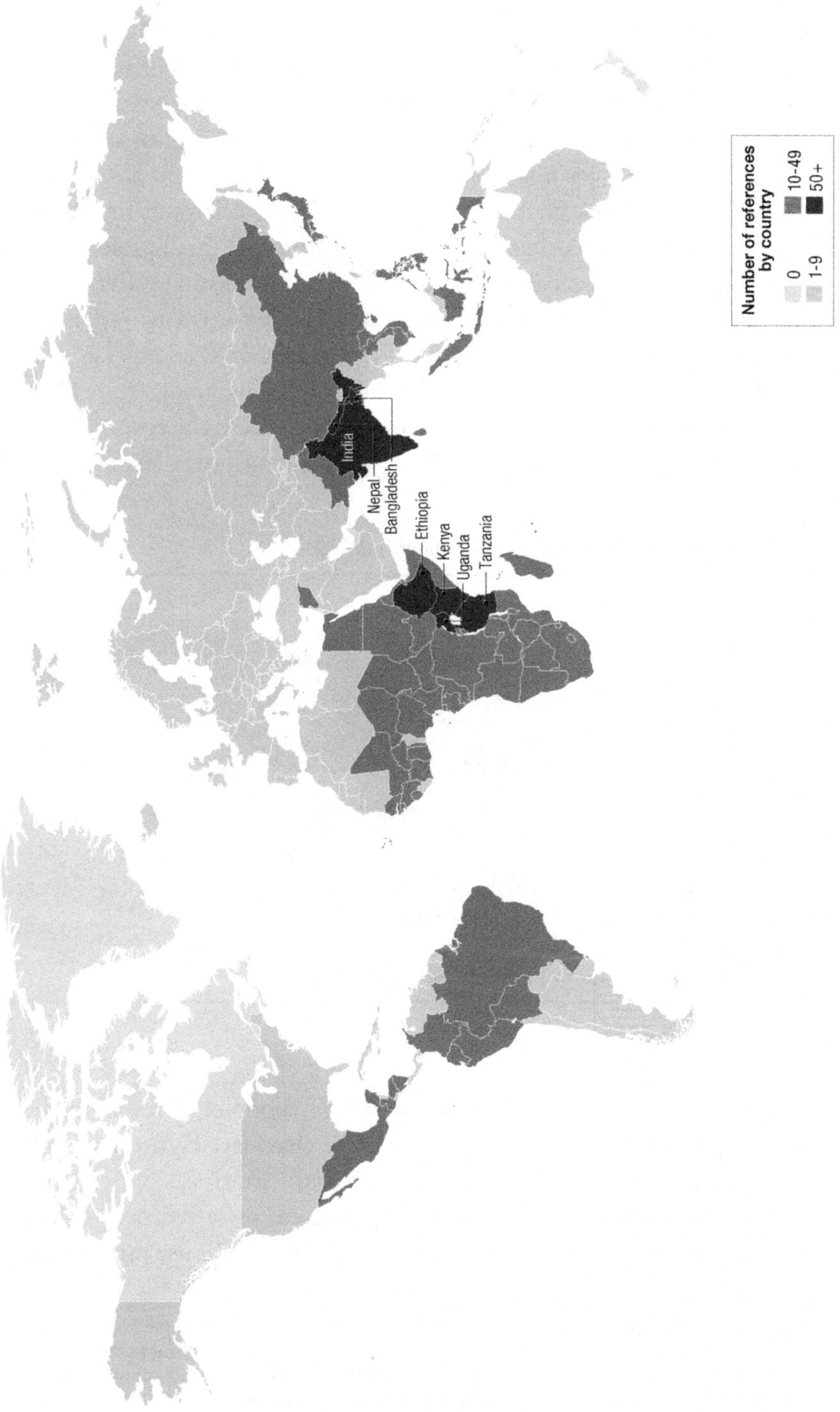

Literature reviews were not intended to be exhaustive but rather a (re-)examination of evidence on key CGIAR gender research themes using a gender equality and women's empowerment lens.

While Figure 1.3 illustrates the overall institutional origins across the book as a whole, it does not show variations across the chapters. Comparing the chapter-specific pie charts, it is evident that some chapters rely more on CGIAR research than others. This speaks to both the methodologies used for the literature reviews and the evidence available within CGIAR and externally on each theme, as well as the maturity of the theme—that is to say, how long it has been studied. In some domains, CGIAR emerges as a clear thought leader; for others, it is one contributor among many.

A third category for consideration of the evidence is the **research methodology** of the studies cited: qualitative, quantitative, mixed methods,

FIGURE 1.3 Number of studies by institutional origin (N=970)

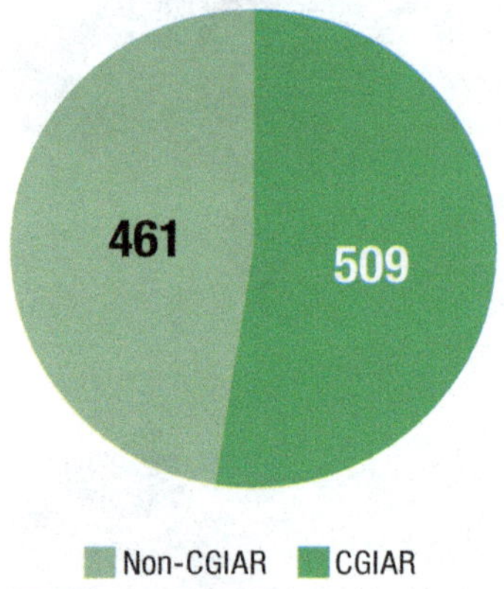

methodological, and "other."[45] The research methodologies for all studies cited in the book are compiled here in one pie chart (see Figure 1.4). Note that Figure 1.4 does not capture the size or scale of the studies.

More than half of the publications (56 percent) reviewed are qualitative studies; about 15 percent are quantitative and 20 percent use mixed methods.

45 The category "methodological" refers to publications that deal with methodological issues or approaches. The "other" category includes guidance documents, presentations, reviews, indices, analytical frameworks, brochures, manuals, webinars, PowerPoint presentations, interviews, and documents that do not fall into the other categories distinguished.

FIGURE 1.4 Number of cited studies by research methodology (N = 970)

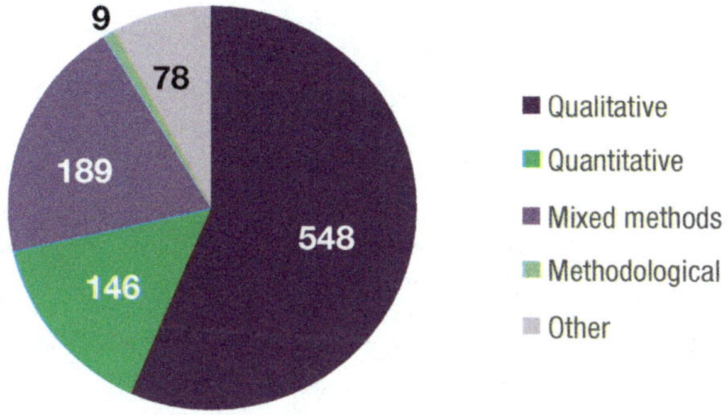

Just a handful are methodological and less than 10 percent fall into the "other" category. Interesting to note is the variation across themes, which can be seen when comparing chapter-specific pie charts. Overwhelmingly, the chapters rely on qualitative publications; however, Chapter 5 (nutrition-sensitive agriculture) and Chapter 8 ("feminization" of agriculture) are exceptions: they refer more to quantitative publications than to qualitative work. Chapter 4 (value chains) has the highest number of mixed-method studies cited (43) and the highest proportion of mixed-methods studies overall (32 percent).

While gender researchers advocate for more mixed-methods research, the studies available seem to be limited for most themes explored in this book. That a limited proportion of studies are mixed methods in character—at best between 21 and 32 percent—is relevant for CGIAR research because it speaks to ongoing discussions on the value of interdisciplinary approaches and mixed methods. Some thematic domains may be more or less amenable to mixed-methods studies. The low number of mixed-methods studies may also be because CGIAR research is most often published in journals, with specific audiences and limited word counts. Accordingly, researchers may send a quantitative paper to one journal and a qualitative paper to another; this may not show if papers are part of a mixed-methods study. There is room for further exploration of opportunities for more mixed-method studies and why they are (or are not) undertaken.

Finally, we consider the **timelines of cited studies** reviewed. The timelines visually represent the publication periods for the citations in each chapter and are compiled for all chapters in one graph (Figure 1.5).

FIGURE 1.5 Timeline of cited studies in this book per chapter

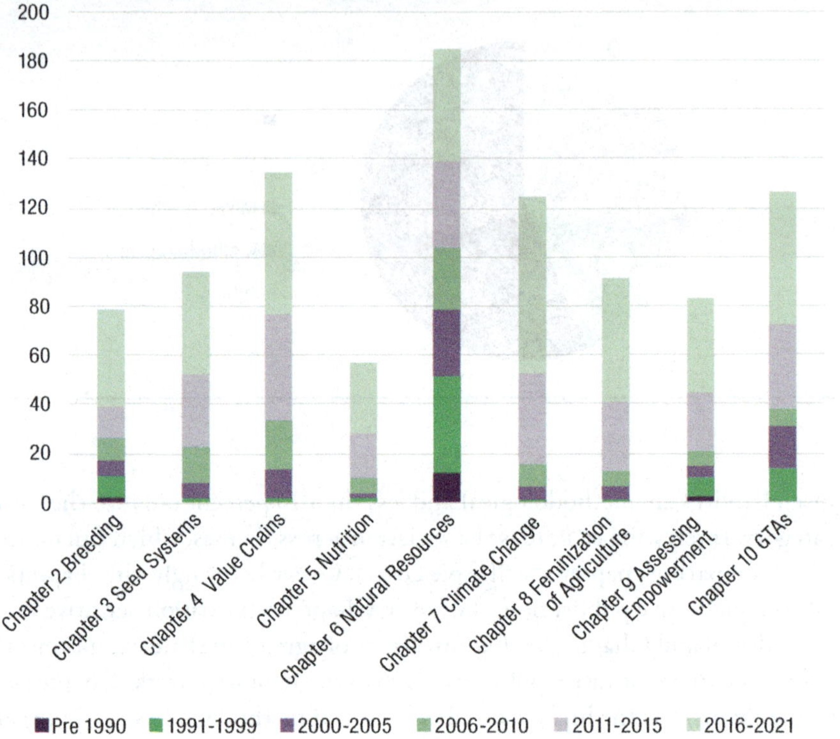

For all chapters, unsurprisingly, a significant proportion of the citations are from the past 10 years (the periods 2011–2015 and 2016–2021). However, some themes have a longer history within CGIAR, with the work on gender and natural resources (Chapter 6) standing out: its citations extend back before 1990. Other themes have emerged more recently (e.g., climate adaptation and mitigation, Chapter 7). Some themes, while more recently emerging, draw significantly from past research, for example that of gender dynamics in seed systems (Chapter 3) or on GTAs (Chapter 10). Figure 1.5 also illustrates the total number of publications reviewed per chapter.

Part 3: Analyzing and reflecting on the book chapters

To begin this part of the chapter, we draw out the main conceptual threads along which the book's nine chapters engage with the key question of **how**

agriculture and natural resource management research and development contribute to women's empowerment and gender equality. We look to the chapters to see what we are learning in relation to the core question of this book, drawing out several conceptual threads. We then turn our attention to a forward-looking gender research agenda and considering what is needed to make it effective, before sharing concluding reflections.

Gender equality through agricultural and environmental initiatives: conceptual threads

First, we synthesize whether and how agricultural and environmental research contributes to gender equality and women's empowerment. We then reflect on three interrelated conceptual threads emerging from the chapters: on conceptualizing beyond the individual, beyond the material; on bringing "bigger picture" trends into focus; and, finally, on inequalities and intentional approaches.

CHANGE HAPPENS, BUT NOT ALWAYS AND NOT AUTOMATICALLY

The chapters show that agricultural and environmental research and development approaches and interventions can, and do, lead to empowerment and gender equality outcomes for women. Women's engagement as seed producers, for example, can enhance their confidence and status in the community and household, contribute to increased income, and enhance their decision-making in the household. Nutrition-sensitive agricultural programs can and do lead to women becoming more able to make decisions in the household, and may increase their control over assets and affect perceptions of their abilities. Climate change adaptation strategies can promote women's participation in household decision-making, on agricultural production as well as income or children's education. Climate-smart agriculture (CSA) technologies, more specifically, can potentially have empowering effects by reducing women's workloads. Value chain participation and value chain development interventions do lead to increases in women's social capital, confidence, leadership, and overall positioning, as well as increased power in household decision-making and enhanced status for women, and sometimes to shifts in gender norms. At the level of communities or rural institutions, women can gain more access to agricultural enterprises and bodies that govern community resources, or the production and management of seed; some women also have more voice in their management and can take on formal and informal leadership positions in such institutions. In short, a range of empowerment and gender equality outcomes are found across the thematic domains covered in this book.

Yet the chapters all underscore that agricultural and environmental developments and interventions also have negative effects on women's empowerment, and, worse, can actually exacerbate gender inequalities. Negative effects include increasing women's workload, in the context of nutrition-sensitive agricultural programs or as a result of climate change adaptation strategies. Such increases to the labor burden, and women's subsequent time poverty, are common because broader shifts in the gender division of labor—including men taking on larger shares of reproductive work—typically do not occur. Other prominent disempowering effects concern women's loss of control over agricultural production processes, in the context of, for instance, value chain development strategies. Critical, yet often undocumented, disempowering effects include increases in household conflict and intimate partner violence. These occur, for instance, if men feel threatened or insecure when patriarchal gender norms—regarding paid work, women's household and reproductive labor, their mobility, and their interactions outside the household—are challenged and start to shift. Also at the community level, changes in women's positions, self-confidence, access, and "power to" can, unfortunately, lead to stigmatization of women who are perceived as having crossed boundaries.

The review of existing evidence not only illustrates the empowering and disempowering effects of agricultural and environmental development but also, strikingly, demonstrates that these two effects can happen *at the same time*. There is a need for explicit attention to these often-contradictory effects. Improvements in one dimension of empowerment can be accompanied by deterioration in another; positive changes at the household level can correspond with negative changes in, for instance, value chain or natural resource governance. Moreover, and maybe even more importantly, across all thematic domains, it is clear that positive and empowering effects *do not happen automatically*. The extensive literature reviewed across the nine thematic chapters confirms the validity of the Reach-Benefit-Empower-Transform framework, and underscores that, "reaching women does not ensure that they will benefit from a project, and even if women benefit [...], that does not ensure that they will be empowered" (Johnson et al. 2018, 5). Increased opportunities to earn can translate into control over that income and a larger say in household decision-making—but often do not. Participation in training can offer women opportunities to use newly acquired knowledge—for instance on the use of new technologies—but women are too often unable to use their new knowledge. Changes in women's involvement and participation in seed production and management, in value chains, in CSA, or in forest or water management can contribute to shifting perceptions and attitudes on appropriate work, roles, and leadership potential for women—but often do not.

ON CONCEPTUALIZING BEYOND THE INDIVIDUAL, BEYOND THE MATERIAL

The simultaneous and often contradictory positive and negative empowerment outcomes of agricultural and environmental development interventions, and approaches, underline the pivotal importance of explicit and comprehensive conceptualizations of gender equality and women's empowerment. This means positioning the type of change underway and articulating a theory of change vis-à-vis how transformative change comes about. We highlight two steps in this direction: (a) look beyond the individual and (b) look beyond the material. The first step requires looking at changes in gender equality and women's empowerment at three levels: the individual,[46] the **relational**, and the **systemic** (Hillenbrand et al. 2015; Lombardini, Bowman, Garwood 2017; van Eerdewijk et al. 2017; Lombardini and McCollum 2018). The second step takes **informal** and ideational aspects into account. This mainly means considering social and gender norms and attitudes (at the systemic and relational levels) and self-confidence and critical consciousness (at the individual level) (Rao and Kelleher 2005; Wong et al. 2019).

The nine thematic chapters reveal that the existing literature tends to conceptualize women's empowerment and gender equality in quite narrow ways, focusing most commonly on individual-level material gains for women. Most of it focuses on what individual women have, do, use, control, gain, or lose (e.g. income, knowledge, skills, assets, time, and so on). The dimensions of empowerment considered often derive from the logic of the field, for example value chain studies looking at market indicators or nutrition studies focusing on nutrition-related ones. The choice of indicators is different if the starting point is a comprehensive understanding of gender equality and empowerment, and exploring how agriculture and natural resource management and interventions contribute to this (or not). Combined, these shortcomings of individual, sector-driven conceptualizations and indicators constrain our ability to understand women's empowerment as a process of social change that goes beyond individual self-improvement and "fixing the women" (see also Chapters 4, 6, 9,

46 Lombardini, Bowman, Garwood (2017) differentiate between three similar level: personal, relational, and environmental. It is worthwhile to note that their understanding of "personal" refers to *immaterial* aspects ("*within the person [...] changes in how a woman sees herself, how she considers her role in society and that of other women, how she sees her economic role, and her confidence in deciding and taking actions that concern herself and other women*", p. 6). Our understanding of the individual level encompasses both material and immaterial dimensions. This is also the case for the relational and systemic levels.

and 10; van Eerdewijk et al. 2017; Wong et al. 2019). In a variety of ways, the nine chapters go "beyond the individual," drawing attention to relational and systemic levels of gender equality and women's empowerment.

At the *relational* level, the role and engagement of men and other influential household or community members come into the frame. In the context of NSAPs (Chapter 5), for instance, recognition of the importance of involving men in nutrition training or in gender awareness interventions, rather than exclusively targeting women, is growing. Similarly, in the context of climate change adaptation (Chapter 7), participatory and collective action approaches engage with men's roles and their responses to changes in women's incomes or resilience. In so doing, these approaches seek to overcome the resistance of men and other power-holders to gender equality by encouraging recognition of women's realities and challenges, and through discussion of patriarchal and constraining attitudes and norms. Engagement of men and boys in change means exploring how they can become allies in transforming power relations toward gender equality, rather than acting as gatekeepers of the status quo in the context of male power and dominant masculinities (Levtov et al. 2014; Edström et al. 2015; van Eerdewijk et al. 2017). Women's relations with partners, relatives, and/or community members also feature prominently in gender transformative approaches (see Chapter 10). GTAs position themselves against an exclusive focus on interventions that engage women only as "atomized units"; they often promote transformative learning within as well as between individuals.

The third level of gender relations—*systemic*—requires looking at the institutional and structural arrangements under which women live, work, and shape their lives. The thematic chapters speak to the systemic level in various ways. Chapter 6 on natural resources explicitly argues for the need to "fix the system." Women's access to and use and control of (irrigation) water, forest resources, or land tenure rights are shaped and constrained, and can be enabled, by informal and formal factors of their governance, at community as well as macro levels. These include legal regulations, policies, customary law, local rules in resource user associations, or social relations that recognize or deny the use of certain resources by certain groups. These systemic factors in rural institutions also affect women's voice in decision-making on such resources, with trends toward privatization and commercialization often contributing to exclusion and marginalization. Chapter 3 highlights how institutional factors in seed governance and management are key to the lack of recognition and further exclusion of women, which directly undermines their benefits and empowerment. On the positive side, stronger engagement of women in seed management, and in particular their voice and leadership in decision-making, carries potential for benefits from seed system and other empowerment indicators, as has been most visible in farmer-managed systems and participatory breeding programs.

The systemic nature of gender relations and inequalities also manifests itself in gender norms. Here, the second step comes into play: *going beyond the material* by taking into account the *informal* and *immaterial*. Social norms on gender are featuring ever-more prominently in gender research in agricultural and environmental domains; recognition is increasing that underlying structural barriers, in particular patriarchal norms, hamper women's ability to adopt and benefit from technological innovations, their expanding engagement in formal employment, or their leadership positions. Gender norms[47] are "collectively held expectations and beliefs as to how women, men, girls and boys should behave and interact in specific social settings and during different stages of their lives" and "are key aspects of institutional structures that shape the empowerment of women and girls" (van Eerdewijk et al. 2017: 35, 40). Despite norms being key dimensions of social practices, they are often largely ignored in AR4D. Chapter 10 discusses how GTAs identify and focus on gender norms as leverage for transformative shifts in underlying structures and barriers, a point that resonates across many other thematic chapters.

The *immaterial* is also at play in how perceptions and stereotypes of what is appropriate for women's work shape whether women's labor is recognized, what (paid) jobs women have access to, how much they are paid (Chapters 4 and 8, for instance), and how perceptions of "who is the farmer" overlook women and act as a barrier to them being approached and included as farmers and farm managers (Chapter 3) or agents of change in climate change adaptation and mitigation (Chapter 7). At an individual level, the immaterial is a core dimension of women's empowerment and change toward gender equality: it relates directly to self-confidence and self-esteem. Shifts toward a critical consciousness are not only at the heart of women's empowerment—as in women

47 Gender norms pertain to, among others, women's behavior and interactions outside the household, assumptions about their reproductive labor and responsibilities, beliefs about male authority and women's submissiveness, and perceptions of women's leadership and decision-making capacities or of appropriate benefits and assets to be owned or controlled by women, and to the acceptability of harassment of and violence against women in public and private spheres. Gender norms are a critical enabling (or disabling) factor affecting agricultural livelihoods and are determinants of the distribution of (material) benefits and resources. Social norms on gender are dynamic and context-specific and present in all domains of social life: in families, communities, the state, private companies, and modern development organizations (Wong et al. 2019; Badstue et al. 2020).

gaining *power within*—but also key to women jointly organizing and engaging in collective action to challenge power hierarchies—women's *power with* (Batliwala 1993, 2007; Cornwall and Edwards 2014; van Eerdewijk et al. 2017). We come back to collective organizing below.

The different chapters highlight different ways of going beyond the individual and the material; together, they illustrate and indeed underscore that an answer to the core question of the book requires a comprehensive conceptualization of gender equality and women's empowerment. Women's empowerment and gender equality occur when shifts happen across different levels (individual, relational, and systemic), and when they encompass both material and immaterial, and formal and informal, dimensions.

ON BRINGING "BIGGER PICTURE" TRENDS INTO FOCUS

The comprehensiveness of the conceptual frameworks pertains not only to the conceptualization of empowerment and gender equality but also to the extent to which the bigger picture comes into view and is interrogated. The chapters point to the commercialization of agriculture (of both large-scale agriculture and smallholder farming), climate change, migration, and conflict and insecurity as larger and fundamental structural processes that affect women and men's lives, and shape processes of gender equality and women's empowerment. These trends, and in particular commercialization and privatization, are manifestations of *neoliberal*[48] logics that have come to dominate economic trends and public policies. In this book, several chapters use the term "neoliberal," for example when discussing the limits of value chains for advancing gender equality (Chapter 4) or in lamenting the development of natural resources thinking (Chapter 6). Neoliberal logics manifest themselves in the context of "markets" but also in relation to challenges to measuring women's empowerment (Chapter 9) and in laying out the limits of individual and instrumental approaches to make the case for GTAs (Chapter 10).

In the seed sector, *privatization* poses challenges to gender equality, as most formal seed systems—private and public—fall short in reaching women smallholder users and producers, let alone meeting their needs and preferences.

---

48 "Neoliberalism" is a term used to refer to "*a macroeconomic doctrine [... that includes] a valorization of the private enterprise and suspicion of the state, along with what is sometimes called 'free-market fetishism*" (Ferguson 2009, 170). It is also used to refer to "*a regime of policies and practices associated with or claiming fealty*" to this doctrine, which in practice have come down to "*loss of public services, and a general deterioration of quality of life for the poor and working class*" (ibid., 170). In addition, "neoliberalism" is used to describe "*the deployment of new, market-based techniques of government within the terrain of the state itself", with new public management techniques applied so that "core functions of the state [...] are 'run like a business'*" (ibid., 170–171; Eyben 2013).

Promising pathways for women's empowerment emerge when women are engaged as seed producers, although gender inequities in the market affect these pathways and outcomes (Chapter 3). The observed finding of simultaneous empowering and disempowering effects surfaces prominently in the context of *commercialization*. Whereas commercialization opens up formal employment or income generation opportunities for women in commercial and smallholder agriculture, the terms under which women participate in value chains or agricultural markets can be biased and gender-unequal. Market logic and inequalities constrain women's empowerment gains in income, professional confidence, or control over productive assets; examples of these constraints include insecure and temporary contracts, gender wage gaps, poor working conditions, and gender segregation of tasks. These constraints hamper the empowering effects of income earning on, for instance, women's bargaining power in the household and in value chains/markets. Both Chapter 4 (on value chains) and Chapter 7 (on the feminization of agriculture) further conclude that the impacts of such trends—commercialization and privatization—are affected by, and contingent on, existing gender relations, inequalities, and patriarchal structures. In addition, empowerment benefits gained through women's participation in value chains or commercial smallholder agriculture do not necessarily mean that these biased market and societal inequalities are being challenged and addressed.

An important driver of rural transformation is *climate change*, and it has both direct effects on agricultural productivity, livelihoods, and natural resources and indirect effects that occur as a result of adaptation responses. Climate change has different impacts on women and men, and can alter gender relations. Climate change adaptation and mitigation responses, in turn, also affect gender inequalities and women's empowerment. *Migration* is a third trend, and has implications for both agricultural and environmental development, and women's empowerment and gender equality. Migration patterns are of particular significance when economic transformations and crises, combined with declining agricultural productivity and landholding sizes, translate into the out-migration of men and young people. Such out-migration directly affects labor patterns in agriculture, and may or may not be accompanied by changes in control over assets and women's say in household and farming decisions. Trends of women being left behind in rural areas and agriculture, with men moving to cities to earn a living, raise questions as to whether agriculture as a sector is actually empowering at all. Finally, *conflict and insecurity* are important drivers of rural transformation and are likely to disrupt the functioning of markets, shift labor patterns, constrain women's mobility and affect their safety, and lead to (gendered) migration. Research on conflict and insecurity in relation to rural transformation and gender (in)equality has been limited and merits more consideration.

These "bigger picture" trends and shifts, and their negative and undermining effects on gender equality and women's empowerment, raise critical questions as to how to interpret specific positive and empowering improvements in women's employment, income, or technology adoption, given that they are at the same time accompanied by declines in resilience and increases in wage or asset gaps. They also raise questions as to what "women's empowerment" means in the context of commercial markets that are driven by neoliberal logic, including the extent to which further incorporation of women into these labor markets is expected. Is it possible to challenge and address the embedded power inequalities and imbalances? It is clear that individual empowerment gains alone are insufficient to effectively challenge the broader inequalities that constrain women's choices and lives and are exposed when we look also at "bigger picture" trends.

ON INEQUALITIES AND INTENTIONAL APPROACHES

The final conceptual thread that merits explicit attention here highlights the importance of recognizing gender as a social relation, thus embedded in power relations and inequalities. To be able to make sense of whether agricultural and environmental research and interventions contribute to gender equality and women's empowerment, it is essential to explicitly consider and address power and inequalities. Agricultural and environmental research and interventions occur in a context that is characterized by structural constraints, power imbalances, and the disempowerment of women. Whether, when, and how they successfully contribute to gender equality and empowerment outcomes is contingent on *pre-existing* inequalities, based on gender and its intersections with other social markers, in the household, the community, and the economic and political domains. This is further shaped by rural transformations that in themselves entail power shifts across many levels, and that interact with efforts to advance women's empowerment and gender equality. In this dialectical interplay and complex context, research and development interventions can either exacerbate inequalities and contribute to disempowerment or have positive empowering effects. Because the desired gendered effects and empowerment outcomes of agricultural and environmental research and interventions do not happen in a vacuum, approaches and interventions need to be *intentional* in recognizing, and then overcoming, prevailing imbalances (see also van Eerdewijk et al. 2017).

From the chapters, two approaches that engage intentionally with power inequalities and dynamics stand out: gender transformative approaches (GTAs) and women's collective action and organizing. Both are proposed as necessary and effective approaches to promote women's empowerment and gender equality in agricultural and natural resource management. GTAs target gender norms as underlying structural barriers to women's empowerment and social change. With this focus on the immaterial and systemic, GTAs explicitly seek to bring about more sustainable and lasting shifts in women's agency and in gender relations, including positive changes in women's decision-making, control over assets, and the division of productive and reproductive labor between women and men. Because gender norms are collective held beliefs and values that transpire in dominant masculine behavior, GTAs not only address attitudes and beliefs but also actively engage with men and masculinities. By focusing on underlying barriers that are limiting women's expression of agency, GTAs challenge watered-down versions of empowerment. Further, their often-participatory tools and methods promote women's—and men's for that matter—reflection and transformative learning, which can strengthen their critical consciousness, "power within," and agency.

Women's organizing and collective action feature in interventions that contribute to women's empowerment in, for instance, seed systems, climate change adaptation, or NSAPs. Women's group-based programming, combined with approaches that engage with masculinities and exclusionary norms, has been a key factor in the realization of empowerment outcomes in NSAPs. In climate change adaptation, strategies that promote women's organizations and groups lead to empowerment outcomes, and in particular strengthen women's voice, shift their sense of self, and contribute to overcoming patriarchal gender norms. The significance of women's organizing and collective action for promoting empowerment underlines the importance of women's agency and voice—rather than a narrow concern with individual choice—so that women can gain control over their lives and futures (see also van Eerdewijk et al. 2017). However, not all interventions promoting women's organizing activate this potential, and some risk "using" women's groups in an instrumental way to facilitate access to resources and services. Moreover, the inclusive character of collective action groups is a key concern: not addressing inequalities between women in women-only groups can lead to the exclusion of poorer, less educated, and marginalized women, or to them benefitting less (Meinzen-Dick et al. 2014b). It is possible to leverage the transformative potential of these strategies only when they are recognized and employed as feminist strategies for change that challenge and overcome power imbalances. As with GTAs, such collective organizing approaches must further engage with macroeconomic dimensions of power shifts and transformation processes, including the disempowering effects of climate change, privatization, and commercialization.

To conclude, we return to and elaborate a figure from earlier in the chapter. Figure 1.6 shows the conceptual threads on gender equality and women's empowerment that emerge out of the gender research in CGIAR, and its shows that the bigger picture trends are (to be) part and parcel of that research. The Figure also seeks to capture that gender research and the emerging insights and threads are, in turn, part of and shaped by, not only the larger research environment, but also these "bigger picture" trends like privatization and migration. Against that background of influences, the conceptual threads emerging from the nine thematic chapters are translated into imperatives that frame a future-oriented research agenda, which is presented in the next and final section of this chapter.

FIGURE 1.6 Situating the content of this book

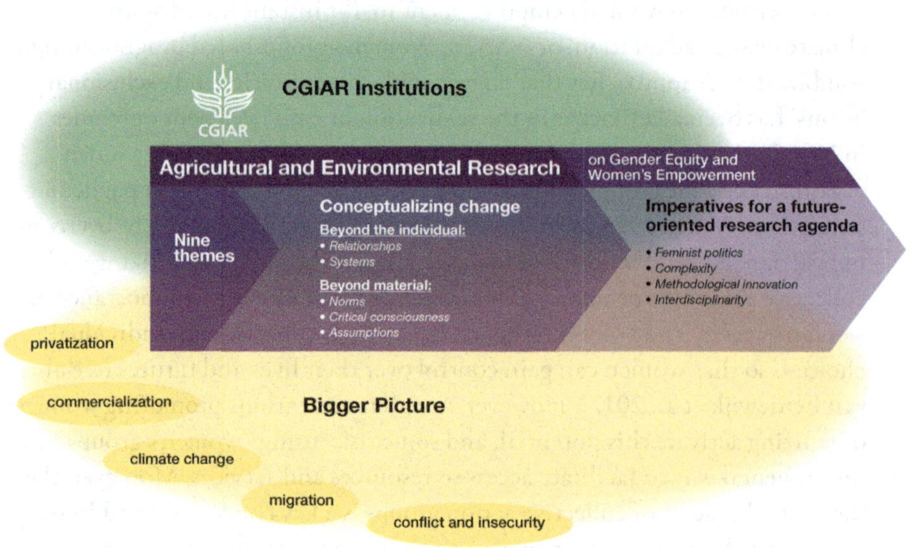

Toward a next-generation gender research agenda

Here we build on the gaps and priorities outlined in the thematic chapters, and set the course for a forward-looking gender research agenda by foregrounding "meta-level" gender research questions that cut across the agricultural and environmental domains. With those research questions in mind, we turn our attention to how to advance this agenda, by articulating imperatives for effective and transformative gender research. We then point to the institutional implications of embracing this research agenda and the imperatives, and call for addressing lingering institutional constraints that persist in hindering progress. We close the chapter with some final observations to further support a next generation of gender equality in agricultural and environmental research.

META-LEVEL GENDER RESEARCH QUESTIONS

Now is the time to be bold and transformative in articulating a next generation of gender research, and to advance and do justice to the fascinating and revealing gender work undertaken since the 1970s. This foundation, coupled with the firm delineation of gender equality as an impact area, can act as a springboard for CGIAR to lead in norm-busting, path-breaking transformative gender research in agricultural and environmental development. Core to a next-generation gender research agenda is a sharper focus on transformative change. This entails further engagement with both the relational and the systemic levels of gender equality and women's empowerment, as well as the ideational, immaterial, and informal dimensions. Another critical element is to move beyond pilot or project-level analysis to understand gender equality and women's empowerment at scale. This speaks to the tough-to-crack relationship between context specificities and, more generalizable conclusions and broader trends. Three sets of research questions that follow start exploring these less explored elements.

Gender research is context-specific, and this context specificity remains critical to understanding local social and gender dynamics, including gender relations, social norms, and power structures. To advance the gender research agenda and enhance the transferability of evidence, we need to develop conceptual frameworks and shared methods that can be implemented across different contexts but that at the same time allow for local contextualization of the learning. It will also be necessary to unpack the relevance of conclusions beyond specific contexts, and the variability between and within different contexts. The mixed evidence emerging from the rich, often case-study based existing body of knowledge calls for research designs and analysis that move **beyond specific settings**, and advance insights into how gender equality and empowerment outcomes vary by context and for different groups of women and men (e.g., by age, socioeconomic status, and other social categories) within the same contexts. Leading research questions that merit further exploration include the following:

- Under what conditions do agricultural and environmental programs and policies have (dis)empowering outcomes?
- How are patterns of empowerment (or disempowerment) outcomes related to specific agricultural and environmental contexts?
- How are these (dis)empowerment outcomes and processes experienced by different groups of women (and men)?

Answers to this set of questions will lead to sharper articulations of the type of change required and will inform interventions to advance gender equality and women's empowerment through agricultural and environmental initiatives.

A second set of research questions seeks to relate what is happening in specific sub-systems—like a value chain or seed system, in nutrition interventions, or related to the governance of natural resources—to the dynamics of **broader rural transformations** within which women and men, their households, and communities live.

- Which broader trends—within and beyond agricultural and environmental domains—affect gender equality and women's empowerment most significantly, and in what ways?
- Specifically, how do these broader rural transformations—economic, institutional, and environmental—enhance or narrow down the options, outcomes, and resilience of women and men?
- Which entry points can harness the potential for positive gender equality and women's empowerment outcomes within these broader rural transformations?
- How can adaptation to (or mechanisms used to cope with) these broader rural transformations sustainably contribute to women's empowerment and gender equality?

Research on how broader processes of rural transformation interlock to produce and reinforce disadvantage is key, as is reflection on how to harness the potential that these transformations may unlock. Interestingly, and perhaps unsurprisingly, "analysis of rural transformations and their gender implications" was identified as a comparative advantage for CGIAR gender work (Baden et al. 2017, 27). More evidence to identify the conditions under which rural transformations increase gender equality is needed, particularly where agrarian change is unfolding in conflict and post-conflict situations or where migration is a factor. This prominent set of questions is about addressing the systemic inequalities that constrain progress on gender equality—looking to the systemic level for insights into transformative change. They connect individual, relational, and systemic change and look at the relationship between specific contexts and more systemic change—and in particular push for future research to explicit position itself in and interrogate macro-level trends of rural transformation.

The third set of questions is about **intentional approaches** to gender equality and women's empowerment, and the challenges involved in scaling them. Building on the premises above, intentional approaches should be explicit in how they offset both the negative effects of broader rural transformations (as well as climate change and other macro-processes) and the potential disempowering effects of interventions and research themselves. This combines comprehensive insight into existing power inequalities with an explicit articulation of gender equality and women's empowerment as core objectives in their own right. A body of work has been developing over the past 10 years within CGIAR and more broadly on gender transformative approaches (see Chapter 10). Work and research on collective action has also been present and evolving across the CGIAR for many years now. More understanding is critical on how approaches that explicitly seek to address systemic inequalities fare (and work).

- How does gender transformative change happen in different contexts?
- How/when can interventions support processes of gender transformative change?
- How to activate the transformative potential of collective action to strengthen women's voice and agency and contribute to overcoming patriarchal gender norms?
- How to address inequalities between women in women-only groups and challenge and overcome power imbalances?
- How can GTAs and collective action challenge systemic gender inequalities and reverse the negative effects of broader trends?

- (How) can inclusive, context-specific, and localized approaches interrelate with more generic "broad-brush" approaches generally necessary for scaling-up, without losing their critical edge and transformative potential?

These sets of meta-level questions offer direction in the move toward a robust next-generation gender research agenda in the agricultural and environmental spheres. Such a research agenda is ambitious: it must meet the expectations of the women and men it is meant to serve, of the dedicated gender researchers working in agricultural and environmental domains within CGIAR and beyond, and of the partners and donors who provide resources. To be impactful, it must be grounded in sincere and profound institutional commitment, including the introspection and self-reflection called for by GTAs.

IMPERATIVES FOR GENDER RESEARCH IN THE AGRICULTURAL AND ENVIRONMENTAL DOMAINS

In consciously "flipping the frame" and positioning women's empowerment and gender equality at the center of agricultural and environmental research and development, this book sharpens analysis to focus on achieving meaningful transformative and empowering change. Reframing inevitably exposes the constraints of prevailing paradigms and approaches. Two cross-cutting imperatives for further research efforts crystallize.

Embrace complexity and feminist politics

Gender relations are complex and messy: "Conflicts of interest between men and women are unlike other conflicts, such as class conflicts. A worker and a capitalist do not typically live together under the same roof, sharing concerns and experiences and acting jointly. This aspect of 'togetherness' gives the gender conflict some very special characteristics" (Sen 1990, 147). Recognition of the complexities and specificities of gender relations in agriculture and in resource governance demands that researchers and their institutes boldly (re-)embrace a feminist research agenda. Adopting empowerment and equality as primary topics of inquiry is critical, as is the assessment of related outcomes in their own right, not merely as a means to an end. This may include more attention for critical consciousness and women's collective action; and a focus on changes in norms and values and in relationships that are often less visible, less tangible, and thus more difficult to measure. Approaches and interventions that address men and masculinities are an integral part of a feminist agenda (Cole et al. 2015; Farnworth et al. 2020)—for example, how to engage with men and power-holders to overcome resistance to gender equality at household, community and organizational/institutional level(s).

The growing hunger for critical approaches that push the boundaries is palpable, yet neither new nor easy: "feminists and some gender advocates have, since the 1970s, fought to retain a transformative agenda" (Okali 2012, 3). To be effective, agricultural and environmental gender research must not shy away from a feminist, transformative agenda that interrogates and confronts power and politics—whether between women and men, within households, communities, and countries, or between different groups in a community. Moving forward, CGIAR gender researchers will do well to engage with feminist research beyond their own agricultural and environmental domains to stay conceptually sharp and effective. Engagements in these broader feminist debates—in particular engaging with feminist thought leaders from the Global South—will be critical to making meaningful, fresh, and up-to-date contributions that advance the field. Exciting, cutting-edge work from Afro-feminists is flourishing, including work exploring decolonization with a feminist lens (e.g, Tamale 2020) as well as an important and innovative "reading of patriarchy… well beyond the exploitation of women"[49] emerging from feminist scholars and activists worldwide. Keeping abreast of such developments, engaging with them, and positioning CGIAR gender research vis-à-vis other feminist research are all critical.

Catalyze methodological innovation and interdisciplinarity

Building a transformative agenda entails making choices on methods and methodologies. The choice and use of quantitative and qualitative methods and measures affect the knowledge generated; they also reflect which voices and perspectives are recognized as valid and knowledgeable (Newton et al. 2019). Making choices on methods and methodologies includes consideration of the paradigms and approaches that emerge, gain space and recognition, and often dominate empirical research on specific themes. Methodological and epistemological preferences and biases reveal the power dynamics at play and the politics of measurement—a point featuring prominently in the thematic chapters calling for greater recognition of qualitative research and, in particular,

49 Renowned development anthropologist, Arturo Escobar commented on this in 2018, in reference to the work of Claudia von Werlhof, Maria Mies, Veronica Bennholdt-Thomsen, Silvia Federici, Barbara Duden, and Frédérique Apffel-Marglin. He noted that related perspectives were found in the work of Vandana Shiva, Carolyn Merchant, Australian ecofeminists Ariel Salleh and Val Plumwood, and political ecologist Wendy Harcourt, and that some of these approaches increasingly dovetailed with Latin American decolonial and autonomous feminists like María Lugones, Rita Segato, Raquel Gutiérrez Aguilar, Silvia Rivera Cusicanqui, Betty Ruth Lozano, Sylvia Marcos, Aura Cumes, Julieta Paredes, Aída Hernández, Yuderkis Espinosa, Diana Gómez, Karina Ochoa, Brenny Mendoza, Rosalba Icaza, Karina Bidaseca, and Ochy Curiel.

of participatory and action learning methodologies (see also Cole et al. 2014; Morgan 2014; Nazneen, Darkwah, Sultan 2014). Many chapters underline and reinforce the need for more mixed-methods research, to take advantage of the strengths of both qualitative and quantitative methods. Action-oriented, participatory methods with multiple actors (such as researchers, diverse local groups, non-governmental organizations, local authorities, government, etc.) may contribute to understanding and addressing the underlying causes of women's disempowerment and gender inequalities, especially to unearth structural barriers (e.g., Newton et al. 2019). An imperative for researchers is to widen the analytical frames, so as to explore non-conventional domains, question common assumptions, and be open to different kinds of evidence on how gender relations are changing. Prioritizing interdisciplinarity, addressing epistemological biases, and promoting the use of mixed methods are integral to invigorating a new wave of methodological innovation.

These two imperatives challenge barriers and biases that hamper or undermine further and meaningful integration of women's empowerment and gender equality into AR4D. They provide a foundational positioning to the robust and innovative directions of a gender research agenda that this moment demands: one with an intentional and explicit focus on gender equality and empowerment. The feminist politics and methodological innovation called for also reveal institutional barriers that constrain (but can also enable) that transformative and robust gender research agenda. Thus, there is a need for institutional change in AR4D, including in how research institutes function, to support and promote this forward-looking research agenda.

INSTITUTIONAL IMPLICATIONS FOR CGIAR: "WALKING THE TALK"

Any serious consideration of the systemic levels involved in gender equality and women's empowerment inevitably puts the spotlight on agricultural and environmental research itself, including the CGIAR Centers and Research Programs where research is undertaken. The final, tricky and perhaps sensitive, though nonetheless critical, aspect of a next generation of CGIAR gender research is that of tackling organizational and institutional change within research institutions themselves. Analysis of the developments in gender research from a feminist perspective shows that, despite significant progress since the 1980s, the challenges to institutionalizing gender research in CGIAR are striking. Gender researchers have struggled to carve out space and have their contributions valued and recognized. Indeed, "... feminist advocates have changed their approaches, but institutional change continues to be elusive (except in a few corners)" (Mukhopadhyay 2016, 77). The many reviews and research and institutional initiatives over time illustrate that ground has been lost and gained. Understanding the dynamics, exposing the mechanisms behind these institutional constraints and trying out strategies to addressing them is a first step to prevent a repetition of these cycles.

One recurring bottleneck relates to the slow and uneven progress vis-à-vis interdisciplinarity. The practice and uptake of gender research in agricultural and environmental fields has been affected by paradigmatic biases of the agricultural and environmental disciplines and their particular understandings of what constitutes "good science." Resistance to gender and feminist analyses—which can be seen in how dominant notions of what constitutes "science" fail to recognize and allow space for feminist approaches, frameworks, data, and insights, and also gender researchers and specialists—undermines the huge potential for stronger synergies (do Mar Pereira 2012; van Eerdewijk and Davids 2014; van Eerdewijk 2016). Indeed, there is a chasm to bridge between feminist analysis and a critical definition of the "gender problem," and the way in which development policy, practice, and research redefine and renegotiate it (Okali 2012, 12). Interdisciplinarity, and addressing disciplinary biases and epistemological differences, is part and parcel of the kinds of institutional change required to further a dynamic, future-oriented research agenda. The compelling example of the iterative developments between natural resources and feminist discourse over the past 30 years or more (discussed in Chapter 6) provides inspiration and optimism as to what is possible on this front.

As CGIAR institutes and portfolios progress towards One CGIAR, with gender equality firmly on the agenda as a priority area, and under the direction of talented gender researchers and the new GENDER Platform, institutional support is needed and organizational change is both inevitable and desirable. CGIAR leadership has an important role to play in setting the tone and in making this happen. Some specific actions are to:

- Ensure strong consultation with gender scientists to thoroughly infuse gender analysis across the One CGIAR portfolio from the start. This will contribute substantially to the quality and value of that work as well as to the effective design of policies and programs through the impact evaluations that will follow: make gender a "spice" permeating all CGIAR work, not a "topping" that can be removed![50]

50 With thanks to Agnes Quisumbing for the analogy.

- Use the Reach-Benefit-Empower-Transform framework to better articulate the gender-related objectives, outcomes and change processes of a project or policy and to better integrate gender equality and women's empowerment.
- Generously resource strategic gender research to advance gender equality through agriculture and environmental research and development where gender research is the "main ingredient": gender research must be well-embedded in core funding streams.
- Investigate higher-level gender research questions that cut across specific agricultural or environmental domains through comparative research and by synthesizing transferable and robust evidence and broader lessons. Gender research cannot be piecemeal: it must both speak to other bodies of gender research as well as to broader agricultural and environmental work.
- Staff permanent, strong, senior gender scientists providing ample resourcing and clear mandates, including that of mentoring junior researchers and supporting them in navigating both the body of work, as well as complex organizational contexts.
- Ensure effective accountability and performance monitoring mechanisms are in place.
- Confirm commitment to gender equality research from CGIAR leadership at all levels (e.g., Centers, platforms, research programs, governance bodies), and coordination on gender research throughout the governance system.
- Facilitate internal reflection on norms and practices as to how research is prioritized, designed, and organized vis a vis gender knowledge, including prevailing beliefs as to what constitutes good science.

Fostering and maintaining an internal culture of gender equality is a foundation for gender transformative change: in this case, we are referring to organizational and institutional change that prioritizes gender research. Addressing the outstanding institutional obstacles—including politics, norms, and practices—is critical for CGIAR if gender equality is truly a priority moving forward: the call is to lead by example, by "walking the talk."

Concluding reflections

Many aspects of this emerging research agenda are not unique to agricultural and environmental research and development: they resonate with broader feminist research. In the broader international development arena, the call to reclaim the transformative potential of gender strategies has been pertinent for over a decade, and has been accompanied by critical concerns about the techno-cratization and depoliticization of core gender and feminist concepts, practices, and strategies. The conceptual positioning of this book gains weight against this backdrop of lessons, critiques, and innovative debates that highlight the institutional and organizational dimensions and complexities of "getting institutions right for women and development," and the politics at play in feminist knowledge transfer (e.g. Goetz 1998; Cornwall, Harrison, Whitehead 2007a, 2007b; Mukhopadhyay 2007; Chant and Sweetman 2012; van Eerdewijk and Davids 2014; Bustelo, Ferguson, Forest 2016; Davids and van Eerdewijk 2016; van Eerdewijk 2016; Harcourt 2016). The threads among the chapters in this book strongly align with existing and recurring feminist and transformative critiques. In the directions for future research, we are taking them to heart to advance them within agricultural and environmental research for development. We hope that this book will not only inspire and provide clear directions for a next generation of CGIAR gender research, but also play a role in catalyzing the institutional change required for sustained advancement toward gender equality. The time is ripe: the question and challenge for One CGIAR leadership and its research portfolio, is whether and how to embrace this golden opportunity.

Acknowledgments

The authors are extraordinarily grateful for the insightful and detailed feedback on the chapter provided by four external reviewers. The substantive input was challenging and thought-provoking. Further, we thank Frank Place, PIM Director, who both coordinated the double-blind review process and provided very practical and strategic feedback. Thanks are owed to Laura Avila who did the leg work in analyzing the evidence across all book chapters (results found in Annex 3). And, finally, profound gratitude goes to the lead and co-authors of the book chapters, who provided input and critique on the chapter, strengthening and refining the formulation of meta-questions for future research and key messages for CGIAR leadership.

Annex 1: CGIAR background

CGIAR is a global research partnership comprising 15 agricultural and environmental research institutes internationally. It was founded in 1971 as a worldwide consultative group and in 2010 repositioned itself as a Consortium. The International Rice Research Institute (IRRI) and the International Maize and Wheat Improvement Center (CIMMYT) were among the first members of the CGIAR network, which then had the goal, responding to Green Revolution demands, *"to end world hunger by increasing food production"* (CGIAR 2015, 4). The CGIAR system expanded throughout the 1970s and 1980s, broadening in terms of both the range of commodities and agro-ecological regions addressed, and its objectives, which came to include "*poverty eradication and protection of the environment*" (ibid.). The 1990s saw the addition of Centers devoted to agroforestry, forestry, and fisheries. Today, the mandate of CGIAR is to produce international public goods to support a food-secure future, "*dedicated to reducing poverty, enhancing food and nutrition security, and improving natural resources.*"[51] The System Management Office, based in Montpellier, France, is the voice of the CGIAR network as a whole, representing all 15 Centers and all 12 CGIAR Research Programs.

Cross-Center collaboration has been fostered through two phases of CRPs, on agri-food systems and global integrating themes, running 2011–2016 and 2017–2021, respectively. These CRPs have a lead Center and contributing Centers, as well as external partners. Agri-food systems CRPs cover wheat; maize; rice; livestock; fish; forests, trees, and agroforestry; grain legumes and dryland cereals; and roots, tubers, and bananas. The global integrating CRPs are on topics that are cross-cutting: agriculture for nutrition and health; climate change, agriculture, and food security; policies, institutions, and markets; and water, land, and ecosystems.

Gender researchers and social scientists have increasingly been a part of CGIAR scientific staff teams. Growing social science contributions within CGIAR began in 1974, with the hiring of recently graduated PhDs through two-year Rockefeller Foundation fellowships—a program that continued until 2001 (Conway et al. 2002; van der Burg 2019). In the 1980s, more headway was made toward gender research, and anthropologists became a part of the research staff at CGIAR Centers (van der Burg 2018). In 2014 and 2015, two cohorts of gender post-doctoral fellows were hired to bolster gender research in phase 1 of the CRPs. Each Center now has a gender focal point and, in phase 2 of the CRPs, each of the 12 CRPs has a dedicated gender research coordinator and

51 https://www.cgiar.org/how-we-work/strategy/

Location of CGIAR Centers worldwide

CGIAR Centers

1. Montpellier - France
   CGIAR System Organization
2. Washington, DC - USA
   International Food Policy Research Institute
3. Texcoco - Mexico
   International Maize and Wheat Improvement Center
4. Cali - Colombia
   The Alliance of Bioversity International and the International Center for Tropical Agriculture (CIAT)
5. Lima - Peru
   International Potato Center
6. Abidjan - Côte d'Ivoire
   Africa Rice Center
7. Ibadan - Nigeria
   International Institute of Tropical Agriculture
8. Rome - Italy
   The Alliance of Bioversity International and the International Center for Tropical Agriculture (CIAT)
9. Beirut - Lebanon
   International Center for Agricultural Research in the Dry Areas
10. Co-hosted by Kenya and Ethiopia
    International Livestock Research Institute
11. Nairobi - Kenya
    World Agroforestry (ICRAF)
12. Patancheru - India
    International Crops Research Institute for the Semi-Arid Tropics
13. Colombo - Sri Lanka
    International Water Management Institute
14. Penang - Malaysia
    WorldFish
15. Bogor - Indonesia
    Center for International Forestry Research
16. Los Baños - Philippines
    International Rice Research Institute

Source: https://www.cgiar.org/research/research-centers/

a team comprising gender researchers across different Centers internationally working on the domain of that Research Program.

The Gender and Agriculture Research Network (2012–2016) supported concerted mobilization of CGIAR gender researchers system-wide. The Network focused on institutional reform, including supporting gender teams in developing gender strategies at Research Program and Center levels that would meet the strategy laid out system-wise in 2011. Other points of discussion among gender researchers in the Network were around the integration of gender into phase 2 CRP proposals, gender budgeting of 10 percent system-wide,[52] and making gender research more visible. The set-up of the network was a Systems Office initiative, with the International Center for Tropical Agriculture (CIAT) coordinating communications (including regular webinars, a website, annual meetings, and so on).

Building on the work of the Network, as part of phase 2 of the CRPs, the CGIAR Collaborative Platform for Gender Research[53] (2017–2019) came into operation. The Platform focus was on gender research and, as part of that focus, it convened the first CGIAR-wide gender research conference in 2017 at KIT Royal Tropical Institute, in Amsterdam, the Netherlands, and another in 2018 at the IRRI campus in Addis Ababa, Ethiopia. By 2019, a jointly convened scientific conference with the Australian Center for International Agricultural Research (ACIAR) and the University of Canberra was held in Canberra, Australia. This third scientific conference opened up participation to a much broader network of gender researchers and practitioners in agricultural and environmental domains.

As of January 2020, the latest iteration of a system-wide gender entity came into being—namely, the CGIAR GENDER—Generating Evidence and New Directions for Equitable Results—Platform.[54] This is housed at the International Livestock Research Institute (ILRI) in Nairobi.

---

52 Up to 14 percent of program budgets (CGIAR Consortium 2012; Russell et al. 2015; CGIAR-IEA 2017; comp. Karlsson and Russell 2017, all cited in van der Burg 2019).

53 The Platform was housed within the CRP on Policies, Institutions, and Markets (PIM), coordinated by KIT Royal Tropical Institute, and had an elected Advisory Committee representing gender researchers from other Research Programs and Centers. Regular communications between Center and CRP gender coordinators were facilitated through virtual meetings as well as an annual face-to-face meeting during the scientific conferences. For more on the achievements of the CGIAR Collaborative Platform for Gender Research, see https://gender.cgiar.org/wp-content/uploads/2020/01/SC_KIT_CGIAR_Ewen_4pager_11.pdf

54 See CGIAR GENDER Platform website for ongoing developments: https://gender.cgiar.org/

Annex 2: Timeline of key CGIAR gender developments

TABLE A2.1 Key gender developments in CGIAR (and beyond)

	1970s	1980s	1990s	2000s	2010-2014	2015-2019	2020
Notable CGIAR gender events	Periodic mentions of women, no analysis	1983 first conference on women in rice farming systems (IRRI) 1985 Inter-Center Seminar (ISNAR) on Women and Agricultural technology: relevance for research 1985 Bellagio conference on user perspectives	1999 Participatory Research for Natural Resource Management: Continuing to Learn Together workshop held at the Natural Resources Institute (NRI), Chatham, UK		2014 Annual Gender meeting at CIAT, Cali, Colombia 2016 Gender, Breeding and Genomics meeting (Nairobi) 2017 Innovation workshop on Gender and Breeding (Nairobi)	2015 Annual Gender meeting at IRRI in Los Banos, Philippines 2016 Annual Gender meeting at CIAT, Cali Colombia 2017 First annual CGIAR-wide gender scientific conference at KIT Royal Tropical Institute, Amsterdam 2018 Second annual CGIAR-wide gender scientific conference at ILRI in, Addis Ababa, Ethiopia 2019 Third annual CGIAR-wide gender scientific conference and first one that is jointly convened with non-CGIAR partners (with ACIAR and University of Canberra): Seeds of Change in Canberra, Australia	
CGIAR gender research initiatives	Late 1960s-early 1970s Farming Systems Research – greater focus on economic and social issues and smallholder farming	IFPRI Commercialization of Agriculture studies	1992 Household dynamics (IFPRI) 1996-2003 IFPRI Strengthening Development Policy through Gender and Intra-household Research 1996-2011 Participatory Research and Gender Analysis (PRGA) Program based at CIAT, co-sponsored by CIMMYT, ICARDA and IRRI Women and Technology Initiative (PRGA-lead) IMWI prolific gender and water research	2001-ongoing CGIAR-wide program on Collective Action and Property Rights (CAPRi) 2009-2013 GAAP 1: Gender, Agriculture and Assets Project (phase 1)	2014-2018 GEN-NOVATE (Enabling Gender Equality in Agricultural and Environmental Innovation) 2016 Gender and Breeding Initiative (RTB-lead) WorldFish initiated research on Gender Transformative Approaches in Agri-culture/Aquaculture 2012 Women's Empowerment in Agriculture Index (WEAI) launched WEAI adaptions for livestock (WELI), fish (WEFI)	2016-ongoing GAAP2: Gender, Agriculture and Assets Project (phase 2) 2018-2020 Gender and Big Data Platform collaboration Research grants via CGIAR Collaborative Platform for Gender Research on: • 2017 gender dynamics in seed systems • 2018 feminization of agriculture • 2019 gender dynamics in value chains 2019 European Commission report on Gender Transformative Approaches in Agriculture (Platform lead, drawing on CGIAR experiences) WEAI adaptations developed: an abbreviated version (A-WEAI); for projects (pro-WEAI); and for market inclusion (pro-WEAI+MI)	

continued

	1970s	1980s	1990s	2000s	2010-2014	2015-2019	2020
Gender evaluations, reviews and assessments	Mentions of women but without analysis or action	1980 First CGIAR Review 1981 Second CGIAR Review 1986 Gender-related impacts study paper (Jiggins) for the 1982-1988 CGIAR-wide assessment	1990 Assessment of WIRFS 1990-91 Gender Assessment of CGIAR (Poats) 1995 CGIAR Gender Research Inventory (Feldstein) 1998 CGIAR Gender Research Inventory (Feldstein) 1996 Annual Report	2005 Assessment of the Strengthening Development Policy through Gender and Intra-household Research (referred to as the Gender and Intra-household Program) (Cecile Jackson) 2008 Stripe Review 2008 Independent Review: Bringing together the best of science and the best of development (Elizabeth McAllister Chair) 2000 PRGA review	2010 CGIAR Gender Scoping Study by the International Center for Research on Women (Kauck et al) 2011 Assessment of Participatory Research and Gender Analysis (PRGA)	2016 CGIAR-wide Gender Evaluation (2017 report Baden et al)	
Institutional developments to support gender re-search		1986-1996 Women in Rice Farming Systems WIRFS Network (IRRI)	1990-1996 First CGIAR Gender Program (both gender staffing and gender research) 1999-2011 Gender and Diversity Program (staffing and workplace issues) led by ICRAF Center liasons on gender (set up by the first Gender Program in the early 1990s and inherited by PRGA in 1998)		2011 CGIAR-wide Gender Strategy. 2011-2013 Gender Strategies put in place in CGIAR Research Programs and Centers. 2012-2016 Gender and Agriculture Network coordinated by CIAT	2016 Gender/inclusion an Intermediate Development Outcome in the SRF 2016 Gender strategies in phase 2 proposals 2017-2019 CGIAR Collaborative Platform for Gender Research (housed in the Policies, Institutions, and Markets Research Program -PIM) 2018 Gender representatives appointed in the CGIAR System Council 2015/2016 Gender and Breeding post-doctoral fellows (2 rounds)	2020 GENDER Platform (ILRI)
Gender and development milestones beyond CGIAR	1970 Esther Boserup book Women in Development (WID) paradigm 1975 First UN Women's conference (Mexico City) 1975 UN International Year of Women 1975-1985 UN Decade for Women 1979 CEDAW Convention on the Elimination of Discrimination Against Women	1980 UN World Conference on Women Copenhagen, Denmark 1985 UN World Conference on Women, Nairobi Kenya 1980s Women and Development (WAD) paradigm 1985 Harvard Gender Framework 1986 Gender Relations Framework	1992 UN Earth Summit, Rio de Janeiro, Brazil (Agenda 21) 1995 Fourth UN Conference on Women in Beijing 1995 Gender and Development (GAD) paradigm 1995 Gender mainstreaming begins 1999 Social Relations Framework (Kabeer) Late 1980s-1990s Ecofeminsm gains momentum (Shiva 1988 and others)	2000 Millennium Development Goals (MDGs) MDG 3: "Promote Gender Equality and Empower Women" 2009 start of major agricultural development organizations' publication profiles on gender	2010-2012 Proliferation of agricultural development publications profile on gender	2015 Sustainable Development Goals (SDGs) SDG 5: "Achieve gender equality and empower all women and girls"	

Source: Adapted and further developed from van der Burg (2019).

Annex 3: Coverage of evidence[55]

This book reviews evidence on different themes to take stock of the gender research agenda and craft a new one within agricultural and environmental domains. At the start of the references for each chapter, four visuals provide the reader with an impression of different aspects of the evidence. First, a map illustrates the geography of case studies in the citations, showing where the evidence for the theme in hand comes from. Second, a pie chart shows the institutional origins of the studies reviewed—the proportion from CGIAR research institutes or programs or with inputs from CGIAR researchers, and that from outside of CGIAR. Third, another pie chart shows the research methodologies used in the publications reviewed—whether qualitative, quantitative, mixed-method, methodological, or other. Fourth and finally, a timeline illustrates when the papers reviewed were published.

In this annex, we reflect on the coverage of the evidence used in the thematic chapters, on which chapter discussions and conclusions are based. We compile the data from the separate thematic chapters to provide an impression of the evidence referred to across the book as a whole.

First, we look at the **geographies** covered in the references for each chapter.[56] Map A3.1 presents the compiled data for all thematic book chapters. Unsurprisingly, coverage is uneven globally, with more studies from Asia and East Africa, with several countries standing out as particularly well cited—namely, **India** (92 publications), **Bangladesh** (83), **Kenya** (69), **Uganda** (67), and **Nepal** (57).

In Asia, after India, Bangladesh, and Nepal, other highly cited countries are Indonesia (34), Viet Nam (30), Syria (22), and the Philippines (21). In Africa, after Kenya and Uganda, other highly studied geographies include Ethiopia (50), Tanzania (50), Ghana (44), Burkina Faso (37), Zambia (33), Malawi (32), Nigeria (25), and South Africa (25). Latin America has much less coverage overall, likely because of reasons related to language: the chapters are written in English, which may have resulted in Spanish language publications being cited less. It may also be explained in part by the decline in development and research funding in Latin America in recent years. That said, the most cited Latin American references are from Bolivia (18), Nicaragua (18), Peru (18), Ecuador

55 Many thanks to Laura Avila, who acted as a research assistant on this section, reviewing all book references and providing the background information needed for the visuals.

56 To determine the countries covered, the abstracts were reviewed, along with tables of contents (where applicable) and any case studies included in the publication. All countries mentioned were noted. Where a region was mentioned rather than specific countries, all countries in that region were noted.

MAP A3.1 Geographical coverage of all studies cited in thematic chapters

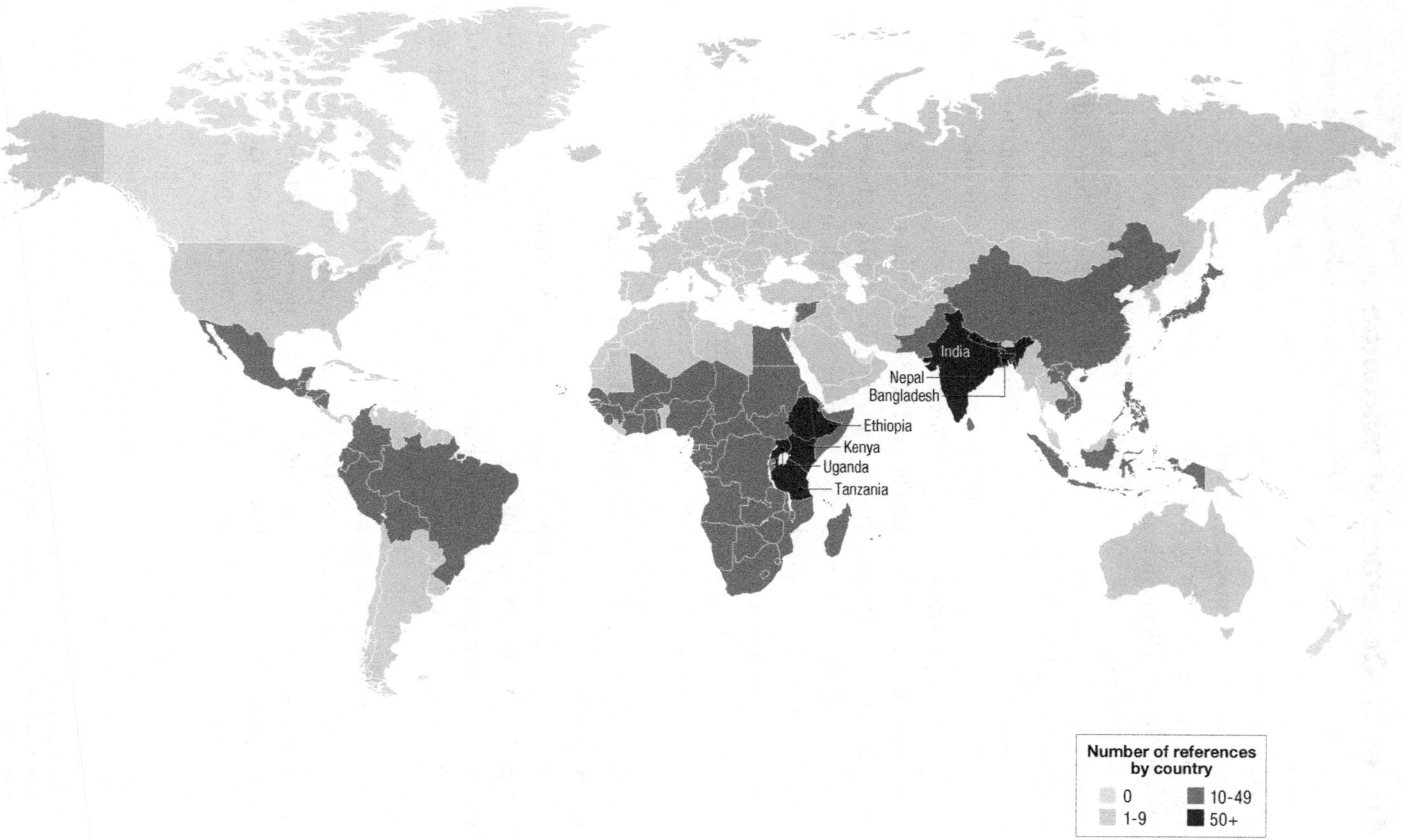

(16), and Guatemala (14). The geographic prevalence of studies cited may loosely correspond with either having a CGIAR Center based there[57] (e.g. Kenya, India, Ethiopia, Bangladesh) or being a priority country for CGIAR research or donor investments. It may also reflect data limitations, with more studies possible in countries where the sex-disaggregated data exists.

While useful and telling as an overview, the compiled map (Map A3.1) blurs significant **variation in geographical referencing** across the thematic chapters. For example, while India is the most cited geography overall and Kenya the third most cited, Chapter 5 (nutrition-sensitive agriculture) has only one publication referring to each. While Tanzania is well cited in most chapters, again Chapter 5 has only one publication from that country, as do Chapter 8 (feminization of agriculture) and Chapter 9 (assessing women's empowerment). Most chapters have multiple publications from Burkina Faso, but Chapter 8 has just one. Likewise, all chapters have several publications from Ethiopia, with the exception of Chapter 6 (natural resources) and Chapters 7 (climate adaptation and mitigation), which both have just one citation each. Finally, all chapters have multiple references on Ghana, with the exception of Chapter 10 (gender transformative approaches), which has just one citation. The compiled map does not capture the differences between the citing of geographies by theme, but this does come out clearly in the chapter-specific maps.

When we compare the chapter-specific maps, it becomes clear that **some themes rely heavily on evidence from specific countries.** For example, a striking 24 of the citations in Chapter 4 (value chains) are from Kenya, and 12 in Chapter 7 (climate adaptation and mitigation) are from that country. Tanzania is the most cited in Chapter 3 (seed systems), with 12 citations, in Chapter 4, with 13 citations, and in Chapter 7, with 10 citations. Uganda, Nepal, and Bangladesh are more evenly cited throughout. While all chapters cite Bangladesh, Chapter 5 (nutrition-sensitive agriculture) and Chapter 9 (assessing women's empowerment) stand out with 17 citations each. Uganda is the most cited in Chapter 4 (12 citations) and Chapter 7 (13 citations), followed by Chapter 2 (breeding), with 8 citations. Nepal is the most cited in Chapter 7 (12 citations). This makes sense as some regions are the focus of multiple studies on specific topics or are more amenable to research on a specific theme. Kenya

57 CGIAR headquarters are based in the following southern countries: Colombia, Côte d'Ivoire, Indonesia, Kenya, Malaysia, Mexico, Nigeria, Peru, the Philippines, and Sri Lanka (see also Map A1.1 in Annex 1). The International Livestock Research Institute also has a campus in Ethiopia, as does WorldFish in Bangladesh. Given the political situation in Lebanon, staff of the International Center for Agricultural Research in the Dry Areas are dispersed internationally.

has for a long time been the location for value chain studies, for example, as the citations for Chapter 4 clearly illustrate.

Striking in comparing the chapter-specific maps is the fact that, while a region may not have abundant citations across all themes of this book, some **theme-specific geographies** emerge. For example, while less cited overall, several clusters of citations from Latin America are noteworthy, including 7 publications from Bolivia on natural resources (Chapter 6); 6 publications from Guatemala on nutrition-sensitive agriculture (Chapter 5); 5 publications from Nicaragua on each of natural resources and climate adaptation and mitigation (Chapters 6 and Chapter 7); and 6 publications from Peru on seed systems (Chapter 3). Likewise, several clusters from Africa stand out, including 4 publications from Burundi on seed systems (Chapter 3); 5 publications on climate adaptation and mitigation from Cameroon and 6 from Senegal (Chapter 7); 10 publications related to gender transformative approaches from Zambia (Chapter 10) and 6 on climate change adaptation and mitigation (Chapter 7); 6 publications from Malawi on breeding (Chapter 2), 7 on seed systems (Chapter 3), and 7 on climate change adaption and mitigation (Chapter 7); 5 publications from Mozambique on value chains (Chapter 4); 10 publications about Nigeria on breeding (Chapter 2); and, from South Africa, 7 on each of value chains (Chapter 4) and climate change adaptation and mitigation (Chapter 7). From Asia, a notable anomaly are the 9 citations from Syria on seed systems (Chapter 3). The strong representation of some countries related to specific themes is noteworthy. In considering the evidence, this variation—the regional specificities—needs to be weighed against more generalizable trends in drawing conclusions.

A second visual in each chapter's references is a pie chart that shows the number of publications reviewed by **institutional origin**: CGIAR versus non-CGIAR. Writing teams reviewed both CGIAR research, from one or more of the 15 CGIAR Centers, and key studies carried out and/or published elsewhere. A publication was considered to be CGIAR if (a) one of the co-authors was based at a CGIAR Center at the time of publication; (b) the publication was funded by a CGIAR Center or Program; or (c) it was published by CGIAR. Figure A3.1 illustrates the institutional origins of evidence and literature reviewed in this book. A narrow majority of the publications are from CGIAR, which is logical, given the purpose of this book: to look at CGIAR gender research using a gender equality and women's empowerment lens.

FIGURE A3.1 Institutional origins (N= 970)

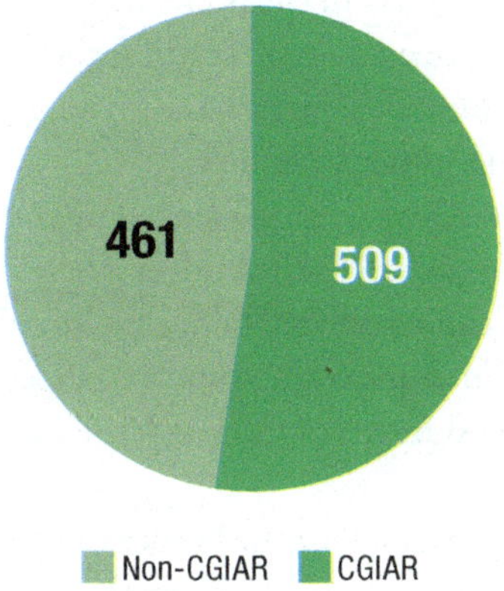

While Figure A3.1 illustrates the institutional origins of publications across the book as a whole, variations come out in the chapters, with some relying more on CGIAR research than others. Chapter teams were asked to begin their reviews with wide outreach across CGIAR and, as such, the visibility of CGIAR publications is not surprising and was, in fact, purposive. Literature reviews were not intended to be exhaustive but rather a (re-)examination of evidence on key CGIAR gender research themes using a gender equality and women's empowerment lens. Notably, 80 percent of the citations (71/89) in Chapter 8 (feminization of agriculture) draw on evidence generated by CGIAR, 74 percent (42/57) in Chapter 5 (nutrition-sensitive agriculture), and 70 percent (87/125) in Chapter 7 (climate change adaptation and mitigation). Chapter 2 (breeding) follows with 56 percent (43/77) of citations from CGIAR publications. The other chapters have between 35 and 48 percent. This speaks to the methodologies used for the literature reviews, the evidence available within CGIAR and externally on each theme, and the maturity of the theme—that is to say, how long it has been studied. In some domains, CGIAR emerges as a clear thought leader; for others, it is one contributor among many.

A third category considered in relation to evaluating the evidence is the **research methodology** of the studies cited. References were reviewed to determine the research methods used or the nature of the publication. The following categories were used: qualitative, quantitative, mixed-method, methodological, and "other."[58] To determine the methodology used, the abstracts,

58 The category "methodological" refers to publications that deal with methodological issues or approaches. The "other" category includes guidance documents, presentations, reviews, indices, analytical frameworks, brochures, manuals, webinars, PowerPoint presentations, interviews, and documents that do not fall into the other categories distinguished.

introductions, and methodology sections were reviewed and cross-checked. As Figure A3.2 shows, for all chapters in the book, more than half of the publications (56 percent) reviewed are qualitative studies; about 15 percent are quantitative; 20 percent use mixed methods; just a handful are methodological; and less than 10 percent fall into the "other" category.

The kind of research undertaken is influenced by the research questions and gender researchers involved, as well as the budget available. What is interesting is that about 35 percent of the studies are quantitative or mixed-method (which includes both quantitative and qualitative) and 76 percent are qualitative or mixed-method. Our review shows that the gender research cited is largely qualitative.

Figure A3.2 is a compiled visual for the book as a whole; the research methodology for publications reviewed for each chapter are illustrated in a similar visual to be found at the start of the reference sections. Interesting to note is the variation across chapters. Overwhelmingly, the chapters rely on qualitative publications, as already discussed. However, there are exceptions. For example, Chapter 5 (nutrition-sensitive agriculture) and Chapter 8 (feminization of agriculture) refer more to quantitative publications than to qualitative work, with 68 and 66 percent (quantitative and mixed-method studies combined) respectively. From this light analysis, we cannot see why this would be the case, but it would be interesting to explore: are these themes more amenable to quantitative research or is there another explanation, for example the literature review methodology?

FIGURE A3.2 Number of cited studies by research methodology (N= 970)

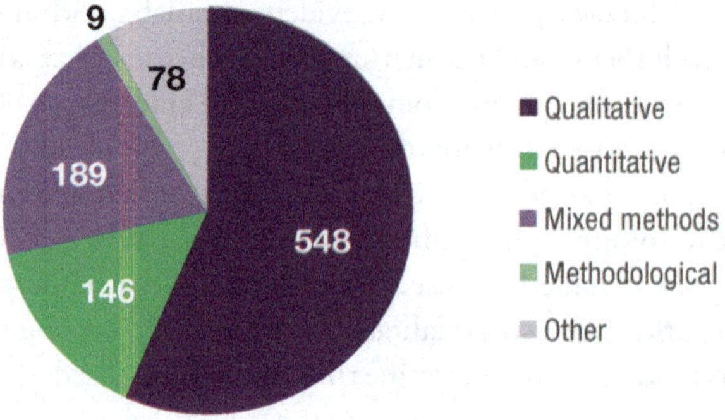

Chapter 4 (value chains) has the highest number of mixed-method studies cited (43) and the highest proportion of mixed-method studies overall (32 percent), followed by Chapter 7 (climate adaptation and mitigation), with 29 mixed-method publications, comprising 23 percent of all mixed-method study references. Proportionally Chapter 5 (nutrition-sensitive agriculture) and Chapter 9 (feminization of agriculture) both have mixed-method studies at 26 percent, with Chapter 2 (breeding) following closely behind with 21 percent. That such a limited proportion of studies use mixed methods—at best between 21 and 32 percent—is relevant for CGIAR research because it speaks to ongoing discussions as to the value of interdisciplinary approaches and mixed methods. While gender researchers advocate for more mixed-method research, the studies available seem to be limited for most themes explored in this book. The low number of mixed-method studies may also be because CGIAR research is most often published in journals, with limited word counts and specific audiences. Accordingly, researchers may send a quantitative paper to one journal and a qualitative paper to another, thus hiding the fact that papers are part of a mixed-method study.

Some thematic domains may be more or less amenable to mixed-method studies. The review of the references in this book shows a lower proportion of mixed-method studies in Chapter 3 (seed systems), at 16 percent, Chapter 6 (natural resources), at 14 percent, Chapter 10 (gender transformative approaches), at 13 percent, and, perhaps most surprisingly, Chapter 9 (assessing women's empowerment), at just 6 percent. While this may make sense for some subjects—for example gender transformative approaches (Chapter 10), with its focus on qualitative issues like gender norms—it is more surprising for publications on the assessment of women's empowerment (Chapter 9), where one might expect the use of mixed methods. A partial explanation for this may be that Chapter 9 refers mostly to tools and measures rather than studies: tools and measures tend to use one methodology rather than a mix. There is an opening here to further explore opportunities for more mixed-method studies and why (or why not) they are being undertaken.

What we do not see from our review of the citations is the size and scale of the studies. For some qualitative research, particularly when it is exploratory, the size of the sample may be of less importance than what the sample offers in terms of insights. For example, earlier studies on the feminization of agriculture or on gender and value chains tended to be smaller in scale, seeking nuance and testing concepts. Generally, there seems to be a move toward larger-scale studies, as exemplified by, for instance, the work on nutrition-sensitive agriculture

(Chapter 5). While Chapter 5 cites more quantitative studies, this shift toward larger studies can also be seen for qualitative research. GENNOVATE—a large-scale qualitative research program running from 2014 to 2018 looking at gender norms and innovation—is perhaps the most striking example of this in recent years. GENNOVATE brought attention to the importance of gender norms research in getting at the less visible aspects of gender inequality and in stimulating normative change (e.g., via gender transformative approaches) (Badstue et al. 2020). GENNOVATE research teams worked with more than 7,500 women and men from 137 communities in 26 countries in Africa, Asia, and Latin America, through both focus group discussions and individual interviews (Petesch et al. 2018). Smaller examples are the clustering of thematic studies to explore broader issues, like grants offered through the CGIAR Collaborative Platform for Gender Research: 2017–2019 on gender dynamics in seed systems; 2018–2020 on the feminization of agriculture; and 2019–2021 on gender dynamics in value chain development, beyond the production node.

Finally, we consider the **timelines of cited studies** reviewed. The timelines visually represent the publication periods for the citations in each chapter and are compiled for all chapters in one graph (Figure A3.3). Figure A3.3 also illustrates the total number of publications reviewed per chapter.

For all chapters, unsurprisingly, a significant proportion of the citations are from the past 10 years (the periods 2011–2015 and 2016–2021). However, some themes have a longer history within CGIAR, with the work on gender and natural resources (Chapter 6) standing out for its citation of work back before 1990. Other themes have emerged more recently (e.g., climate adaptation and mitigation, Chapter 7). Some themes, while more recent, draw significantly from past research, for example that of gender dynamics in seed systems (Chapter 3) and that of gender transformative approaches (Chapter 10).

FIGURE A3.3 Compiled timeline of references cited in this book per chapter

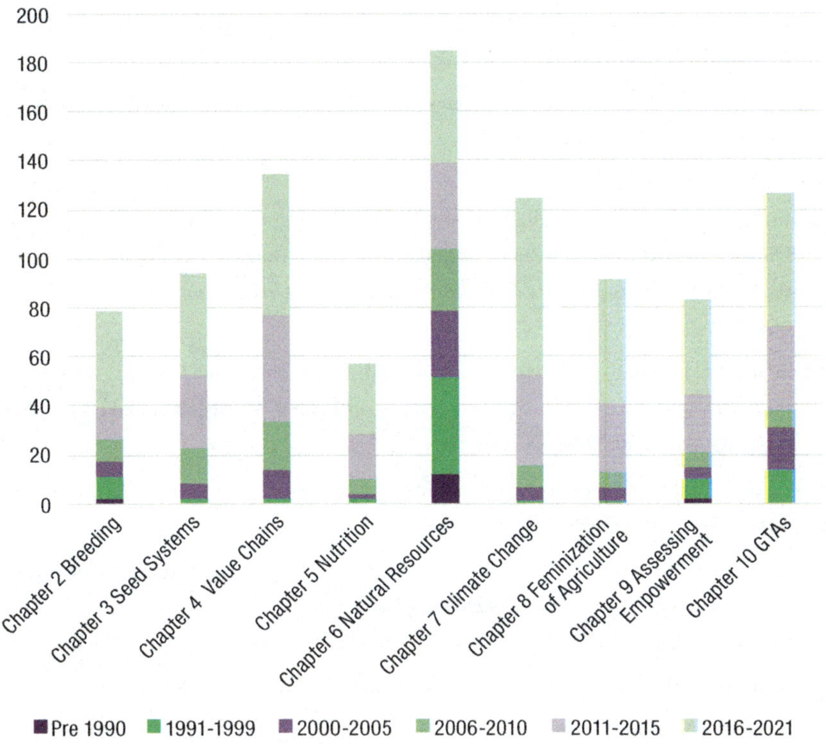

References

Anderson, J.R., R.W. Herdt, and G.M. Scobie. 1988. "Science and Food: The CGIAR and Its Partners." *Agricultural Systems* 31 (3): 315–316.

Baden, S., L. Brown, D. Merrill-Sands, R. Percy, and F. Coccia. 2017. *Evaluation of Gender in CGIAR – Volume I, Evaluation of Gender in Research.* Rome: Independent Evaluation Arrangement of CGIAR.

Badstue, L., M. Elias, V. Kommerell, P. Petesch, G. Prain, R. Pyburn, and A. Umantseva. 2020. "Making Room for Manoeuvre: Addressing Gender Norms to Strengthen the Enabling Environment for Agricultural Innovation." *Development in Practice* 30 (4): 541–547.

Batliwala, S. 1993. "Empowerment of Women in South Asia: Concept and Practice." Report for Asian-South Pacific Bureau of Adult Education and FAO's Freedom from Hunger campaign/Action for Development.

———. 2007. "Taking the Power Out of Empowerment: Experiential Account." *Development in Practice* 17 (4–5): 557–565.

Batliwala, S., and D. Dhanraj. 2007. "Gender Myths that Instrumentalize Women: A View from the Indian Frontline." In *Feminisms in Development: Contradictions, Contestations and Challenges*, edited by A. Cornwall, E. Harrison, and A. Whitehead, 21–34. London: Zed Books.

Boserup, E. 1970. *Woman's Role in Economic Development.* London: Allen & Unwin.

Bustelo, M., L. Ferguson, and M. Forest. 2016. *The Politics of Feminist Knowledge Transfer: Exploring Gender Training and Gender Expertise.* New York: Palgrave MacMillan.

CGIAR. 1981. "Second Review of the CGIAR." Report of the Review Committee.

———. 2015. "Strategy and Results Framework 2016-2025." Final draft for consultation.

CGIAR Consortium Board. 2011. "Consortium-Level Gender Strategy." Montpellier, November.

CGIAR Independent Review Panel. 2008. "Bringing Together the Best of Science and the Best of Development." Independent Review of the CGIAR System. Report to the Executive Council. Washington, DC., chaired by Elizabeth McAllister, November.

CGIAR Research Program on Fish Agri-Food Systems. 2017. "CGIAR Research Program on Fish Agri-Food Systems (FISH): Gender Strategy." Penang: CGIAR Research Program on Fish Agri-Food Systems.

———. 2020. *FISH Gender Integration Guidelines.* Penang: CGIAR Research Program on Fish Agri-Food Systems.

CGIAR Science Council 2007. *Report of the First External Review of the Systemwide Program on Participatory Research and Gender Analysis (PRGA).* Rome: Science Council Secretariat.

———. 2009. *Stripe Review of Social Sciences in the CGIAR*. Rome: Science Council Secretariat.

Chant, S., and C. Sweetman. 2012. "Fixing Women or Fixing the World? 'Smart Economics', Efficiency Approaches, and Gender Equality in Development." *Gender & Development* 20 (3): 517–529.

Chater, S., and V. Carangal. 1996. *On Farmers' Fields: Portrait of a Network*. Los Baños: International Rice Research Institute (IRRI).

Cole, S.M., P. Kantor, S. Sarapura, and S. Rajaratnam. 2014. "Gender-Transformative Approaches to Address Inequalities in Food, Nutrition and Economic Outcomes in Aquatic Agricultural Systems." Working Paper AAS-2014-42, CGIAR Research Program on Fish Agri-Food Systems, Penang.

Cole, S.M., R. Puskur, S. Rajaratnam, and F. Zulu. 2015. "Exploring the Intricate Relationship between Poverty, Gender Inequality, and Rural Masculinity: A Case Study from an Aquatic Agricultural System in Zambia." *Culture, Society and Masculinities* 7 (2): 154–170.

Colfer, C.J.P., B. Basnett, and M. Ihalainen. 2018. "Making Sense of 'Intersectionality': A Manual for Lovers of People and Forests." Occasional Paper 184. Bogor: Center for International Forestry Research.

Conway, G., A. Adesina, J. Lynam, and J. Moock. 2002. "The Rockefeller Foundation and Social Research in Agriculture." In *Researching the Culture in Agriculture: Social Research for International Development*, edited by M. Cernea and A. Kassam, 373–381. Wallingford: CABI.

Cornwall, A. 2018. "Beyond 'Empowerment Lite': Women's Empowerment, Neoliberal Development and Global Justice." *cadernos pagu* 52: e185202.

Cornwall, A., and A.M. Rivas. 2015. "From 'Gender Equality' and 'Women's Empowerment' to Global Justice: Reclaiming a Transformative Agenda for Gender and Development." *Third World Quarterly* 36 (2): 396–415.

Cornwall, A., and J. Edwards. 2014. *Feminisms, Empowerment and Development: Changing Women's Lives*. London: Zed Books.

Cornwall, A., E. Harrison, and A. Whitehead. 2007a. *Feminisms in Development: Contradictions, Contestations and Challenges*. London/New York: Zed Books.

———. 2007b. "Gender Myths and Feminist Fables: The Struggle for Interpretive Power in Gender and Development." *Development and Change* 38 (1): 1–20.

Danielsen, K., F. Wong, D. McLachlin, and S. Sarapura. 2018. *Typologies of Change: Gender Integration in Agriculture and Food Security Research*. Amsterdam: KIT.

Davids, T., and A. van Eerdewijk. 2016. "The Smothering of Feminist Knowledge: Gender Mainstreaming Articulated through Neoliberal Governmentalities." In *The Politics of Feminist Knowledge Transfer: Exploring Gender Training and Gender Expertise*, edited by M. Bustelo, L. Ferguson, and M. Forest, 80–95. New York: Palgrave MacMillan.

Do Mar Pereira, M. 2012. "'Feminist Knowledge Is Proper Knowledge, But?...' The Status of Feminist Scholarship in the Academy." *Feminist Theory* 13 (3): 283–303.

Doss, C., R. Meinzen-Dick, A. Quisumbing, and S. Thies. 2018. "Women in Agriculture: Four Myths." *Global Food Security* 16: 69–74.

Doss, C., and C. Kieran. 2014. *Standards for Collecting Sex-disaggregated Data for Gender Analysis: A Guide for CGIAR Researchers*. January 1. https://hdl.handle.net/10568/76974.

Edström, J., A. Hassink, T. Shahrokh, and E. Stern. 2015. *Engendering Men: A Collaborative Review of Evidence on Men and Boys in Social Change and Gender Equality*. Brighton, UK: EMERGE Evidence Review, Promundo-US, Sonke Gender Justice, and Institute of Development Studies.

Elson, D. 1991. *Male Bias in the Development Process*. Manchester/New York: Manchester University Press.

Escobar, A. 2018. "An Exchange on the Essay Feminism and Revolution: Looking Back, Looking Ahead." Contribution to GTI Roundtable on Feminism and Revolution, June.

Eyben, R. 2010. "Subversively Accommodating: Feminist Bureaucrats and Gender Mainstreaming." *IDS Bulletin* 41 (2): 54–61.

Eyben, R. 2013. "Uncovering the Politics of 'Evidence' and 'Results'. A Framing Paper for Development Practitioners." Politics of Evidence Conference, Brighton, April 23–24.

FAO (Food and Agricultural Organization of the United Nations). 2011. *The State of Food and Agriculture 2010–2011: Women in Agriculture—Closing the Gender Gap for Development*. Rome.

FAO, IFAD (International Fund for Agricultural Development), and ILO (International Labour Organization). 2010. *Gender Dimensions of Agricultural and Rural Employment: Differentiated Pathways Out of Poverty: Status, Trends and Gaps*. Rome.

Farnworth, C.R., L. Badstue, and S.M. Cole. 2020. "Engaging Men in Gender-Equitable Practices in Maize Systems of Sub-Saharan Africa." GENNOVATE Resources for Scientists and Research Teams. Mexico City: CIMMYT.

Feldstein, H.S. 1995. "Inventory of Gender-Related Research and Training in the International Agricultural Research Centers 1990-95." CGIAR Gender Program Working Paper 8, CGIAR.

———. 1998. "An Inventory of Gender-Related Research and Training in the Consultative Group on International Agricultural Research (CGIAR) Centers 1996-1998." Cali, Colombia: CGIAR.

Feldstein, H. S., and J. Jiggins. 1994. *Tools for the Field: Methodologies Handbook for Gender Analysis in Agriculture.* London: Intermediate Technology Publications.

Feldstein, H. S., and S. Poats. 1989a. *Working Together—Gender Analysis in Agriculture, Volume 1: Case Studies.* Bloomfield, CT: Kumarian Press.

———. 1989b. *Working Together—Gender Analysis in Agriculture, Volume 2: Teaching Notes.* Bloomfield, CT: Kumarian Press.

Ferguson, J. 2009. "The Uses of Neoliberalism." *Antipode* 41 (S1): 166–184.

Firetail. 2014. "Gender, Agriculture, and Assets Project." End of Project Evaluation, March 27.

Gibbs, C. 2008. "Draft Meta-Analysis of the External Program and Management Reviews." Input to the CGIAR Independent Panel Review.

Gomez, J. 2009. "CGIAR Systemwide Program on Participatory Research and Gender Analysis." Cali: Center for Tropical Agriculture.

Goetz, A.-M. 1998. *Getting Institutions Right for Women in Development.* London: Zed Books.

Harcourt, W. 2016. *Palgrave Handbook on Gender and Development: Critical Engagements in Feminist Theory and Practice.* New York: Palgrave MacMillan.

Hillenbrand, E., N. Karim, P. Mohanraj, and D. Wu. 2015. *Measuring Gender-Transformative Change: A Review of Literature and Promising Practices.* Washington, DC: CARE USA.

IGWG (Interagency Gender Working Group). 2017. *Gender Integration Continuum 2017.* Washington, DC: United States Agency for International Development.

Jiggins, J. 1986. "Gender-Related Impacts and the Work of the International Agricultural Research Centers." *CGIAR Study Paper* 17. Washington, DC: World Bank.

Johnson, N., C. Kovarik, R. Meinzen-Dick, J. Njuki, and A.R. Quisumbing. 2016. "Gender, Assets, and Agricultural Development: Lessons from Eight Projects." *World Development* 83: 295–311.

Johnson N., M. Balagamwala, C. Pinkstaff, S. Theis, R. Meinzen-Dick, and A. Quisumbing. 2018. "How Do Agricultural Development Projects Empower Women? Linking Strategies with Expected Outcomes." *Journal of Gender, Agriculture and Food Security* 3 (2): 1–19.

Kabeer, N. 1994. *Reversed Realities: Gender Hierarchies in Development Thought.* London/New York: Verso.

Kabeer, N. 1999. "Resources, Agency, Achievements: Reflections on the Measurement of Women's Empowerment." *Development and Change* 30 (3): 435–464.

Kabeer, N. 2017. "Economic Pathways to Women's Empowerment and Active Citizenship: What Does the Evidence from Bangladesh Tell Us?" *Journal of Development Studies* 53 (5): 649–663.

Kleiber, D., P. Cohen, C. Gomese, and C. McDougall. 2019. "Gender-Integrated Research for Development in Pacific Coastal Fisheries." Program Brief: FISH-2019-02. Penang: CGIAR Research Program on Fish Agri-Food Systems.

Kauck, D., S. Paruzzolo and J. Schulte (2010). *CGIAR Gender Scoping Study*. Washington, DC: International Center for Research on Women (ICRW).

Levtov, R., G. Barker, M. Contreras-Urbina, B. Heilman, and R. Verma. 2014. "Pathways to Gender-Equitable Men: Findings from the International Men and Gender Equality Survey in Eight Countries." *Men and Masculinities* 17 (5): 467-501.

Lombardini, S., and K. McCollum. 2018. "Using Internal Evaluations to Measure Organisational Impact: A Meta-Analysis of Oxfam's Women's Empowerment Projects." *Journal of Development Effectiveness* 10 (1): 145–170.

Lombardini S., K. Bowman, and R. Garwood. 2017. "A 'How To' Guide for Measuring Women's Empowerment: Sharing Experiences from Oxfam's Impact Evaluations." Methodological Paper. Oxford: Oxfam.

Meinzen-Dick, R., C. Kovarik, and A.R. Quisumbing. 2014a. "Gender and Sustainability." *Annual Review of Environmental Resources* 39: 29–55.

Meinzen-Dick, R., N. Johnson, A.R. Quisumbing, J. Njuki, J.A. Behrman, D. Rubin, A. Peterman, and E. Waithanji. 2014b. "The Gender Asset Gap and Its Implications for Agricultural and Rural Development." In *Gender in Agriculture: Closing the Knowledge Gap*, edited by A.R. Quisumbing, R. Meinzen-Dick, T.L. Raney, A. Croppenstedt, J.A. Behrman, and A. Peterman, 91–115. Amsterdam: Springer.

Miller, C., and S. Razavi. 1995. "From WID to GAD: Conceptual Shifts in the Women and Development Discourse." Occasional Paper 1. Geneva: UNRISD.

Morgan, M. 2014. "Measuring Gender Transformative Change." Program Brief. Penang: CGIAR Research Program on Aquatic Agricultural Systems.

Moser, C. 1993. *Gender Planning and Development: Theory, Practice and Training*. New York/ London: Routledge.

Moser, C., and C. Levy. 1986. "A Theory and Method of Gender Planning - Meeting Women's Practical and Strategic Needs." DPU Gender and Planning Working Paper 11, Development Planning Unit, London.

Moser C., and A. Moser. 2005. "Gender Mainstreaming since Beijing: A Review of Success and Limitations in International Institutions." *Gender and Development* 13 (2): 11–22.

Mukhopadhyay, M. 2007. "Mainstreaming Gender or 'Streaming' Gender Away: Feminists Marooned in the Development Business." In *Feminisms in Development: Contradictions, Contestations and Challenges*, edited by A. Cornwall, E. Harrison, and A. Whitehead, 135–149. London: Zed Books.

Mukhopadhyay, M. 2016. "Mainstreaming Gender or 'Streaming' Gender Away: Feminists Marooned in the Development Business." In *Palgrave Handbook on Gender and Development: Critical Engagements in Feminist Theory and Practice*, edited by W. Harcourt, 77–91. New York: Palgrave MacMillan.

Newton, J., A. van Eerdewijk, and F. Wong. 2019. "What Do Participatory Approaches Have to Offer the Measurement of Empowerment of Women and Girls." Working Paper 2019-1, KIT, Amsterdam.

Nazneen, S., A. Darkwah, and M. Sultan. 2014. "Researching Women's Empowerment: Reflections on Methodology by Southern Feminists." *Women's Studies International Forum* 45: 55–62.

North, D. 1990. *Institutions, Institutional Change and Economic Performance.* Cambridge/New York: Cambridge University Press.

Njuki, J., E. Waithanji, B. Sakwa, J. Kariuki, E. Mukewa, and J. Ngige. 2014. "A Qualitative Assessment of Gender and Irrigation Technology in Kenya and Tanzania." *Gender Technology and Development* 18 (3): 303–340.

Okali, C. 2012. "Gender Analysis: Engaging with Rural Development and Agricultural Policy Processes." Working Paper 26. Future Agricultures Consortium, Brighton, UK.

———. 2017. "The Relevance and Contribution of CGIAR Gender Research (2011-16)." In *Evaluation of Gender in CGIAR*, by S. Baden, L. Brown, D. Merrill-Sands, R. Percy, and F. Coccia. Annex F Issue Paper. Rome: Independent Evaluation Arrangement of CGIAR.

Paris, T.R. 1989. "Philippines: Women in Rice Farming Systems, Crop-Livestock Project, Sta. Barbara Pangasinan." In *Working Together – Gender Analysis in Agriculture, Volume 1: Case Studies*, edited by H. Sims Feldstein and S. Poats, Chapter 8. Bloomfield, CT: Kumarian Press.

Petesch, P., L. Badstue, L. Camfield, S. Feldman, G. Prain, and P. Kantor. 2018. "Qualitative, Comparative and Collaborative Research at Large Scale: The GENNOVATE Field Methodology." *Journal of Gender, Agriculture and Food Security* 3 (1): 28–53.

Prain, G., H. Hambly, M. Jones, W. Leppan, and L. Navarro. 2000. "CGIAR Program on Participatory Research and Gender Analysis." CGIAR Internally Commissioned External Review, December.

Quisumbing, A.R., D. Rubin, C. Manfre, E. Waithanji, M. van den Bold, D. Olney, N. Johnson, et al. 2015. "Gender, Assets, and Market-Oriented Agriculture: Learning from High-Value Crop and Livestock Projects in Africa and Asia." *Agriculture and Human Values* 32: 705–725.

Rao, A., and D. Kelleher. 2005. "Is There Life after Gender Mainstreaming?" *Gender & Development* 13 (2): 57–69.

Rathgeber, E. 1990. "WID, WAD, GAD: Trends in Research and Practice." *Journal of Developing Areas* 24 (4): 489–502.

RF (The Rockefeller Foundation) and ISNAR (International Service for National Agricultural Research). 1985. "Report of a Seminar on Women and Agricultural Technology: Relevance for Research. Volume 1 – Analyses and Conclusions." Report from the CGIAR Inter-Center Seminar on Women and Agricultural Technology, Bellagio, March 25–29, 1985.

Sachs, C.E. 1996. *Gendered Fields: Rural Women, Agriculture and Environment.* Rural Studies Series of the Rural Sociological Society. Boulder, CO: Westview Press.

Sachs, C.E. 2019. *Gender, Agriculture and Agrarian Transformations: Changing Relations in Africa, Latin America and Asia.* New York: Routledge.

Sen, A. 1990. "Gender and Cooperative Conflicts." In *Persistent Inequalities: Women and World Development*, edited by I. Tinker, 123–149. New York: Oxford University Press.

Tamale, S. 2020. *Decolonization and Afro-feminism.* Ottawa: Daraja Press.

Tangka, F.K., M.A. Jabbar, and B.I. Shapiro. 2000. "Gender Roles and Child Nutrition in Livestock Production Systems in Developing Countries: A Critical Review." Socioeconomics and Policy Research Working Paper 27. International Livestock Research Institute, Nairobi.

Van der Burg, M. 2018. "'Change in the Making': Building on the Past Gender Trends in CGIAR Agricultural Research." International Livestock Research Institute Conference, Addis Ababa, September 27.

———. 2019. "'Change in the Making': 1970s and 1980s Building Stones to Gender Integration in CGIAR Agricultural Research." In *Gender, Agriculture and Agrarian Transformations: Changing Relations in Africa, Latin America and Asia*, edited by C.E. Sachs, 35–57. New York: Routledge.

Van Eerdewijk, A. 2016. "Gender Mainstreaming: Views of a Post-Beijing Feminist." In *Palgrave Handbook on Gender and Development: Critical Engagements in Feminist Theory and Practice*, edited by W. Harcourt, 117–131. New York: Palgrave MacMillan.

Van Eerdewijk, A., and T. Davids. 2014. "Escaping the Mythical Beast: Gender Mainstreaming Reconceptualised." *Journal of International Development* 26 (3): 303–316.

Van Eerdewijk, A., F. Wong, C. Vaast, J. Newton, M. Tyszler, and A. Pennington. 2017. *White Paper: A Conceptual Model of Women and Girls' Empowerment.* Amsterdam: KIT.

Wong, F., A. Vos, R. Pyburn, and J. Newton. 2019. "Implementing Gender Transformative Approaches in Agriculture." Discussion Paper for the European Commission. Amsterdam: CGIAR Collaborative Platform for Gender Research.

World Bank. 2012. *World Development Report 2012: Gender Equality and Development.* Washington, DC.

World Bank and IFPRI (International Food Policy Research Institute). 2010. *Gender and Governance in Rural Services: Insights from India, Ghana and Ethiopia.* Washington, DC: World Bank.

World Bank, FAO, and IFAD (International Fund for Agricultural Development). 2009. *Gender in Agriculture: Sourcebook.* Washington, DC: World Bank.

Yuval-Davis, N. 2006. "Intersectionality and Feminist Politics." *European Journal of Women's Studies* 13 (3): 193–209.

Chapter 2

EXAMINING CHOICE TO ADVANCE GENDER EQUALITY IN BREEDING RESEARCH

Vivian Polar, Rohini Ram Mohan, Cynthia McDougall, Béla Teeken, Annet Abenakyo Mulema, Pricilla Marimo, and Jummai Othniel Yila

Breeding is a technical pillar of CGIAR research: the animal/fish breeds, and plant varieties developed are international public goods that contribute to agricultural development for low-income contexts worldwide. Gender equality and women's empowerment are critical social dimensions underpinning agricultural development in these contexts. Progressing toward gender equality in agriculture requires that women, as well as men, have equal capabilities to make decisions about agricultural innovation, and specifically technology choice. Current evidence, however, suggests the situation here is not yet equal. Nevertheless, despite ongoing inequalities, there is a dearth of literature on the connection between gender and breeding in agricultural research.

This chapter critically examines what has been done to address gender dynamics in (current) breeding structures and processes, and what more can be done so that breeding programs contribute to advancing gender equality. We are specifically concerned with **technology choices in relation to the plant variety or animal/fish breed** by resource-poor smallholders in low-income countries. The chapter explores how CGIAR and public breeding programs generate options based on user needs, preferences, and constraints, and the institutional requirements needed to develop them in such a way that they contribute to gender equality and women's empowerment.

We begin with a discussion on why gender matters for breeding. Here, we also position this chapter within a technology, society, and gender frame, and briefly look at how the incorporation of gender dimensions in breeding processes have evolved, starting with the Green Revolution.

We next introduce our conceptualization of how breeding relates to gender equality, emphasizing, in particular, technology options and the power to choose. We deepen this by presenting an impact pathway that can be used to identify intervention and information gaps, and potential new areas of research that can foster a better understanding of the linkages and relations between breeding for agricultural development and gender equality.

We go on to review a process of incorporating gender dimensions into different stages in the breeding cycle, looking at *when* and at what stages of the breeding cycle we need to take gender into account, and *how* this can concretely be done. This allows us to delve into examples of how technology options relate to gender equality by contributing to the generation of "real choices" that fit the needs, preferences, opportunities, and constraints of men and women.

Finally, we look forward, laying out new research opportunities and challenges in advancing breeding toward gender equality. We focus specifically on how to ensure the inclusion of gender equality dimensions at the level of breeding objectives. We also further articulate how monitoring, evaluation, and learning can strengthen feedback loops in the breeding cycle, and how this can be conducive to the integration of gender dimensions.

Why gender matters in breeding

Agricultural innovations such as new crop varieties and animal/fish breeds have great potential to contribute to agricultural production and development if these innovations meet users' needs and demands. Plant and animal breeding aims to improve the genetics or traits of breeding products to produce desired characteristics with the goal of achieving food security and better livelihoods. In meeting these objectives, understanding the priorities that women and men assign to genetically determined traits becomes critical (Orr et al. 2018). A socially inclusive, and thus gender-inclusive, process in product development can enhance gender equality when women and men's voices are heard and effectively inform the breeding process.

Although there is growing recognition of the vital role women have in informal (farmer-led) breeding processes, knowledge on how and when to involve men and women farmers and how gender-responsive breeding can advance gender equality is limited. Furthermore, the pathway from biophysical research to gender equality is complex and requires careful attention to multiple factors.

In the case of breeding for agricultural development, gender-differentiated access to and control over assets and resources can influence the technology, crop, and/or variety selected for production (Njenga and Gurung 2011, Kawarazuka and Prain 2019, Olaosebikan et al. 2019). Studies on gender-differentiated trait preferences show that varietal choice is related to resources, rights, and responsibilities shared differentially by men and women who are differently engaged in production, processing, and marketing (Fisher and Carr 2015, Christinck et al. 2017, Bentley et al. 2018, Isaacs et al. 2018, Teeken et al. 2018, Ashby and Polar 2019, Marimo et al. 2019, Olaosebikan et al. 2019). Similarly, formal and informal social structures and social relations shape men and women's innovation experiences and choices of technology (Bullock and Tegbaru 2019, Kawarazuka and Prain 2019).

In the broad context of technology and innovation in agriculture, **lower adoption of modern varieties among women producers** (Wale and Yalew 2007, Ashby and Polar 2019) emerges as a significant trend, reflecting unequal access to technology. Unequal access may imply that technology is *physically* not accessible equally. Alternatively, it may imply that the technology developed has not considered or does not respond (equally) to the needs and demands of gender-differentiated segments of the population (Mulema et al. 2019, Polar et al. 2021). This speaks to the need for institutional and structural innovations that revitalize the way new varieties are developed and disseminated for uptake, to ensure the consideration at multiple levels and stages of factors such as gender-differentiated control over assets and resources, and normative climate. Crop and animal breeding programs need to consider gender differences when setting priorities and targets for breeding, since overlooking traits important to women farmers and consumers may lead to women's disempowerment and aggravated household food insecurity and poverty (Tufan et al. 2018), thus increasing the gender gap and inequality.

Before proposing and further exploring ways to integrate gender dimensions into breeding processes and practices, we take two steps back: one step to reflect on the relationship between technology, society, and gender and a second step to briefly reflect on the history of breeding since the Green Revolution.

Framing technology, society, and gender

Technology is not neutral. For one thing, it tends to be associated with masculinity, not only in popular assumptions but also, in some cases, in academic "truths" (Gill and Grint 1995). The gendered nature of technology reflects that technology and society co-constitute one another (Bijker 1995, MacKenzie and Wajcman 2011) and co-evolve (see Chapter 10, this volume). Social and cultural factors condition ideologies, while policies shape the development and endurance of technology, and vice versa (Johnson 2010). Technologies are thus inherently political as they can be designed, consciously or unconsciously, to open certain social options and close others (MacKenzie and Wajcman 2011).

As such, in agricultural research for development, it is important that technology be framed not as an artifact, technique, system of knowledge, or expertise but rather as a sociotechnical system (Hughes 1994) produced by the interactions of the technology and society. Technology change is then one factor among many others—political, economic, cultural—in the dynamics of social change (MacKenzie and Wajcman 2011). Adopting a technology may have far more effects than are first evident. Technology matters not only physically and biologically but also to human relations and social processes. This shifts the frame of technology adoption from passive adoption by users/receivers of a technological innovation toward a more active role of users in shaping technological change (ibid.).

This shift is crucial to advancing toward gender equality. The material features of a technology are necessary but not sufficient conditions for gender-equitable relations (Johnson 2010). Two aspects required in understanding the potential contribution of technology to gender equality are (a) the features of the technology and how they may (or may not) carry a deliberate gender bias; and (b) the sociocultural context surrounding the access to and use of the technology (Gill and Grint 1995, Johnson 2010, Polar et al. 2017).

The philosophical and actual separation between women and technology in western culture is linked to changes that took place during the Industrial Revolution. The separation of public and private spheres and the move toward factories for manufacturing resulted in a gendered division of labor that fostered male dominance of technology (Gill and Grint 1995). Similarly, the Green Revolution marks a breaking point between women and technology resulting from a drive toward specialization in agricultural production.

Toward gender integration in breeding processes

Between the 1950s and the late 1960s, when food shortage was one of the world's major challenges, the Green Revolution endeavors, advanced by Norman Borlaug, leveraged agricultural research and technology to increase productivity in the developing world (Hazell 2009). The introduction of packages of new high-yielding plant varieties with improved practices, fertilizers, and other improved inputs was crucial to increase food production in many countries (Farmer 1986, Zaidi et al. 2019).

Decades after Borlaug won the Nobel Peace Prize as the "Father of the Green Revolution," credited with saving over a billion people from starvation, breeding programs continue to increase smallholder farmers' yields, reduce pesticide use, improve nutrition and health, and contribute to poverty reduction (Qaim et al. 2007, Osei et al. 2014) through new varieties and animal breeds. However, the drive to reach the maximum number of farmers and generate the greatest production improvements has also led to lack of attention to the diversity of needs, challenges, and preferences differentially faced by men and women. Furthermore, although many studies show the positive contribution of breeding for agricultural development, there is a gap in knowledge on the outcomes or impact of the new technologies on women and gender equality.

Male and female farmers in marginal ecologies have not benefited from the outstanding yield increases that Green Revolution endeavors obtained in environments that are naturally favorable or that can be made favorable profitably by using inputs (Ceccarelli et al. 1996). Consequently, adoption of new crop varieties by resource-poor farmers has been limited. To enhance the adoption of breeding products developed by breeding programs with social welfare and development goals, a key consideration in product design and advancement is the potential impact of a new breeding product on the welfare of end users. The drive toward more "client-focused" breeding processes has grown and evolved over the past decades (Ashby 1996, Persley and Anthony 2017, Ragot et al. 2018).

Unequal power relations and differentiated roles and/or control over assets and resources imply that men and women engage in agriculture with different means of production and face different constraints. As a result, women frequently develop different strategies for farming than men, based on systematic differences in their rights, opportunities, and resource endowments (Ashby and Polar 2019, Olaosebikan et al. 2019). Whenever poor men and women on small farms produce for direct household consumption as well as for the market, gender inequalities often translate into gender disparities in the adoption of new technologies (Peterman et al. 2014, Fisher and Carr 2015, Mehar et al. 2017). These adoption disparities are aggravated when technologies are developed to address "common denominator" traits such as yield or disease resistance (Teeken et al. 2018) and not necessarily gender-differentiated needs and preferences.

Because of this, we argue that it is necessary to deliberately address gender dimensions in the development of new technologies, including new varieties and animal breeds, to level the playing field toward gender equality among women and men belonging to different age and ethnic groups. This involves providing farmers with **real technology choices** that better address their needs, preferences, and constraints (Polar et al. 2021). With new technological advances in breeding such as genomic selection and high throughput processing, there are also more opportunities for breeding to expand the range of traits prioritized and to tailor new products to specific end-user preferences. This opens the door to the formulation of breeding objectives that actively integrate gender considerations (Ashby and Polar 2019) that advance breeding endeavors beyond the Green Revolution by developing new crop varieties and animal breeds that foster inclusion, equality, and sustainability.

How do technology options relate to gender equality?

The power to choose!

Gender inequality relates directly to power. One aspect of power is the ability to make choices (Kabeer 2005). Women and men may not have the same possibilities to make choices, and gender-related disparities often intensify the effects of poverty, creating cycles of greater inequality. Individual preferences are an important dimension of choice;[1] they are not so much features of individuals but rather also reflect internalized inequalities from the wider social context (Kabeer 2002). Moreover, individual preferences also embody the extent to which individuals seek to challenge such societal inequalities.

People are not free when they cannot make choices about their lives (Sen 1994, 2004). The power to make such choices refers to human agency, which creates new possibilities and actions (Rowlands 1997), but also to social structures (Akram 2010) that enable or restrain choices and choice-making. Empowerment relates to the existence of real choices, the exercise of choice, and the outcomes that result from the process (Kabeer 1999, Alsop et al. 2006).

For there to be "real choices," two basic conditions need to be met (Kabeer 2005):

> *There must be alternatives to choose from that make meaningful choices possible.*
>
> *Alternatives must not only exist but they must also be seen to exist.*

Both the existence of choice and users' perception of its existence stem directly from the relationship between individuals and society, or human agency and social structures (Akram 2013). Human agency and social structures are assumed as interdependent processes that shape the way culture,

1 Collective and individual choice-making are both part of human agency. However, for the purpose of exemplifying the existence of real choices, this chapter addresses only individual aspects of choice-making.

institutions and values, norms, beliefs, and behaviors of humans co-evolve (Musolf 2003).

While issues of human agency and social structures have multiple dimensions, we are interested in how they are at play in the existence and perception of practical interests and choice **related to the adoption of a new plant variety or animal/fish breed.** We are looking at the breeding processes and structures involved in the generation of options (plant varieties or animal breeds) that do respond to the needs, preferences/priorities, and constraints facing men and women. This calls for a dualistic perspective: (a) the identification of gender-differentiated needs, priorities, and constraints related to plant varieties and animal breeds; and (b) the institutional and organizational frameworks of breeding that enable or restrain the generation of options for meaningful choice-making. We consider these two issues across the different stages of breeding that we discuss further below.

There is a growing body of research and insights on the needs, priorities, and constraints facing men and women, but only just emerging are examples of how this information is and can be incorporated meaningfully into breeding programs. Decisions made by breeders about which traits to incorporate in a new plant variety or animal breed often involve tradeoffs about whose preferences among different end users are prioritized (Ragot et al. 2018, Tufan et al. 2018, Polar et al. 2021).

For technology users, gender inequalities in the availability of options for meaningful choice-making occur when (a) breeding programs do not develop products with traits that women value positively, whether or not men producers also value those traits; or (b) the new breeding products incorporate traits that men producers value highly but that are detrimental to women. The latter is the case, for example, when the new, higher-yielding variety increases women's unpaid labor in threshing or requires the use of inputs of unequal access for women. The integration of gender dimensions in breeding hence entails both "doing good" and "doing no harm."

A prospective impact pathway linking breeding and gender equality

Gender equality refers to equal rights, responsibilities, and opportunities of women and men, implying that the interests, needs, and priorities of both are taken into consideration (Fredman et al. 2015; see also Chapter 1, this volume). Advancing toward gender equality and women's empowerment requires transformative shifts, integrated approaches, and new solutions through innovations in policies, management, finance, science, and technology (Waezi 2017). It is increasingly clear that science and technology can create new, unforeseen problems and that they may not benefit all equally (UN Women, 2019)—but can also create new opportunities if purposefully approached.

In this definition of gender equality, the elements for biophysical research and breeding for agricultural development are providing and ensuring that the interests, needs, and priorities of both women and men are taken into account when setting priorities and targets for breeding. Based on these premises, Figure 2.1 presents an impact pathway describing the prospective changes expected from setting breeding priorities toward gender equality, including a summary of intermediate steps. Innovations in science and technology that disrupt "business as usual" are increasingly being recognized as preconditions (Waezi 2017) in the path toward gender equality. The process described in Figure 2.1 presents a disruption of business as usual by changing traditional technology development structures (breeding programs) through institutional innovations that incorporate gender analysis to set breeding objectives that contribute to equality in opportunities.

This proposed impact pathway focuses only on interventions based on formal breeding systems. While there may be multiple mechanisms to achieve gender equality that do not necessarily address formal structures of technology development and breeding, the pathway presented below describes the steps from breeding processes and products to socioeconomic change and gender equality. The simplified logic holds a series of assumptions, mainly associated with social and behavioral change, that are fundamental to equality outcomes.

The impact pathway presented in Figure 2.1 is subdivided into three phases. Each phase has specific outputs and outcomes; the three are linked to each other sequentially but with some degree of overlap. Within and between phases there are a number of important assumptions that need to be intentionally addressed to make sure they are actually in place in reality. For example, the first assumption will require a great deal of biophysical research to translate the basic components of preferred traits (for example taste, smell, other traits) in terms of genes, heritability, and breeding potential. A second assumption is the need for flexible structures that enable an inclusion of users' vision in priority-setting. Other assumptions also are not automatic and require an intentional approach to ensure they are in place.

Phase 1 includes the first steps of the breeding cycle. Gender in this phase may be included in multiple ways and through different tools and mechanisms. Essentially, the expected outputs are:

FIGURE 2.1 Impact pathway from breeding for development to gender equality

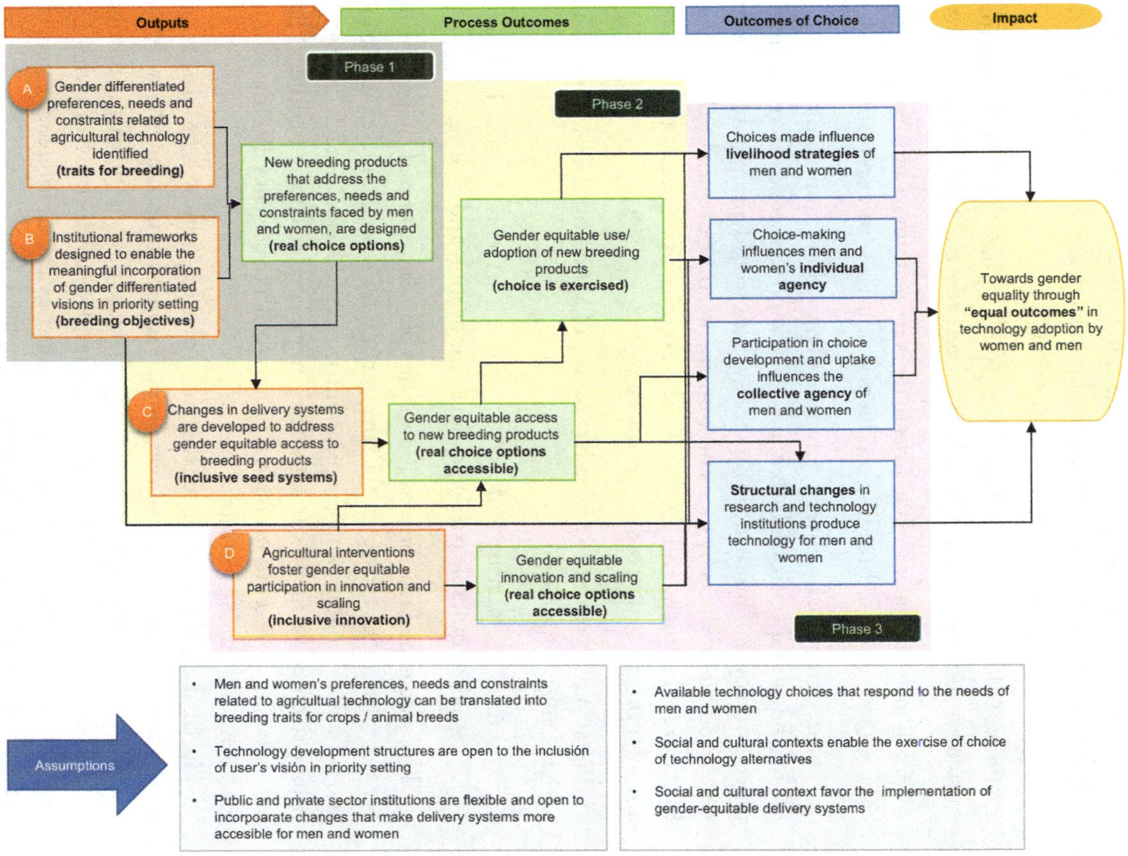

Source: Developed by the authors.

(A) Identify the preferences, needs, and constraints related to agricultural technology from the experiences of men and women; and
(B) Design institutional frameworks to enable meaningful incorporation of gender-differentiated visions in priority-setting for breeding objectives.

The process outcome of this intervention is the existence of **"real choice options"** in the form of new breeding products that respond to the preferences, needs, and constraints facing men and women.

Phase 2 deals more with the delivery system. Once **"real choice options"** exist, men and women must be able to access them equitably. This requires:

(C) The development of changes in delivery systems to address gender-equitable access to breeding products (Inclusive seed systems).

The design of such delivery systems will probably require multiple approaches and specific tools that may go beyond the scope of agriculture. The process outcome of more inclusive seed systems would be equitable access to new breeding products by men and women, meaning essentially the **accessibility of real choices** and potentially the **exercise of choice** in the form of use and adoption of new breeding products.

Phase 3 reaches out to a broader context of agricultural development and social change. This requires:

> Gender-responsive and transformative interventions that (D) foster gender-equitable participation in innovation and scaling.

The outcome of this process is also the accessibility of real choices through behavioral change. However, the stronger emphasis on inclusive innovation and behavioral change, and the prior outcomes of Phases 1 and 2, can contribute to empowerment, through outcomes related to choice-making. These outcomes are changes in individual agency, collective agency, livelihood strategies, and social and institutional structures.

The final expected impact of this three-phased pathway is progress toward gender equality through equal outcomes. It is important to highlight that, as the pathway advances, the outcomes are influenced by an increasing number of variables and actors and thus may yield results that deviate from what is expected. This process is non-linear: it comprises a thick interconnected network of variables that may require multiple feedback loops.

In the next section, we take an in-depth look at Phase 1 of the impact pathway. That means we focus on outputs (A) and (B) in the impact pathway.

We explore in more detail the changes in the breeding process, both methodological and institutional, that need to be made in the breeding research process to generate **real choice options** for men and women. In this, we keep in mind that technologies have a political dimension, and how this can and must be addressed as early as the design phase.

Inclusion of gender considerations in the breeding cycle

In assessing the place of gender considerations in critical breeding decisions, two aspects need to be considered:

When, along the breeding decision process, is gender a consideration?

How are gender-differentiated needs, preferences, and constraints incorporated and what does this imply for institutional/organizational structures?

This section presents some of the products from the CGIAR Gender and Breeding Initiative (GBI), which focused on these two aspects.

GBI emerged from the conclusions of a workshop held in Nairobi in late 2016, with the objective of bringing together plant and animal breeders, and social scientists to develop a strategy for gender-responsive breeding. GBI started in 2017 through a one-year grant from CGIAR System Management Office, coordinated by the CGIAR Research Program on Roots, Tubers and Bananas. In order to enhance the inclusion of gender considerations in breeding programs, GBI identified seven critical decision points along the breeding process where gender must be included (Ashby et al. 2018). Based on these decision points, a set of questions were developed to trace activities and information needed to make decisions, including also the expected results. Table 2.1 presents a summary of the questions formulated in the "decision checklist."

The first four decision points seek to incorporate gender in **customer segmentation and targeting,** and in the definition of the **product profile or package of traits** for the target group of customers, considering also the breeding feasibility of these traits. Decision point 5 takes place iteratively during breeding and early testing. Decision points 6 and 7 are part of the product delivery process.

Using the guiding questions of these decision points, Figure 2.2 presents a simplified breeding cycle overlaid with concrete INTERVENTIONS and expected OUTPUTS. The checklist can be used at any stage in the breeding program but is likely to be most effective if used during the early stages, when decisions are made about whom to target and what desired package of traits is

TABLE 2.1 Guiding questions in the "decision checklist for gender-responsive breeding"

Point	Questions that guide actions and decisions	Focus area
1	Who are the potential customers when gender is considered?	Segmenting and targeting
2	What customers to target? What is the justification for targeting one segment of the user population versus another, considering differences in gender equality?	
3	Which trait preferences could the program potentially breed for? Which existing or new-bred plant or animal traits could potentially satisfy some aspects of identified demand?	Understanding trait preferences
4	What product meets the needs of a gendered target customer? What product can feasibly be developed to meet the priority demand of the most important customer group?	Changing priority-setting
5	How is the program going to breed for the traits needed to reach the gender-responsive product profile? Is new variation needed to meet the specifications of the product profile?	
6	How will selection of bred genotypes meet the specifications of the gender-responsive product profile?	Testing and selection
7	What gendered constraints should be included in the design of delivery systems for the breeding products?	

Source: Adapted from Ashby et al. (2018).

to be prioritized. Therefore, we take a much closer look at the decision points in Phase 1 by discussing four areas: (a) segmenting and targeting, (b) understanding trait preferences, (c) changing priority-setting, and (d) selection and testing. For each of these areas, we discuss ways to integrate gender dimensions, and share concrete examples of how that has been done.

Segmenting and gender targeting when breeding for the poor

A first challenge for the incorporation of gender considerations in public sector breeding programs is a methodological one. For public breeding programs focused on breeding for the urban and rural poor, gendered dimensions of demand and market signals are often obscured or not readily discerned (Orr et al. 2018). The challenge is compounded by the fact that, unlike private entities,

FIGURE 2.2 Gender-related interventions and outputs along the breeding cycle

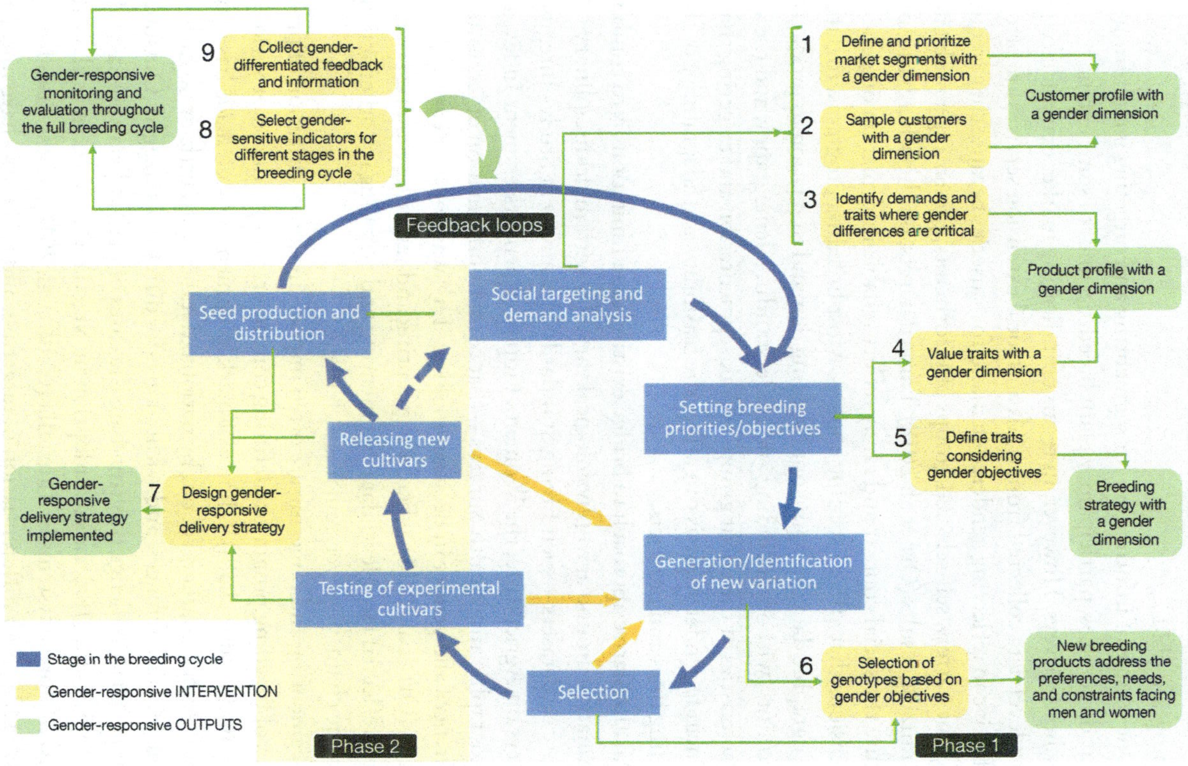

Source: Developed by the authors as a representation of how "Decision Points" (Asby et al. 2018) integrate with the Breeding Cycle.

public breeding programs, especially programs for resource-poor farmers, do not have the wherewithal to conduct extensive market research to understand their clients, and to assess and evaluate the findings of the research to redevelop the program (ibid.).

One source of information for segmenting and targeting is data from extension services, participatory plant breeding, and, to some extent, household surveys. However, in terms of indicating market demand accurately, micro-level data collected from trials and local extension services are often ineffective in reflecting gendered demand for varieties because they frequently represent narrow, self-selected groups of informants, presenting a problem for generalizing results (Orr et al. 2018). Large datasets, on the other hand, often obscure the social context and rarely reflect reliable information on gender dimensions that affect breeding choices. Most breeding programs, to date, rely on geographic data, mapping out breeding demand in relation to production constraints, without including and understanding demography, and the social characteristics of the client groups. Weak segmentation and targeting that does not include gender analysis can affect the feasibility and adaptability of the breeding lines and products.

An alternative approach to bridge existing data limitations is to combine information around agro-physiological variables with multidisciplinary large-scale datasets—such Living Standards Measurement Study–Integrated Surveys on Agriculture, the Women's Empowerment Assessment Index, and Demographic and Health Surveys—that contain information on consumer preferences and gender in decision-making. This was done for cassava in Nigeria (Orr et al. 2018), where a segmenting and targeting exercise predominantly reflected farm-level data. Segmenting and targeting for other actors in the value chain, however, remains a challenge.

Segmenting and targeting are often implicit in the organization of priorities of breeding programs (defining consumers and producers at various stages of the value chain). Choice of market segment is often made based on agro-ecological markers set out by national datasets with little or no inclusion of consumer preferences, and without gender differences flagged. Outstanding examples that challenge this trend are the cases of market beans in East Africa (Katungi et al. 2018a), cassava adoption in Nigeria (Olaosebikan et al. 2018), and *ololili* forage systems in Tanzania (Galiè et al. 2018). These cases have considered gender differentiation in targeting, and the results have influenced other stages of the breeding cycle.

In the case of market beans in East Africa, the International Center for Tropical Agriculture conducted different studies, including household surveys, choice experiments, and participatory varietal selection, to gather socioeconomic data from male and female respondents. Results revealed that farmers' preferences for bean traits were influenced by landholding size, age, household size, sex, and wealth of the household (Katungi et al. 2018b). These are variables that also influence technology adoption. An important finding was that traits cannot be labeled as men or women's, since often both prefer the same traits but with varying intensity or for different reasons.

The above study also revealed a shared preference for reduced cooking time, and this trait was further explored in a complementary market study with consumers, which revealed the significance of selling precooked beans, given the lower cost per person of boiling beans and the fundamental importance of color for buyers (Aseete et al. 2018).

These findings have two significant implications for setting breeding objectives: (a) if precooking beans is a viable marketing option, fast cooking may not need to be included in the package of traits for biofortified beans; and (b) if color is a main driver for buyers, this trait should be included in the development of new breeding products.

Incorporating gender into social targeting can take a functional or transformative approach, especially when formulating breeding objectives (Orr et al. 2018, 2).[2] Figure 2.3 depicts four different options in deciding on what market segment to target, taking gender into account. The approach taken to incorporating gender into targeting will determine gender equality outcomes. On the other hand, both functional and transformative approaches to targeting have the potential to influence breeding product and program design, thus supporting the generation of **real choice options** for women as a basic step toward empowerment and gender equality.

Understanding gender dynamics in trait preferences: simplifying a complex picture

Historically, decision-making in breeding programs has been economic, focusing on the value of a preference linked to the genetic gain of a specified trait (Hazel et al. 1994, Mehar et al. 2019). With the twofold incentive of more inclusive agricultural research for development and increased adoption of breeding products, the opportunity to incorporate potential users' preferences

2 "A functional approach takes gender differences among small producers into account only when the delivery of relevant breeding products to both men and women users is essential for achieving desired levels of adoption and impact" (Orr et al. 2018).

FIGURE 2.3 Decisions that incorporate gender into targeting through a functional or transformative approach to gender equality

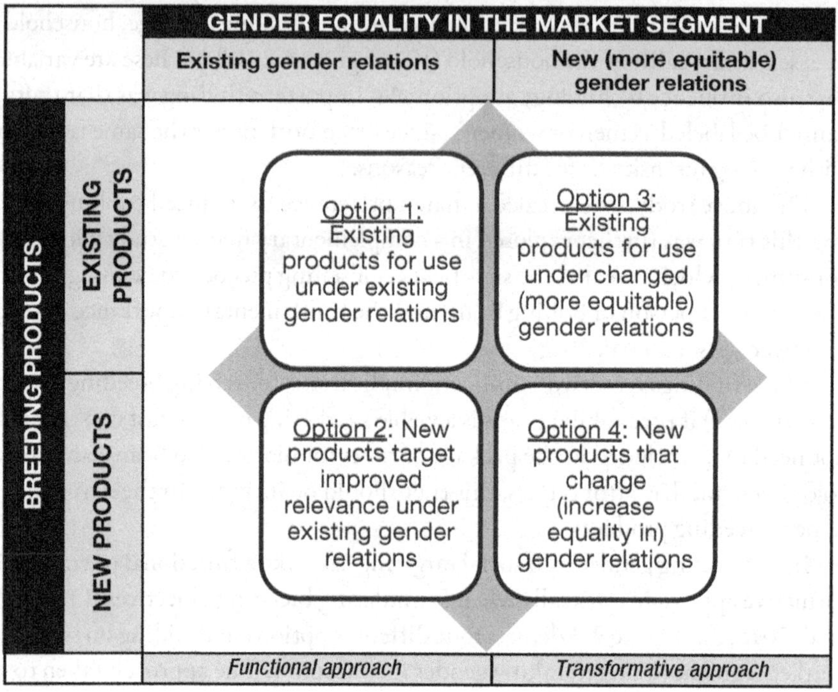

Source: Orr et al. (2018).

has gained traction in recent years. A critical foundation for this is clear and accurate data regarding women and men clients' preferences.

To this end, CGIAR carried out a series of foundational literature reviews and empirical research studies to systematically assess knowledge regarding gender differentiation in trait preferences, with the aim of identifying options for breeding programs to better address gender-specific needs (Christinck et al. 2017, Ramasawmy et al. 2018, Mehar et al. 2019, Mulema et al. 2019, Murphy et al. 2020). Overall, the studies highlighted that gendered trait preference information was relatively scarce and reliable methods were still a work in progress in public sector agricultural research for development in the Global South.

A multiorganization, multicommodity initiative with post-doctoral researchers to unpack methodological challenges and opportunities related to gender analysis in trait preferences surfaced several methodological challenges that need further work. These challenges include accurate assessment when there are notable differences between what people mean, what they say, and what they do (for example what they actually purchase as opposed to what they say they purchase); and the challenge of translating information about preferences into traits that are potentially actionable by breeding programs. The latter also involves sifting through preferences to determine which can be included in a breeding program versus those that can be addressed through the development of best management practices and extension (see Mehar et al. 2019).

To understand gender dimensions in trait preferences it is important to look at how these latter reflect underlying gender differences in assets, markets, information, and risk, and the ways institutions and policies condition these. Unpacking the links between trait preferences, social differences, and gender asset inequalities can provide indications as to the expected potential impact of a breeding product. For example, a participatory breeding program in Mali identified that women expressed preferences for varieties tolerant to low soil fertility. Further analysis reflected underlying structures of gender inequality in land rights, land access, and access to inputs for soil fertility (Rattunde et al. 2018). Although challenging and changing gender norms and access to land would be an ideal intervention, it may be outside the sphere of control of agricultural researchers working on breeding. In this case, breeders proactively acted to decrease gender inequality by developing varieties with improved tolerance to low phosphorus, which were particularly beneficial for women producers (Ashby 2018).

Similarly, the study of trait preferences with the *ololili*[3] in Tanzania showed that men gave higher importance to livestock fattening whereas women gave higher importance to milk production and this reflected unequal asset distribution: men own the animal while women control other assets like milk production (Galiè 2013). In this situation, breeders can actively make a choice to support gender equality through breeding for a composite of traits that benefit both men and women.

In relation to livestock, the African Chicken Genetic Gains Project tested different strains of chickens to improve their performance under different agroecologies in Africa. At its onset, the project assessed the traits preferred by men and women chicken farmers from more than 3,500 households through a baseline survey in Ethiopia. Following this, a qualitative study was conducted to triangulate the results. Although men and women showed similar preferences for physical traits, the reasoning behind these preferences differed. Male

3 *Ololili* is a traditional dry season forage reserve maintained by Tanzania's pastoral Maasai communities to feed their animals. In this system, a portion of land is fenced to let the natural pasture grow during the wet season.

respondents focused on the productivity, health, and marketing of chickens while women elicited behavior and consumption traits as well (Mulema 2018, Ramasawmy et al. 2018).

Although the project has not gone through the entire breeding and dissemination process, the results will guide breeders as to which traits to integrate into the second phase of the breeding program, and how to involve men and women in the selection, testing, and delivery of preferred chickens. The project aims at setting up a Long-Term Chicken Genetic Gains Program for sub-Saharan Africa, to produce chicken breeds that are more relevant for women and their households, providing options for different contexts.

An often-observed trait preference of women and food-insecure producers is for early-maturing varieties, despite the tradeoff of lower productivity. This may be because women and the poor are often land-scarce and cash-poor and face food insecurity, trying to meet household subsistence food needs year-round. Early-maturing varieties are one way to manage asset scarcity early in the growing season. Dependence on rain, vulnerability to climatic risks, availability of labor, and priorities in time allocation may also be factors affecting women and men differently and influence the preference for early-maturing varieties. Going beyond trait preferences as such—and into the underlying factors that shape them—can help breeders **set breeding priorities** that more effectively address the needs of the target population (Ashby 2018, Weltzien et al. 2019, Mudege et al. 2020) to provide them with **real choice options**.

Changes in how breeding priorities are set

Before the formalization and specialization of breeding programs, both men and women farmers, depending on their role in seed management and status, were involved in breeding. This included selection of naturally emerging crosses (Mokuwa et al. 2014), testing, cross-breeding, conducting varietal trials across seasons from material sourced from neighbors and those in their social circles, and evaluating the merits and demerits of these varieties in their specific sites (Farnworth and Jiggins 2003). With formal (current) breeding, processes, and decision-making largely shifted to breeding scientists, in terms of managing and controlling the gene flow and deciding what genetic qualities should be valued and for what markets, the operational implications of the differentiated roles of men and women have often been unwittingly ignored (ibid.). Nevertheless, some initiatives have targeted a bridging process to incorporate user perspectives, and more specifically, gender perspectives in breeding objectives.

An early example of changes in breeding priorities, influenced by gender-differentiated information and/or preferences, is the case of bean breeders in Colombia during the 1980s. Inspired by new evidence from participatory research, breeders learned about the multiple uses of beans in households and the key role of women in making the choice on what varieties to grow. As a result, breeders' fields kept varieties that they would not otherwise have selected (Ashby 1990). Nearly 40 years later, bean breeding teams in Africa are using segmentation and targeting tools and techniques combined with thorough socioeconomic analysis to make more fundamental decisions to consistently incorporate gender considerations in breeding priorities (Nchanji 2018).

At the turn of the century, the Green Revolution breeding paradigm began to shift toward more participatory breeding schemes. Animal and plant breeders started to acknowledge the need for gender analysis in breeding (Ceccarelli et al. 2007, Christinck et al. 2017, Katungi et al. 2018a). The barley breeding program in Syria is an example of an intervention that decided to carry out participatory diagnosis early in the design of new breeding products. This allowed a focus on the strategic needs of women for income-generating opportunities and the co-development of new varieties with women producers to expand their participation in seed marketing (Galiè et al. 2018).

The case of cassava in Nigeria is a more recent example of changes in the operational structure of breeding programs and how they set priorities to include gender considerations. Cassava in Nigeria is a major smallholder crop primarily for local processing and home consumption. Trait preference studies on cassava underscored the extent to which food product quality and processing traits were more important for women, including ease of peeling and swelling ability in *gari* and *fufu*[4] (Bentley et al. 2017, Olaosebikan et al. 2018). These findings, and additional gender and monitoring information, have led the breeding unit to (a) include a focus on women farmers and processors; (b) integrate social science and food science as breeding team competencies; and (c) include information on social and food quality variables for decision-making processes.

This example is critical to understand two important aspects in addressing gender in breeding to advance toward gender equality. The first aspect is that of tools, methods, and procedures that enable a better understanding of the interplay between traits and gender. Such an understanding can drive concrete actions and suggestions that can be addressed through breeding. The second aspect is the incorporation of these suggestions and the traits selected through

---

4 *Gari* and *fufu* are traditional West African foods based on flour made from cassava roots.

deliberate attention to gender. Important here is that the incorporation of gender dimensions is not by chance, or the result of fleeting opportunities, but rather a consequence of formal and systematic processes embedded in breeding structures.

Selection and testing experimental cultivars and new animal breeds

Improving access to and adoption of improved crop varieties and livestock breeds that are adapted to specific environmental conditions is an important approach to increasing production, productivity, and food and nutrition security. However, improving access is a challenge that requires rethinking approaches to mechanisms and market systems that can advance access to new breeding material. Conventional breeding programs that focus narrowly on high input use to minimize environmental risks have limited success. Environmental variations at both the landscape and the field level, limited access to resources by smallholder farmers, poor infrastructure, poor governance, and limited risk mitigation measures further hinder the success of conventional breeding (Charles et al. 2010). Social structures, including gender norms and gender division of labor, that inform farmer selection criteria are often unknown to the breeders or not part of breeders' standards for selection (Mulatu and Zelleke 2002).

Most experiences of gender integration in breeding are related to the later stages of breeding decision-making, with women involved in evaluating advanced material or released varieties and in their distribution. Mother and baby trials, participatory varietal selection, and other participatory research appraisal tools are frequently used to conduct evaluations and extract information to refine breeding products and enhance gender responsiveness in delivery mechanisms.

In response to low rates of adoption of improved released cultivars in rain-fed rice environments in eastern India, a participatory plant breeding project was implemented in the late 1990s. Male and female farmers in the drought-/submergence-prone villages agreed that grain yield and crop duration were the most important traits when choosing varieties for upland and low-lying areas. However, women gave more importance to traits related to tasks that they conducted, such as weed competitiveness and post-harvest qualities (ease of de-husking and threshing, high milling recovery, and suitability for different food preparations, for example puffed rice).

This led to the program revising the methods for evaluating rice varieties on farmers' fields. Farmers' selection criteria were included in rice varietal selection (mother trials) and farmers were also included in the early evaluation of new rice lines under their own management (baby trials). The women were able to select lines with good eating qualities and suitable for making other rice products. The approach described in the rice case contributed to equitable access to cultivars with traits that responded to the general needs of men and women (Paris et al. 2008). It is during the evaluation processes of this later stage that valuable information is generated to support further changes in earlier stages of the breeding cycle.

These experiences have contributed to growing recognition that gender analysis is not something that should come at the end of the research process, essentially to enhance adoption of the new breeding products generated with so-called "gender-neutral" objectives. The findings support the need to develop feedback loops to learn from experiences of early and late testing of material, in a way that these can feed into developing gender equality objectives in early stages of the (next) breeding process. Developing effective feedback loops requires systematically and cyclically incorporating gender equality concerns throughout the different stages of the breeding cycle. The continuous monitoring and evaluation of outputs and outcomes is essential to enhance learning and to redirect the process toward the desired outcomes.

Toward a next generation agenda for gender research and breeding-related development outcomes

The incorporation of gender in breeding processes has gained attention from biophysical researchers over time, mainly as an opportunity to tackle low adoption rates of new breeding products. However, it is important to point out that higher adoption rates do not necessarily mean progress toward gender equality. Outputs and outcomes achieved are determined by the objectives set in breeding priorities. The three possible outcomes of a breeding process depend on the objectives pursued and the measurement indicators established. Figure 2.4 describes these.

A gender and breeding research agenda should carefully articulate gender objectives in breeding to advance toward gender equality. The big question is: **How do we ensure gender equality objectives are incorporated at the level of breeding objectives?**

Taking this question to heart, with differentiated objectives as a starting point, a future-oriented research agenda that builds on the existing and emerging gender and breeding work can be envisioned. Such a research agenda includes both methodological and institutional innovations. Further **methodological** innovation is needed for segmentation, targeting, trait identification, and breeding selection procedures at each stage, and gender-inclusive product

FIGURE 2.4 Process and outcomes of different gender objectives in breeding

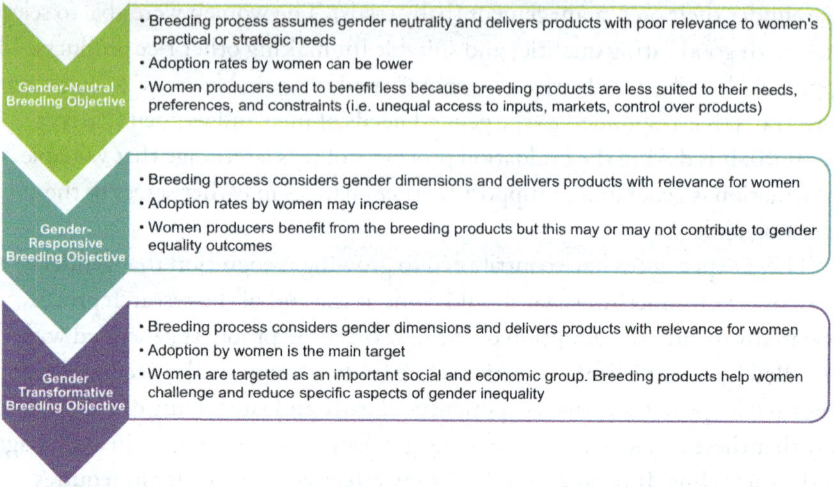

Source: Orr et al. (2018).

evaluation. Innovative mixed methods approaches and big data need to be further used and adapted to integrate data from gender relations and the underlying factors that shape preferences, needs, and constraints.

New tools and methods, at all stages of the breeding cycle, should include in their design a prospective analysis of positive or negative effects of the new breeding products vis-à-vis gender equality. They can build on emerging insights into the usefulness of taking a gender relations perspective and looking at the underlying factors of gender differences and disparities. These tools will need to collect data *from* women and men, and *on* women and men, to shed light on gender dynamics and the underlying factors that shape them. In order to formally incorporate the use of new methodological innovations in a systematic and consistent way across breeding programs, **institutional** innovations will also be needed. The CGIAR Excellence in Breeding Platform, currently working to change the breeding mindset toward interdisciplinary and client-oriented breeding, offers great potential to formally incorporate gender dimensions across the breeding process.

At this point, we want to focus our attention on a third core element of the future research agenda: the key role that **monitoring, evaluation and learning** can play in advancing gender equality through breeding. Evaluation and learning are central in the complex scenario of understanding the potential and actual contribution of technology development to gender equality.

Monitoring and evaluation plays a key role in strengthening the impact pathway, presented in Figure 2.1, and the possible outputs aligned with the breeding cycle, presented in Figure 2.2. These provide a robust starting point for further monitoring, evaluation and learning innovations that can play a critical role in assessing and strengthening contributions toward gender equality. This section presents specific suggestions for evaluation and learning loops in four specific areas: (a) segmentation, targeting, and the definition of trait preferences; (b) changing breeding priorities; (c) selection and testing of experimental cultivars; and (d) creating a new network of feedback loops.

Evaluation and learning in segmentation, targeting, and the definition of trait preferences

Research should focus on **understanding when, where, and why gender relations and inequalities influence beneficiary or user groups**. At this early stage, it is also important to anticipate how design decisions may affect and be influenced by gender-differentiated constraints and access to resources and opportunities (Ashby et al. 2018). This will enable the breeding program to better understand the gender dimensions in each social segment the program decides to target, and in relation to the potential package of traits. The final outcome of Stage "0" should be the definition of a product profile that holds breeding objectives (Kotch n.d.) to benefit specific beneficiary groups taking into account gender differences.

As discussed in earlier sections, there are concrete examples of incorporating gender dimensions at the stage of product design (Paris 2001, in Farnworth and Jiggins 2003, Ceccarelli et al. 2007, Galiè 2013, Nchanji 2018). However, all cases have used different approaches and methods, and have collected information on different variables. Looking ahead, it is important to formally incorporate a systematic and replicable process of data collection, aggregation, and analysis in existing breeding structures.

Evaluation and learning in changing breeding priorities

Experiences with participatory plant breeding demonstrate that engaging farmers in the breeding process and in early testing can yield positive results, both in terms of breeding products with traits that are useful for women and men and for achieving women's empowerment (Galiè 2013, Nchanji 2018). In Rwanda, women were identified as bean experts to work with breeders for better cultivars—a revolutionary move in a patriarchal context (Nchanji 2018). Participatory plant breeding emerged as a means of decentralizing breeding processes, increasing the responsiveness of breeding processes to the gender roles of men and women, yielding empowering outcomes for women (Galiè et al. 2018). While multiple factors contributed to the successful integration of gender, an element emerging as a common denominator is that the breeding objectives were set and defined by the **end users**, therefore generating **real choice options.**

Looking ahead, the challenge is to replicate the success of these cases in defining breeding objectives that respond to the needs, preferences, and constraints of both men and women, and that promote structural changes that allow breeding to produce real choice options. This should be accompanied by a thorough assessment of the social and economic impact of breeding in terms of gender equality, creating and dynamically incorporating feedback loops across different stages of the breeding cycle to enhance the learning process.

Evaluation and learning in selections and testing of breeding material

An essential component for incorporating gender equality dimensions in the selection and testing of breeding material is the definition and implementation of evaluation criteria with, and for, gender-differentiated target segments of the population. Experience includes tools such as participatory varietal selection (Agboh-Noameshie et al. 2013, Misiko 2013, Mudege et al. 2015, 2017), mother and baby trials (Paris et al. 2008), and other participatory research appraisal tools for evaluation (Paris et al. 2008, Misiko 2013, Mudege et al. 2017) applied to intentionally capture gender-differentiated perceptions.

For the most part, gender has been considered only at the later stages of breeding, with women involved in evaluating advanced material or released varieties, and in their distribution. Insufficient consideration of gender-responsive or gender transformative dimensions in data collection processes over varietal adoption and impact is reflected in inadequately described product profiles (Thiele et al. 2020), and this makes the creation of feedback loops a real challenge. Lessons need to be harvested to make it possible to develop and institutionalize feedback loops, which can contribute to breeding priorities that promote equality of opportunity in accessing meaningful technology options for choice-making.

Evaluation and learning: creating a network of feedback loops

Gender dimensions and equality are rarely identified as a priority consideration in breeding decisions at the beginning of a process (Ashby 2018). Interestingly, however, learning loops and feedback emerging from the inclusion of gender dimensions in the later stages of the process prompt breeders to reconsider the gender analysis upstream. An outstanding example of this is in sorghum in Mali, where a fundamental shift in product definition occurred in response to research on gender undertaken during the testing of experimental varieties. The program found that, while sorghum is locally considered a "man's crop", women also grow it for their own specific uses. This led to a rethink of the product profile and acknowledgment that gender dimensions must be included at earlier stages (ibid.).

Stakeholder consultations and socioeconomic surveys are critical to integrate gender into decision-making. Examples from programs like the *ololili* forage system in Tanzania and Matooke in Uganda have successfully integrated insights from gender-responsive methods, complementing meaningful qualitative research with surveys. A more nuanced approach should (a) integrate gender considerations in mixed methods and large datasets; (b) use gender-disaggregated data with reference to socioeconomic indicators; and (c) not rely solely on "women's participation" in activities, but more on their specific roles. This will shape a more disciplined, less anecdotal approach that can be formally institutionalized.

Moreover, some programs, as with groundnuts in Malawi and barley in Syria, have proactively sought to alter procedures to suit the convenience of women participating in various consultations. Examples from some of these programs have evolved to transform the composition of breeding research teams, with the inclusion of seed system actors, gender and social scientists, and traders, in addition to breeders. They thus systematically embedded these actors in the breeding decision-making process (Ashby and Polar 2019). Others, like the Maize program in Africa, have seen the inclusion of manuals for gender-responsive breeding (Mulema 2018, Adam et al. 2019) for a more thorough institutionalization of approaches.

Over the years, the formal inclusion of gender analysis in breeding has been attempted across CGIAR institutions and in many national agricultural research and extension system networks (Farnworth and Jiggins 2003), as well as through initiatives like GREAT[5], which focused specifically on integrating gender into the biophysical sciences. Breeding programs have evolved in terms of changing some protocols and considerations as to how they evaluate demand. However, with few exceptions, gender analysis in breeding is still at a formative stage, evolving "from ad hoc discovery of gender-differentiated traits" but more

5 Gender-Responsive Researchers Equipped for Agricultural Transformation (GREAT) is a Gates Foundation–funded five-year collaboration between Cornell University, in the United States, and Makerere University (www.greatagriculture.org).

often benefiting from lessons encountered at the later stages of the breeding process (Ashby 2018).

The objective ahead is to identify and institutionalize systematic entry points to create feedback loops linking gender analysis into the procedures, protocols, and practices of the breeding cycle and breeding structures. While current market trends visualize breeding as a pipeline, to foster learning and benefit from feedback loops, the process needs to be internalized as a cycle. Feedback should be formally established through a definition of entry points (moments and frequency of data analysis). While the specific moments of critical analysis have been identified (see Table 2.1 on the decision checklist and Figure 2.2 on the breeding cycle entry points), the iterative generation of data and their collective analysis as part of the breeding process are yet to be tested, moving from ad hoc processes to systematic, replicable, and cyclical ones. This type of analysis will provide evidence to support further institutional innovations and structural change to advance toward engendering breeding processes that disrupt the status quo and create equal opportunities for men and women to benefit from agricultural science and technology development.

Acknowledgments

We would like to thank Jacqueline A. Ashby for comments and suggestions during the preparation of this chapter. We are grateful to Alessandra Galiè, Eileen Nchanji, and Graham Thiele for input, comments, and ideas during early stages of its preparation. We would also like to thank Claudio Proietti for his contribution to the construction of the impact pathway. Finally, we also with to thank two anonymous reviewers and the book editorial committee, who provided insightful comments and suggestions to redirect the analysis of this chapter.

This research was coordinated by the CGIAR Gender and Breeding Initiative and the CGIAR Research Program on Roots, Tubers and Bananas, and supported by CGIAR Trust Fund contributors. https://www.cgiar.org/funders/

Geography of study sites for publications cited in Chapter 2

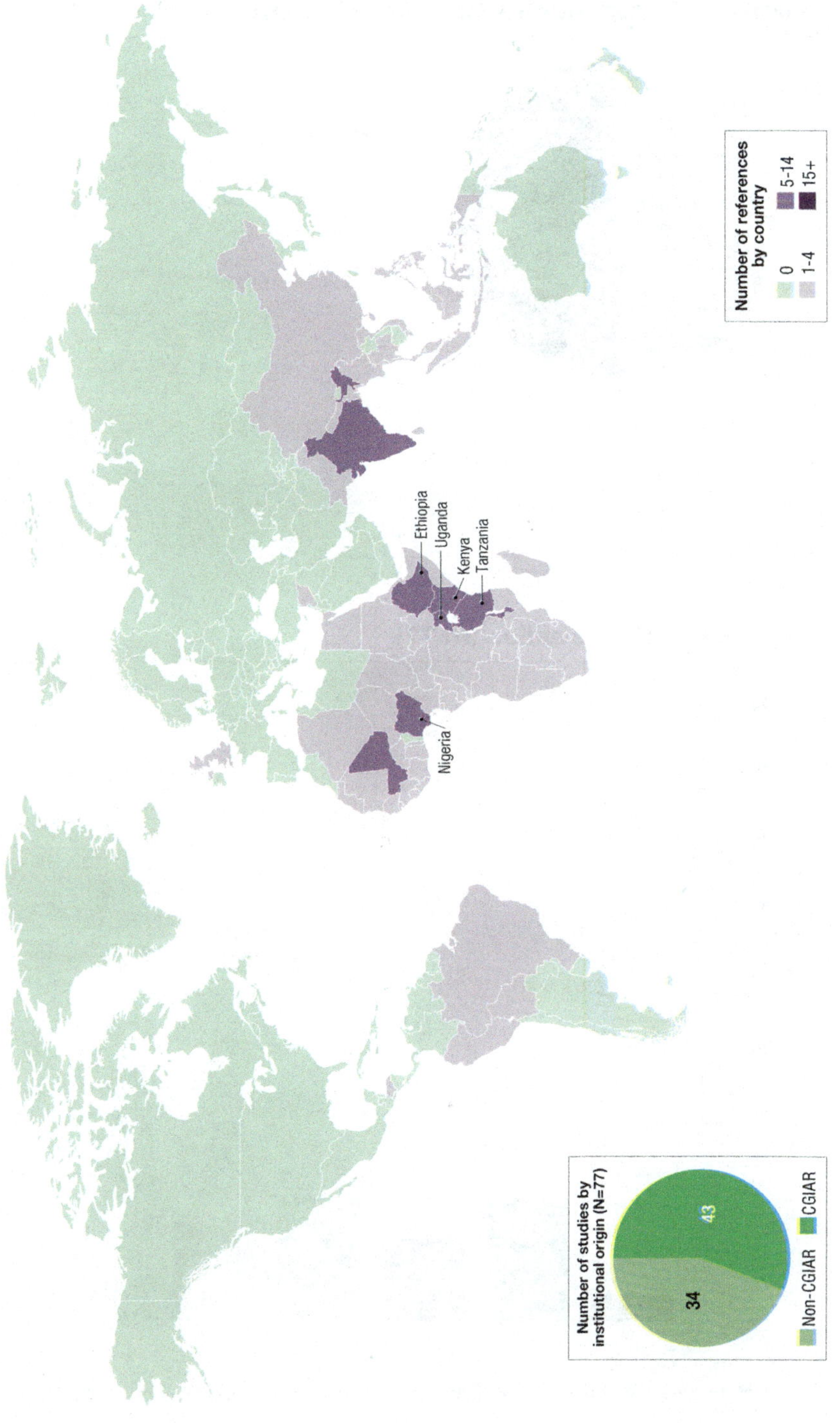

Number of cited studies by research methodology (N=77)

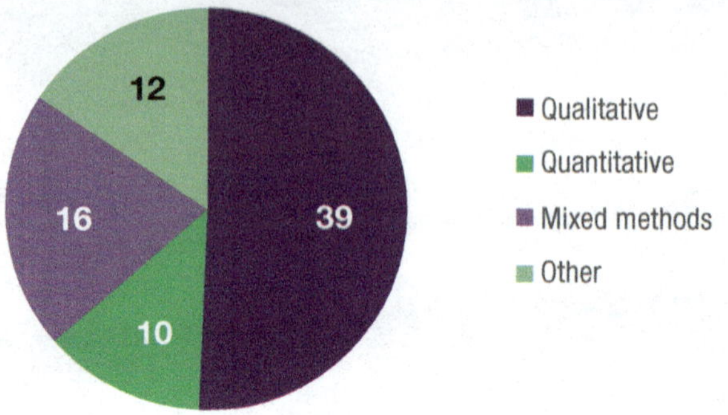

Timeline of cited studies (N=77)

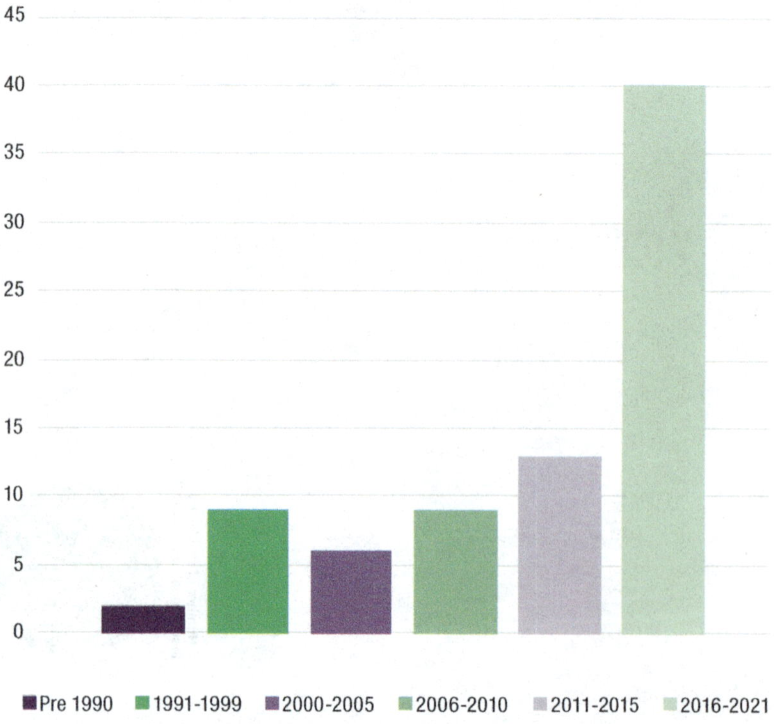

References

Adam, R. I., V. Kandiwa, S. David, and P. Muindi. 2019. *Gender-Responsive Approaches for Enhancing the Adoption of Improved Maize Seed in Africa: A Training Manual for Plant Breeders and Technicians*. Mexico City: CIMMYT.

Agboh-Noameshie, A. R., A. Kaboré, and M. T. Misiko. 2013. "Integrating Gender Considerations in Rice Research for Development in Africa." In *Realizing Africa's Rice Promise*, edited by M. Wopereis, D. Johnson, N. Ahmadi, E. Tollens, and A. Jalloh, 343–344. Oxfordshire, UK: CABE.

Akram, S. 2010. "Re-Conceptualising the Concept of Agency in the Structure and Agency Dialectic: Habitus and the Unconscious." PhD Thesis, University of Birmingham, UK.

Akram, S. 2013. "Fully Unconscious and Prone to Habit: The Characteristics of Agency in the Structure and Agency Dialectic." *Journal for the Theory of Social Behaviour* 43 (1): 45–65.

Alsop, R., M. F. Bertelsen, and J. Holland. 2006. *Empowerment in Practice: From Analysis to Implementation*. Washington, DC: World Bank.

Aseete, P., E. Katungi, J. Bonabana-Wabbi, E. Birachi, and M. A. Ugen. 2018. "Consumer Demand Heterogeneity and Valuation of Value-Added Pulse Products: A Case of Precooked Beans in Uganda." *Agriculture & Food Security* 7: 51.

Ashby, J. A. 1990. "Small Farmers' Participation in the Design of Technologies." In *Agroecology and Small Farm Development* , edited by M. A. Altieri and S. B. Hecht, 245-253. Boca Raton, FL: CRC Press Inc.

Ashby, J. A. 1996. "What Do We Mean by Participatory Research in Agriculture?" In *New Frontiers in Participatory Research and Gender Analysis*, 14–22. Proceedings of International Seminar on Participatory Research and Gender Analysis for Technology Development, Cali, September 19–14.

Ashby, J. A. 2018. "Lessons Learned." In *State of the Knowledge for Gender in Breeding: Case Studies for Practitioners*, edited by H. A. Tufan, S. Grando, and C. Meola, 133–147. Gender and Breeding Initiative. Lima: CGIAR.

Ashby, J. A., and V. Polar. 2019. "The Implications of Gender for Modern Approaches to Crop Improvement and Plant Breeding." In *Gender, Agriculture and Agrarian Transformation*, edited by C. Sachs, 11-34. London: Routledge.

Ashby, J. A., V. Polar, and G. Thiele. 2018. "Critical Decisions for Ensuring Plant or Animal Breeding is Gender-Responsive." GBI Brief 1. Lima: CGIAR.

Bentley, J. W., A. S. Olanrewaju, T. Madu, O. Olaosebikan, T. Abdoulaye, T. Assfaw Wossen et al. 2017. "Cassava Farmers' Preferences for Varieties and Seed Dissemination System in Nigeria: Gender and Regional Perspectives." Cali: CIAT.

Bentley, J. W., J. L. Andrade-Piedra, P. Demo, B. Dzomeku, K. Jacobsen, E. Kikulwe, P. Kromann et al. 2018. "Understanding Root, Tuber, and Banana Seed Systems and Coordination Breakdown: A Multi-Stakeholder Framework." *Journal of Crop Improvement* 32 (5): 559–621.

Bijker, W. E. 1995. "Sociohistorical Technology Studies." In *Handbook of Science and Technology Studies*, edited by S. Jasanoff, G. E. Markle, J. C. Peterson, and T. Pinch, 229–256. London: SAGE.

Bullock, R., and A. Tegbaru. 2019. "Women's Agency in Changing Contexts: A Case Study of Innovation Processes in Western Kenya." *Geoforum* 105: 78–88.

Ceccarelli, S., S. Grando, and R. H. Booth. 1996. "International Breeding Programmes and Resource-Poor Farmers: Crop Improvement in Difficult Environments." In *Participatory Plant Breeding*, edited by P. Eyzaguirre, and M. Iwanaga, 99–116. Proceedings of Workshop on Participatory Plant Breeding, Wageningen, July 26–29.

Ceccarelli, S., S. Grando, and M. Baum. 2007. "Participatory Plant Breeding in Water-Limited Environments." *Experimental Agricultural* 43 (4): 411–435.

Charles, H., J. Godfray, J. R. Beddington, I. Crute, H. Lawrence, L. Haddad et al. 2010. "Food Security: The Challenge of Feeding 9 Billion People." *Science* 327 (5967): 812–818.

Christinck, A., E. Weltzien, F. Rattunde, and J. A. Ashby. 2017. "Gender Differentiation of Farmer Preferences for Varietal Traits in Crop Improvement: Evidence and Issues." Gender and Agriculture Research Network Working Paper 2, CIAT, Cali.

Farmer, B. H. 1986. "Perspectives on the 'Green Revolution' in South Asia." *Modern Asian Studies* 20 (1): 175–199.

Farnworth, C. R., and J. Jiggins. 2003. "Participatory Plant Breeding and Gender Analysis." PPB Monograph 4. Cali: PRGA Program Coordination Office.

Fisher, M., and E. R. Carr. 2015. "The Influence of Gendered Roles and Responsibilities on the Adoption of Technologies that Mitigate Drought Risk: The Case of Drought-Tolerant Maize Seed in Eastern Uganda." *Global Environmental Change* 35: 82–92.

Fredman, S., and B. Goldblatt. 2015. *Gender Equality and Human Rights*. New York: UN Women.

Galiè, A. 2013. "The Empowerment of Women Farmers in the Context of Participatory Plant Breeding in Syria: Towards Equitable Development for Food Security." PhD Thesis, Wageningen University, Netherlands.

Galiè, A., J. Kizima, W. Mengesho, and B. Lukuyu. 2018. "Gender-Responsive Forage Intensification in the Ololili System of Tanzania." In *State of the Knowledge for Gender in Breeding: Case Studies for Practitioners*, edited by H. A. Tufan, S. Grando, and C. Meola, 44–54. Gender and Breeding Initiative. Lima: CGIAR.

Gill, R., and K. Grint. 1995. "Introduction." In *The Gender-Technology Relation: Contemporary Theory and Research*, edited by K. Grint, and R. Gill, 1-28. London: Taylor & Francis.

Hazel, L. N., G. E. Dickerson, and A. E. Freeman. 1994. "The Selection Index—Then, Now, and for the Future." *Journal of Dairy Science* 77 (10): 3236–3251.

Hazell, P. 2009. *The Asian Green Revolution*. Washington, DC: IFPRI.

Hughes, T. P. 1994. "Technological Momentum." In *Does Technology Drive History? The Dilemma of Technological Determinism*, edited by L. Marx, and M. R. Smith, 101–114. Cambridge, MA: MIT Press.

Isaacs, K., E. Weltzien, C. Diallo, M. Sidibe, B. Diallo, and F. Rattunde. 2018. "Farmer Engagement in Culinary Testing and Grain-Quality Evaluations Provides Crucial Information for Sorghum Breeding Strategies in Mali." In *State of the Knowledge for Gender in Breeding: Case Studies for Practitioners*, edited by H. A. Tufan, S. Grando, and C. Meola, 74–85. Gender and Breeding Initiative. Lima: CGIAR.

Johnson, D. 2010. "Sorting Out the Question of Feminist Technology." In *Feminist Technology*, edited by L. L. Layne, S. L. Vostral, and K. Boyer, 36–54. Chicago, IL: University of Illinois Press.

Kabeer, N. 1999. "Resources, Agency, Achievements: Reflections on the Measurement of Women's Empowerment." *Development and Change* 30 (3): 435–464.

Kabeer, N. 2002. *The Power to Choose: Bangladeshi Women and Labor Market Decisions.* London and Dhaka: Verso.

Kabeer, N. 2005. "Gender Equality and Women's Empowerment: A Critical Analysis of the Third Millennium Development Goal 1." *Gender & Development* 13 (1): 13–24.

Katungi, E., P. Aseete, C. Mukankusi, and S. Nkalubo. 2018a. "Towards a More Gender-Responsive Bean Breeding Program: Lessons from East Africa." In *State of the Knowledge for Gender in Breeding: Case Studies for Practitioners*, edited by H. A. Tufan, S. Grando, and C. Meola, 23–34. Gender and Breeding Initiative. Lima: CGIAR.

Katungi, E., S. Louise, J. Chianu, F. Andrew, and S. Beebe. 2018b. "Common Bean in Eastern and Southern Africa." Situation and Outlook Analysis. Cali: CIAT.

Kawarazuka, N., and G. Prain. 2019. "Gendered Processes of Agricultural Innovation in the Northern Uplands of Vietnam." *International Journal of Gender and Entrepreneurship* 11 (3): 210–226.

Kotch, G. P. n.d. "Applying Stage-Gates to Better Manage Public Breeding Programs." Excellence in Breeding Platform. https://excellenceinbreeding.org/blog/applying-stage-gates-better-manage-public-breeding-programs

MacKenzie, D. A., and J. Wajcman. 2011. *The Social Shaping of Technology*. Maidenhead: Open University Press.

Marimo, P., D. Karamura, R. Tumuhimbise, M. M. Shimwela, I. van den Bergh, M. Batte, and C. Massawe. 2019. "Post-Harvest Use of Banana in Uganda and Tanzania: Product Characteristics and Cultivar Preferences of Male and Female Farmers." RTB Working Paper 3, CGIAR, Lima.

Mehar, M., T. Yamano, and A. Panda. 2017. "The Role of Gender, Risk, and Time Preferences in Farmers' Rice Variety Selection in Eastern India." *Asian Journal of Agricultural Development* 14 (1): 17–36.

Mehar, M., W. Mekkawy, C. McDougall, and J. A. Benzie. 2019. "Fish Trait Preferences: A Review of Existing Knowledge and Implications for Breeding Programmes." *Reviews in Aquaculture.* 12 (3): 1273–1296.

Misiko, M. 2013. "Dilemma in Participatory Selection of Varieties." *Agricultural Systems* 119: 35–42.

Mokuwa, A., E. Nuijten, F. Okry, B. Teeken, H. Maat, P. Richards et al. 2014. "Processes Underpinning Development and Maintenance of Diversity in Rice in West Africa: Evidence from Combining Morphological and Molecular Markers." *PLoS ONE* 9 (1): e85953.

Mudege, N. N., E. Mukewa, and A. Amele. 2015. "Training on Gender Integrated Potato Participatory Varietal Selection (PVS) in Ethiopia." Report on Workshop, Addis Ababa, 12–14 January. Lima: CIP.

Mudege, N. N., E. Salas, S. de Haan, S. Bonierbale, T. Mendes, and A. Amele. 2017. "Experience and Guidelines for PVS: Integrating Gender and Potato Breeding." Gender and Breeding Meeting, Nairobi, October 5–6.

Mudege, N. N., S. Sarapura Escobar, and V. Polar. 2020. "Gender Topics on Potato Research and Development." In *The Potato Crop: Its Agricultural, Nutritional and Social Contribution to Humankind*, edited by H. Campos and O. Ortiz, 475–506. Cham: Springer International.

Mulatu, E., and H. Zelleke. 2002. "Farmers' Highland Maize (Zea mays L.) Selection Criteria: Implication for Maize Breeding for the Hararghe Highlands of Eastern Ethiopia." *Euphytica* 127: 11–30.

Mulema, A. A. 2018. "Gender Strategic Research for the African Chicken Genetic Gains (ACGG) Project: A Qualitative Research Guide." Report. Nairobi: ILRI.

Mulema, A., J. Hassen, S. Belay, B. Tekleyohannes, M. Lakew, S. Abegaz, and T. Dessie. 2019. "Can Chickens Empower Women? Perceptions from Chicken Producers in Peri-Urban and Rural Ethiopia." Project Report. Nairobi: ILRI.

Murphy, S., H. Charo-Karisa, S. Rajaratnam, S. Cole, C. McDougall, A. Nasr-Allah et al. 2020. "Selective Breeding Trait Preferences for Farmed Tilapia among Low-Income Women and Men Consumers in Egypt: Implications for Pro-Poor and Gender-Responsive Fish Breeding Programmes." *Aquaculture* 525: 735042.

Musolf, G. R. 2003. "Social Structure, Human Agency, and Social Policy." *International Journal of Sociology and Social Studies* 23 (6/7): 1–12.

Nchanji, E. 2018. "The History of Gender and Plant Breeding in CIAT/PABRA: Where Are We?" Gender and Breeding Initiative 1. Cali: CIAT.

Njenga, M., and J. Gurung. 2011. "Enhancing Gender Responsiveness in Putting Nitrogen to Work for Smallholder Farmers in Africa (N2Africa)." WOCAN Report.

Olaosebikan, O., P. Kulakow, T. Madu, C. Egesi, and B. Teeken. 2018. "A Case Study of Cassava Trait Preferences of Men and Women Farmers in Nigeria: Implications for Gender-Responsive Cassava Variety Development." In *State of the Knowledge for Gender in Breeding: Case Studies for Practitioners*, edited by H. A. Tufan, S. Grando, and C. Meola, 35–43. Gender and Breeding Initiative. Lima: CGIAR.

Olaosebikan, O., B. Abdulrazaq, D. Owoade, A. Ogunade, O. Aina, P. Ilona, A. Muheebwa et al. 2019. "Gender-Based Constraints Affecting Biofortified Cassava Production, Processing and Marketing among Men and Women Adopters in Oyo and Benue States, Nigeria." *Physiological and Molecular Plant Pathology* 105: 17–27.

Orr, A., C. M. Cox, Y. Ru, and J. A. Ashby. 2018. "Gender and Social Targeting in Plant Breeding." Gender and Breeding Initiative Working Paper 1, CGIAR, Lima.

Osei, M. K., M. D. Asante, A. Agyeman, M. A. Adebayo, and H. Adu-Dapaah. 2014. "Plant Breeding: A Tool for Achieving Food Sufficiency." In *Sustainable Horticultural Systems: Issues, Technology and Innovation, Sustainable Development and Biodiversity*, edited by D. Nandwani, 253–274. Cham: Springer International.

Paris, T. R., A. Singh, A. Cueno, and V. Singh. 2008. "Assessing the Impact of Participatory Research in Rice Breeding on Women Farmers: A Case Study in Eastern Uttar Pradesh, India." *Experimental Agriculture* 44 (1): 97–112.

Persley, G. J., and V. M. Anthony. 2017. *The Business of Plant Breeding: Market-Led Approaches to New Variety Design in Africa*. Wallingford: CABI.

Peterman, A., J. A. Behrman, and A. R. Quisumbing. 2014. "A Review of Empirical Evidence on Gender Differences in Nonland Agricultural Inputs, Technology, and Services in Developing Countries." In *Gender in Agriculture: Closing the Knowledge Gap*, edited by A. R. Quisumbing, R. Meinzen-Dick, T. Raney, A. Croppenstedt, J. A. Behrman, and A. Peterman, 145–186. Dordrecht: Springer Netherlands.

Polar, V., C. Babini, C. Velasco, P. Flores, and C. Fonseca. 2017. "Technology Is Not Gender Neutral: Factors That Influence the Potential Adoption of Agricultural Technology by Men and Women." RTB Report. La Paz: CIP.

Polar, V., J. A. Ashby, G. Thiele, and H. Tufan. 2021. "When Is Choice Empowering? Examining Gender Differences in Varietal Adoption through Case Studies from Sub-Saharan Africa." *Sustainability* 13: 3678.

Qaim, M., A. J. Stein, and J. V. Meenakshi. 2007. "Economics of Biofortification." *Agricultural Economics* 37 (1): 119–133.

Ragot, M., M. W. Bonierbale, and E. Weltzien. 2018. "From Market Demand to Breeding Decisions: A Framework." Gender and Breeding Initiative Working Paper 2, CGIAR, Lima.

Ramasawmy, M., A. Galiè, and T. Dessie. 2018. "Poultry Trait Preferences and Gender in Ethiopia." Gender and Breeding Initiative Working Paper 3, CGIAR, Lima.

Rattunde, F., M. Sidibe, B. Diallo, E. van den Broek, H. Some, K. vom Brocke, A. Diallo et al. 2018. "Involving Women Farmers in Variety Evaluations of a 'Men's Crop': Consequences for the Sorghum Breeding Strategy and Farmer Empowerment in Mali." In *State of the Knowledge for Gender in Breeding: Case Studies for Practitioners*, edited by H. A. Tufan, S. Grando, and C. Meola, 95–107. Gender and Breeding Initiative. Lima: CGIAR.

Rowlands, J. 1997. *Questioning Empowerment: Working with Women in Honduras*. Oxford: Oxfam.

Sen, A. K. 1994. "Well-Being, Capability and Public Policy." *Giornale degli Economisti e Annali di Economia* 53 (7/9): 333–347.

Sen, A. 2004. "Capabilities, Lists, and Public Reason: Continuing the Conversation." *Feminist Economics* 10 (3): 77–80.

Teeken, B., O. Olaosebikan, J. Haleegoah, E. Oladejo, T. Madu, A. Bello, E. Parkes et al. 2018. "Cassava Trait Preferences of Men and Women Farmers in Nigeria: Implications for Breeding." *Economic Botany* 72 (3): 263–277.

Thiele, G., D. Dufour, P. Vernier, R. Mwanga, M. Parker, and E. Schulte Geldermann et al. 2020. "Review of Varietal Change in Roots, Tubers and Bananas: Consumer Preferences and Other Drivers of Adoption and Implications for Breeding." *International Journal of Food Science and Technology* 56 (3): 1076–1092.

Tufan, H. A., S. Grando, and C. Meola. 2018. "State of the Knowledge for Gender in Breeding: Case Studies for Practitioners." Gender and Breeding Initiative. Lima: CGIAR.

UN Women. 2019. *Innovation for Gender Equality*. New York: UN Women.

Waezi, F. 2017. "Making Innovation and Technology Work for Women." Background Paper. New York: UN Women.

Wale, E., and A. Yalew. 2007. "Farmers' Variety Attribute Preferences: Implications for Breeding Priority Setting and Agricultural Extension Policy in Ethiopia." *African Development Review* 19 (2): 379–396.

Weltzien, E., F. Rattunde, A. Christinck, K. Isaacs, and J. A. Ashby. 2019. "Gender and Farmer Preferences for Varietal Traits." In *Plant Breeding Reviews*, edited by I. Goldman, 243–278. New York: John Wiley and Sons.

Zaidi, S., H. Vanderschuren, M. Qaim, M. Mahfouz, A. Kohli, S. Mansoor, and M. Tester. 2019. "New Plant Breeding Technologies for Food Security." *Science* 363 (6434): 1390–1391.

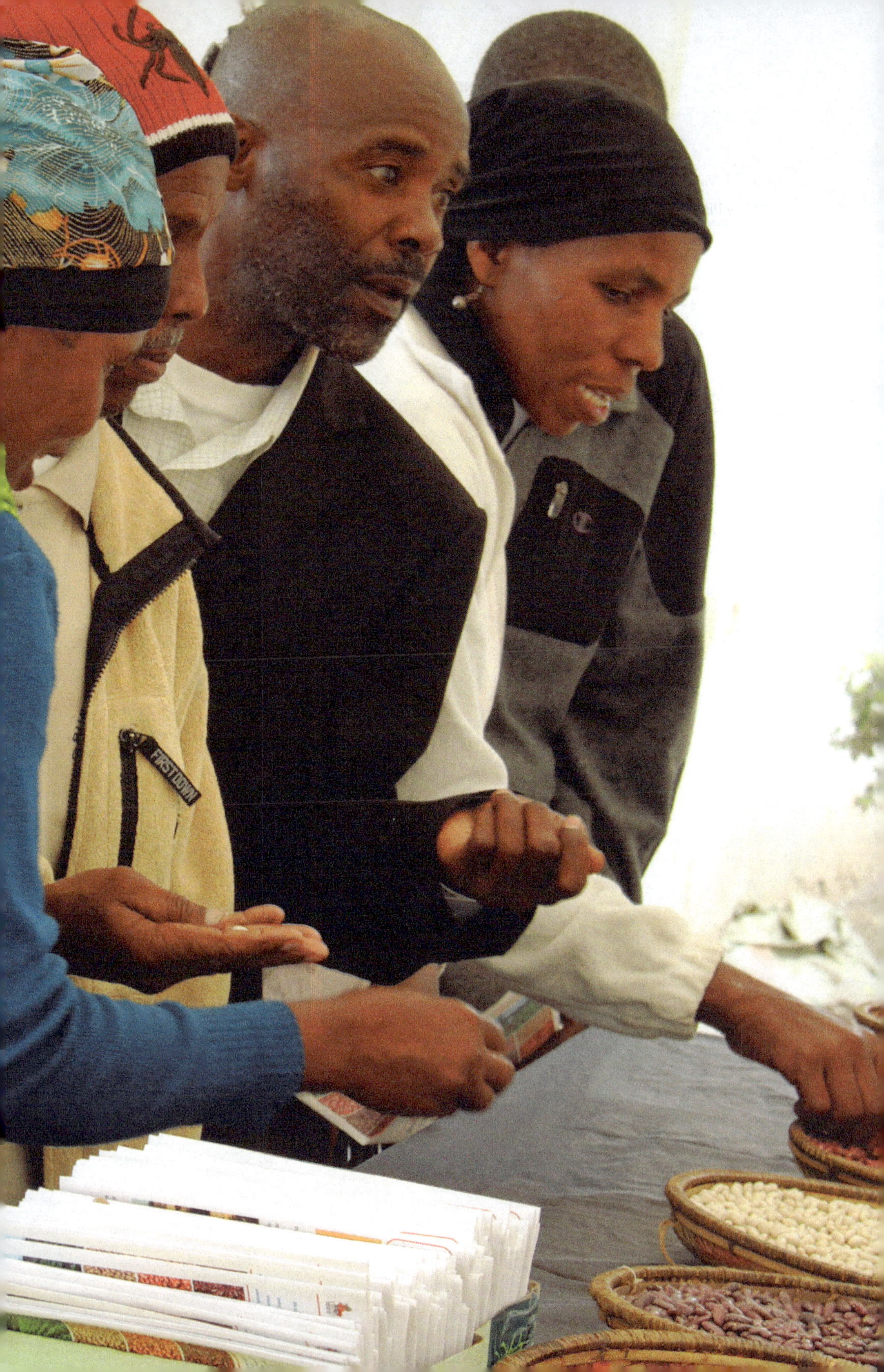
FIRSTDOWN

Chapter 3

MOVING BEYOND *REACHING* WOMEN IN SEED SYSTEMS DEVELOPMENT

Ranjitha Puskur, Netsayi Noris Mudege, Esther Njuguna-Mungai, Eileen Nchanji, Ronnie Vernooy, Alessandra Galiè, and Dina Najjar

Seed is critical to food security as the first link in the food value chain (Galiè 2013) and can be a powerful agent of change (Reddy et al. 2007). Similarly, women's empowerment and gender equality are key to food and nutrition security (Agarwal 2018). The interplay between the two is becoming increasingly important: socioeconomic and gender differences in seed and food security must be understood to target seed interventions effectively (FANRPAN 2011). However, the importance of seed systems to empower women has so far been neglected. This chapter contributes toward closing this gap. Gender analysis is important for a comprehensive understanding of seed systems and to shape effective and inclusive interventions that go beyond reaching women to benefit and empower them.

Gender relations shape seed access, use, and outcomes. In developing countries, seed is often managed by men and women on family farms and sourced mainly from farm-saved seed and the local market (GRAIN 2000, Sarapura 2012). Gender and other socioeconomic factors mediate how farmers access seed sourced through local and formal market channels (Thuo et al. 2014). These factors influence access to information about seed (e.g. origin, quality, price) and ability to purchase seed (e.g. access to cash, negotiating power), which in turn influence utilization (e.g. who plants which seed and where). Applying gender analysis to improve seed systems and reduce/overcome existing biases in access to and availability and use of quality seed of local and improved varieties is an essential first step toward empowering women (FANRPAN 2011).

This chapter reviews existing literature and evidence to answer the question: *How can seed system interventions advance women's empowerment and gender equality?* The study of seed systems does not have a very long history, compared with that of breeding, value chain development, or natural resource management—which likely explains why gender analysis of seed systems is at an early stage. Relevant literature is emerging slowly but still limited. That made this chapter's task not an easy one. It therefore starts with a more typical framing—of how gender dynamics affect seed system outcomes. It then explores how seed system interventions can benefit and empower women. The central premise of this chapter is that seed system interventions can contribute to women's empowerment when designed intentionally to be gender-responsive.

This chapter is based on a systematic review of the seed system literature in English available in CGIAR repositories, using Mendeley and Google Search. The key words used included seed systems + gender roles; formal and informal systems and networks; seed banks and cooperatives; production, processing, storing, saving, exchange, marketing; entrepreneurship; access to seed and information; affordability; willingness to pay; and actors, policies and institutional landscapes. The initial categorization of documents was complemented by identification of parameters and gender frameworks used for analysis. From an initial set of 86 relevant references, 49 were selected for detailed review. Material about Africa was dominant, thus an attempt was made to include more references about Asia and the Middle East. While acknowledging the contextual specificities of gender dimensions and seed systems in these studies, the geographical coverage provided a sufficient base from which to identify and analyze gender asymmetries prevalent across seed systems in different locations.

Most of the literature reviewed lacks sex-disaggregated data; it focuses on how "farmers" as a unit access and use seed. The scarce sex-disaggregated literature that was available was based on household surveys, likely using responses from women heads of household and overlooking the majority of women who are living in households headed by men. These latter likely face different limitations and constraints compared with women household heads. The literature that considers roles of women and men and interactions in the seed system looks largely at informal systems.

We first introduce key gender and seed system concepts, to provide a basis for the discussion. We then take four key seed system outcomes related to seed users—availability, quality, access, and use and control—and review the available evidence through a gender lens. We next reflect on women seed producers and empowerment outcomes, and then summarize insights on the impact of gender dynamics on seed system outcomes and how seed system interventions can contribute to women's empowerment. We conclude by presenting key components of a forward-looking research agenda that support women's empowerment through seed systems.

Gender and seed systems concepts

Here, we present two sets of concepts that guide our analysis and discussion: seed system concepts; and gender equality and empowerment pathway concepts. For the first, we draw on seed security and associated outcomes (Remington et al. 2002, Sperling 2008, McGuire and Sperling 2011, Sperling and Boettiger 2013, FAO 2015, Subedi and Vernooy 2019) and on multi-stakeholder institutional frameworks for intervening in seed systems (RTB 2016, Bentley et al. 2018).

A brief introduction to seed system outcomes, delivery channels, and enabling environment

There are four main seed system outcomes: availability, quality, access, and use and control of seeds.

- **Availability** means having enough seed physically present at the right time and in the right place, especially seed of preferred crops and varieties.
- **Quality** is a standard of excellence in seed attributes that determine the potential performance of a seed lot. It includes both physical qualities (e.g. size, weight, color) and appropriateness (i.e. genetic quality) that result in good seed viability.
- **Access** concerns the capacity to obtain reliable information about how and where to obtain quality seed, its price, and how best to use it. Access is influenced by the mobility/networks of seed users, which is partly reflected in the delivery mechanisms they are able to tap into, such as available transport and logistics of getting seed from point A to B. Affordability is defined as the ability of farmers to buy seed at a reasonable price; it can be affected by the market.
- **Use and control** of seed and the benefits arising from its use are influenced by the resources to which women and men have access and the power relations in the household and community.

Seed delivery channels, a key aspect of seed access, fall broadly into three categories, often referred to as types of seed system: the national public seed system (formal), commercial seed distribution networks (formal), and farmer-managed or community-based seed systems (informal). **National public seed systems** (formal) produce and market certified seed through national/state/parastatal (extension) agencies, cooperatives, and private sector dealer networks. In **commercial seed distribution networks** through wholesale and retail agrodealers (formal), private sector seed companies play a major role; seed of hybrid varieties (cereals and vegetables) that farmers demand every planting season are popular in this channel, and the seed is certified and branded. In **farmer-managed or community-based systems** (informal), seed is exchanged through social and local networks, often moral economy-based (Bates et al. 2011, Schöley et al. 2017, Suma and Großmann 2017) and/or obtained from local markets; seed is predominantly uncertified, truthfully labeled, or quality declared.

The institutional structures/factors that are key in determining the enabling environment for seed systems are seed policy, seed system governance, and seed system management. **Seed policy** refers to a statement of principles that guides government action and explains the roles of relevant stakeholders in a seed system. It can have a significant influence on all seed system outcomes. **Seed system governance** is the process whereby a group of individuals work as a collective to assure the health of a seed system. It usually includes moral, legal, political, and financial aspects. **Seed system management** includes the day-to-day coordination, execution, and monitoring of key tasks required to maintain a seed system in the short and long terms. It usually involves human resources, as well as technical, administrative, organizational, and financial elements.

A brief introduction to gender analysis and women's empowerment pathways

In this chapter, we look at gender dynamics in seed systems for women seed users and producers. Our analytical lens is informed by the Social Relations Framework, which argues that underlying causes of gender inequality are not confined to the household and family but are reproduced across a range of institutions, including the community, state, and market (Kabeer 1994). The use of this Framework enables the examination of power issues, gender dynamics, and social change in relationships, communities, and institutions, and assessment of the potential for gender transformation through seed systems. The Framework conceptualizes gender in terms of social relations that define the ways in which a culture or society defines rights, responsibilities, and the identities of men and women in relation to one another (Bravo-Baumann 2000). These gender relations have strong institutional dimensions, and encompass recognition, gender division of labor, access to and control over resources, decision-making, and social and gender norms (Galiè 2013, Bezner Kerr 2013, Sen et al. 2007). Each is elaborated below.

- **Recognition** is the acknowledgment of the identities and associated roles individuals freely choose or aspire to take in society.
- **Gender division of labor** refers to the allocation of different jobs, responsibilities, or types of work to women and men. This is often influenced by social norms and what is considered suitable and appropriate for each sex. These roles could vary over time and space and are constantly under negotiation. They are also affected by societal, economic, and historical changes. This division of labor influences the economic and other opportunities men and women have.
- **Access to resources** refers to the opportunities and rights to use a resource as per one's need and **control over resources** refers to the rights and power to decide on the use of resources.
- Women express agency in **decision-making** when they influence and make decisions and establish and act on goals. Key decisions that affect women's lives and futures occur in both private and public spheres and often entail a process that includes negotiation and compromise.
- **Gender and social norms** refer to rules and shared social expectations that define acceptable and appropriate actions for women and men in a given group or society.

To deepen the analysis and look at empowerment pathways, we draw on the Reach-Benefit-Empowerment Framework (Johnson et al. 2018) introduced in Chapter 1 of this volume. In short, in the context of seed systems, seed interventions need first to *reach* women and address barriers to reaching them; once reached, women can potentially *benefit* from the use of seed and this could contribute to their *empowerment*. We can consider women's involvement, benefits, and empowerment in seed systems from two angles: women as seed users and seed producers.

A seed system that recognizes the differentiated needs and preferences of women and men farmers as seed users ensures the availability of good quality seed of preferred crops and varieties through appropriate and effective delivery channels. It provides the first step on the pathway toward empowerment by **reaching** women farmers. Women farmers' access to quality seed can be enhanced through creating awareness and providing access to information about the seed and its availability. For seed systems to reach women, quality seed also needs to be affordable and measures need to be in place to support women to overcome the barriers to mobility and networking (Bezner Kerr 2013). These actions have the potential to expand women's access to resources and opportunities and result in positive **benefits** in the form of increased productivity, higher incomes, and food security. Only when women have access to and control over inputs, resources, knowledge, skills, and decision-making can seed systems contribute to women's **empowerment**.

Empowerment through seed systems first requires social and institutional structures that create enabling conditions where women's contributions are valued and rewarded. Gender transformative approaches are those that aim to influence the social context, creating an enabling environment for gender equality and women's empowerment (Razavi 2009, Chant and Sweetman 2012, Kabeer 2012, Okali 2012, Galiè and Kantor 2016; see also Chapter 10, this volume). When seed systems are empowering, the equal rights of men and women to seed are acknowledged and opportunities are equally available to men and women as seed users and as producers. Second, empowerment requires that both men and women farmers are represented when decisions affecting operations of seed systems are discussed and implemented. Third, it requires that women farmers are able to make strategic decisions related to their ability to access, utilize, and benefit from seed.

While this chapter focuses mainly on women seed users, we also examine evidence related to women's empowerment through their engagement as seed producers. Women seed producers challenge stereotypes that narrow women's roles to that of farm laborer without any decision-making power. Seed production is inherently associated with better access to inputs, including knowledge, land, fertilizer, machinery, and credit. This means that women seed producers usually have access to good quality seed to multiply and to associated inputs and resources through networks and institutions. By controlling seed production, it is expected that women producers will generate benefits. When they also have agency over the use of the seed they produce and these benefits (e.g. higher incomes that they control), women seed producers are empowered. An interesting question to explore, for which not much evidence is available now, is whether seed production by women leads to enhanced access to quality seed for women seed users in informal and formal seed systems.

Women seed users and gender dynamics in seed systems

This section focuses on women as seed users and reviews the existing knowledge on gender dynamics in seed systems. We look at seed availability, quality, and access, and use of and control over seed.

Availability

Availability of seed is influenced by the type of delivery channel or seed system (as described above). The formal seed sector generally sells certified seed through a limited number of officially recognized seed outlets. It often focuses on seed from formal breeding programs and a few crops of high commercial value. This sector has generally been less successful in marginal, more variable, low-potential areas; seed often does not reach certain social groups (Shiferaw et al. 2008). As a result, crops and crop varieties with low market value, that are important for smallholder farmers' household food security, are left out. These crops and crop varieties often are under women's control, sometimes referred to as "women's crops." Sometimes, the varieties available may not be relevant (Ceccarelli and Grando 2007, Galiè 2012) or well suited to the complex and gendered needs and preferences of smallholder farmers. Another constraint is the high(er) price of formal sector seed, which makes it unaffordable for many smallholder farmers (Galiè 2013). In sum, the formal seed sector—both private and governmental—faces challenges in reaching women smallholder seed users and meeting their preferences and needs.

Participatory plant breeding (PPB) addresses some of the shortcomings of the formal system to make crop diversity better available to smallholder farmers (Vernooy 2003). PPB brings farmers and scientists together on equal terms to assess and improve varieties (from the national breeding system and local sources) under local farm conditions, based on the selection of locally preferred traits. Varieties created through PPB can then be multiplied locally, and, where government rules and regulations allow, be submitted for formal registration and release. PPB contributes to a seed system that is better able to address supply- and demand-side issues (Almekinders et al. 2007). It is also influenced by gender dynamics. For example, in Syria, women farmers faced difficulties getting involved in a PPB program because male government extension workers were accustomed to working only with men farmers (Galiè 2013). Communities did not appreciate women's involvement in agriculture and did not support them. Gender norms discouraged women's interaction with unrelated men, including breeders and extension workers. In addition, men farmers involved in the sale of seed produced by the program obstructed women's involvement, fearing potential competition.

Smallholder farmers continue to rely to a large extent (70–90 percent) on informal seed sector seed. Medium-scale and better-off farmers also rely to some extent on seed from these sources. The literature on farmer-managed (informal) seed systems is rich and varied and offers good insights into gender roles, community dynamics, and kinship structures. Women play a central role in these systems in seed exchanges, selection, production, and storage, contributing to enhancing nutrition and maintaining crop diversity. Seed is obtained through family, kinship, and friendship relationships, including from saved seed and local markets. An advantage of these systems is that they provide seed of varieties preferred by local communities and that respond to women's needs and preferences. Farmer-managed systems reach women more easily as they circumvent the barriers of mobility and cash to buy seed (Galiè et al. 2017).

Farmer-managed systems are critical in maintaining agrobiodiversity and contribute to resilience and diversity (Subedi and Vernooy 2019), but they are not always able to meet all of the seed needs of the farming community. An emergent new group of seed producers (sometimes referred to as the intermediary seed sector), made up of seedbanks and seed cooperatives, has started to bridge gaps (Box 3.1). In the Gumbu community seedbank, women are key actors, and are provided with varieties that meet their demands. They benefit (from earning extra cash) and are empowered through their active involvement in seed governance and management. While there are examples of women playing a critical role in community seedbanks (World Bank 2009, Vernooy et al. 2015, Mudege et al. 2020), systemic gender inequities, including biased community attitudes, tend to prevent women from realizing the full benefits of such initiatives. For example, a seedbank initiative set up by a nongovernmental organization in 15 villages in northern Ghana saw lower participation of women in seedbank governance as well as constraints to women's membership and leadership, as a result of unrealistic criteria set by male village leaders, hindering availability of seed, information, and skills to women (Nyantakyi-Frimpong et al. 2019).

While there are differences in the way formal and informal systems reach women seed users to make preferred seed available, they share important points of integration. Farmers use both systems to access seed for different crops from both channels. An increasing number of (formal) breeding programs involve farmers in variety selection; farmers sit on variety release committees; and some improved varieties are diffused through local channels.

BOX 3.1 Women farmers ensure availability of preferred crops and varieties through their seedbank in South Africa

The Gumbu community seedbank in Limpopo province, South Africa, operated by women, is located in a remote dryland area with limited market access and low crop diversity. Farming and seed exchange are mostly conducted by poor smallholder women farmers. The seedbank prioritizes the maintenance of nutritious crops and varieties needed for the preparation of traditional dishes. The women farmers of Gumbu contend that the community seedbank allows them to maintain the range of crop species and varieties inherited from their parents. This supports their household food supply, gives them a sense of satisfaction, and allows them to earn some extra cash. They say that the community seedbank is a great new space to meet, talk, decide, and work together.

Source: Vernooy et al. (2015).

It is critical that both systems coexist and together help meet the needs of women seed users.

Quality

Seed quality is defined by genetics, physical appearance, physiological response, and health (Tripp et al. 1997, Weltzien and von Brocke 2001). Use of quality seed alone could increase productivity by 15–20 percent (Yapa 2015). In the formal seed system, quality is defined by trueness to variety, germination percentage, vigor, appearance, and freedom from disease. In a farmer-managed seed system, seed quality is defined more loosely to reflect the performance of seed in a given context. It is "measured" based on the expectations of the seed user.

Men and women farmers often have their own subjective parameters for determining quality. A case study on rice in India, for example, revealed that women and men farmers often related seed quality to yield and thereby conflated varietal traits and seed quality parameters. This kind of "measurement" influences their trust in seed sources (R. Puskur, pers. comm.). A study conducted with bean and cowpea farmers in Tanzania and Ghana to gauge the relative demand for three types of seed products that differ in price and quality (certified, quality-declared, recycled) indicated that farmers were willing to pay significantly more for the higher-rated relative to the lower-rated seeds.

However, for a majority of farmers, the magnitude of the premium they are willing to pay for higher-quality seed is less than the current price differential between certified seed and grain (Maredia et al. 2019). Unfortunately, the literature on seed quality generally fails to capture whether or not there are differences in criteria women and men use to judge or measure seed quality, and what women might value most.

It is often assumed that the seed produced in farmer-managed systems (uncertified) is of low quality. However, a study in Ethiopia and Syria of wheat seed found that a large number of seed samples produced by farmers met the minimum physical purity and germination standards for certified seed, comparable with those of seed from the formal sector (Bishaw et al. 2012). A study in Tanzania found that more than 90 percent of farmer rice seed samples met national seed quality standards. Only about a fifth of the samples from formal and informal systems met the minimum standard for genetic purity (Gebeyehu et al. 2019).

While formal seed system actors often remain skeptical about the quality of seed from farmer-managed systems, most farmers tend to trust the seed produced and provided by fellow farmers more than they trust seed bought from commercial actors (Box 3.2). However, trust is also gendered. For example, in Malawi, men were comfortable obtaining potato seed from outside the village while women concentrated their efforts on accessing seed from farmers they knew in their communities. This was in part because the women

BOX 3.2 What is good quality seed and whom can we trust to provide this?

Women rice farmers in two eastern states of India and Nepal have very high levels of trust in the quality of seed from women's seed producer groups. When probed about quality, they mentioned physical purity (cleanliness of seed) and mixture of varieties as important. They rated the quality of seed produced by seed producer groups more highly than they did seed from private companies or seed dealers, and preferred it. They also felt they would not be cheated, as peer pressure is at play and the seed producer groups are locally accountable. The women seed producers have become true seed experts. They do the quality testing themselves, which enhances users' trust in their seed.

Source: Puskur et al. (forthcoming).

were afraid of being blamed by their husbands if the crop failed as a result of poor-quality seed (Mudege et al. 2016b).

Access to seed

Seed access is determined by access to information and mobility and networks of women and men, and by affordability. We look at the gender dynamics in these two aspects.

ACCESS TO INFORMATION, NETWORKS, AND MOBILITY

Access to information creates awareness of and demand for seed and is mediated by gender norms and relations. Key information sources include extension services, other farmers, family members and relatives, and religious and peer groups (Akca et al. 2008, Poudel et al. 2015). Seed flows tend also to follow these channels. A small proportion of farmers obtain information through information and communication technologies and mass media sources, and from seed companies and agrodealers. Women may not always be targeted (effectively) to receive information and may not have equal access to information and knowledge about seed (Galiè 2013, Mudege et al. 2016, Njuguna-Mungai et al. 2016).

Seed information channels in communities are anchored in local social networks and community processes that are often gendered. In Laos and Indonesia, men have larger seed information networks, and this influences the flow of seed-related information to women negatively (Tatlonghari et al. 2012). In Bangladesh, women depend on family members and neighbors to obtain information on seed, while men obtain this information from both local and external sources, such as extension agents. Older women, who generally have higher mobility, are more able to access information from the public arena (Aktar et al. 2010). While women's low literacy levels do not affect access to informal services, women are less confident about interaction with agrodealers (Lamontagne-Godwin et al. 2018).

For open-pollinated varieties and vegetatively propagated planting materials, awareness-raising and information-sharing can be conducted very effectively through participatory processes, such as field demonstrations, farmer field schools, and participatory varietal selection. While such efforts are often small in scale, the information and experience gained is high as farmers interact intensively with other seed system actors. For women, these processes are often very beneficial: they interact freely, unconstrained by social norms of interaction with men, who are non-family members. Because of time and financial constraints, participatory methods are often implemented with a

BOX 3.3 ***Alea*** **—a social labor-sharing arrangement as an entry point for sharing seed information in Uganda**

In northern and eastern Uganda, farmers replace groundnut and sorghum seed every two/three seasons, with women farmers replacing their seeds more often sourcing it from other farmers. The only external actors with whom women interact are local grain stockists. Women farmers have a traditional labor-sharing mechanism called *Alea*: a common interest group. They take turns working on members' farms at critical times in the crop production cycle, during which they share information about groundnut varieties. At harvest time, they share seed for the next season. *Alea* is a key entry point for sharing new information and seed by the groundnut breeding program in Uganda.

Source: Njuguna-Mungai et al. (forthcoming).

small number of farmers (sometimes referred to as lead farmers or trainers), who tend to be mostly, or only, men. Participatory approaches may thus not reach women.

Innovations are emerging to improve women's access to seed information. Box 3.3 presents a case in which a groundnut breeding program leveraged a unique social arrangement in Uganda through which women farmers access information and seed. Another example is Victoria Seeds' use of local transportation (*tuk-tuks* in Uganda) to function as mobile seed shops in remote villages. Private seed companies have started using videos featuring local farmers to enhance the efficiency and speed of information reach to women (Aktar et al. 2010). This technique borrows from gendered advertising, used to target niche areas for users of a specific gender and age for certain products based on preferences (Njuguna 2009). This approach is used mainly for selected crops and varieties for which profits are assured.

AFFORDABILITY

Seed affordability is not gender-neutral. Lack of money to purchase seed is a major constraint for women, affecting both the quality and volume of seed use (Mudege and Torres 2017). It has impacts on women's ability to buy seed sold by formal systems and increases their dependence on farmer-managed systems. Women use other means to access quality seed in communities, such as seed exchange, casual labor supply, labor exchange, gifts, seed loans or money loans, and deferred payments (Lukonge et al. 2015, Mudege et al. 2015, McGuire and Sperling 2016). However, purchasing seed has a higher prestige than saving one's own in some contexts (Bates et al. 2011).

Farmers are less likely to buy seeds for open or self-pollinated crops such as rice and wheat. The yield potential of open or self-pollinated crops diminishes less between generations, so farmers tend to carry over farm-produced seed. In some cases, both men and women may be willing to pay for new seed, but not once it becomes ubiquitous in the community and they can obtain it from their neighbors and friends. As women and men have overlapping as well as different needs from the same crop, willingness to pay is guided by different drivers (Khan et al. 2016, McGuire and Sperling 2016). The social determinants and gendered differences of farmers' willingness to pay for particular traits and actual ability to pay often remain unaddressed in studies (Kassie et al. 2017).

Women commonly can only afford seed that is relatively cheap (Mudege et al. 2015). In Malawi, women purchased low-quality potato seed that they could afford, while being aware of the risk of pests and disease that could spread to other plots (Mudege et al. 2016b). Large- and medium-scale farmers often opt for more expensive, superior quality, seed of high-yielding varieties obtained from commercial producers (Bogale et al. 2018).

Small packages, "mom and pop" stores, and presentations and promotions in local markets all positively influence the purchase of seeds by women and young farmers (McGuire and Sperling 2016, Kandiwa et al. 2018). For maize in Malawi or sweet potato in Tanzania, subsidies and vouchers are provided (Mudege et al. 2018), although this approach is criticized for creating artificial markets and increasing aid dependency (Mudege and Walsh 2016, Bentley et al. 2018). Little gendered research has been done to understand the influence of incentives and subsidies on affordability.

Use of and control over seed

We now discuss how gender dynamics affect women's use of and control over seed mediated by women's access to resources, including knowledge and skills, and by social norms that influence decision-making.

KNOWLEDGE AND SKILLS

Women's limited access to new knowledge and skills affects their ability to use new seed technologies. Extension services often fail to adequately reach women and provide the services they need (Manfre et al. 2013). Extension officers in Malawi who trained men heads of household on seed marketing

BOX 3.4 Training is necessary but not sufficient

An Irish Aid-funded project, Rooting Out Hunger in Malawi through Orange Fleshed Sweet Potato, targeted women farmer groups for training on conserving vines. A woman participant mentioned that she was conserving her vines using a traditional method by digging a hole and planting the vines in it. This technique conserves moisture and is easier than what is taught in the training (line-planting of vines and irrigating them). She said that she did not have money to purchase irrigation equipment or clean planting material. Women in her community did not control money from the sale of most crops and thus could not save enough to invest in new technologies and practices.

Source: Mudege et al. (2016b).

and profit maximization excluded women farmers (Mudege et al. 2015). This also perpetuates the perception that women are not farmers and therefore need not be consulted by men for decision-making related to seed and farming in general (Mudege et al. 2017).

Some research and development programs do target women to provide information, training, and good quality seed of the varieties they prefer (Mudege and Torres 2017). In Eastern Africa, regional seed companies outperform their global peers in many areas, including in addressing women farmers' needs. Programs for women are conducted by Seed Co. in Kenya and Victoria Seeds in Kenya and Uganda (Access to Seeds Index 2019). While intentional targeting may be necessary to reach women, it is not sufficient. Women farmers' participation in training sessions is usually contingent on negotiating gender norms enforced by kin and community. Moreover, when women's socioeconomic environment is not conducive to implement what has been learned, women may not apply the knowledge and skills they gain. Women's use of seed technologies may be further constrained when they negatively affect their labor burden or when women lack resources to buy additional inputs (Box 3.4).

DECISION-MAKING

Women's decision-making power to acquire and use seeds is important given their roles as household managers and custodians of seed, in particular in farmer-managed systems (Mudege et al. 2015, Khan et al. 2016). In Africa and Asia, planting, selecting, cleaning, and drying seeds are often in the female domain, giving women exclusive access to seed at the sowing and postharvest stages. Similar patterns regarding seed selection are reported for potato in the Andes (GRAIN 2000, Sarapura 2012) and Uganda (Mudege et al. 2016a).

There is, however, little systematic evidence on gendered decision-making and control with regard to seed access and use within households. Gender differences in control of seed in the household appear to vary with the crop and variety and are related to market orientation. In Tanzania and Ethiopia, women manage seed selection, processing, and storage for food crops but not for cash crops (Amri 2010). In some contexts, women do not control the income of crop sales; this in turn affects their ability to purchase seed (Mudege et al. 2018).

The question on women's decision-making power extends beyond a narrow focus on the household, in particular in farmer-based systems. Farmer-managed systems and well-designed participatory breeding programs, in which women play key roles in seed management and production, offer women significant access to and control and use of the desired crops and varieties (Bezner Kerr 2013, Galiè 2013, Mudege et al. 2016a, 2018). In South India, for example, farmers tend to rely on saved seed, which gives them self-reliance in seed, crop diversity, and nutrition. These are three realms that are largely under women's control (Pionetti 2006).

Being able to save their own seed means women can ensure diversity in crops and food, both now and in the future. They can also ensure crop characteristics meet their own specific needs, including sowing at the optimal time. Women's engagement in seed management, including their decision-making, is a factor that positively affects seed system outcomes for women. In addition, women are able to accumulate seed capital and bargaining power within the household. This is discussed in more detail below on empowerment pathways. First, we take a closer look at women as seed producers and the related gender dynamics.

Gender dynamics and women seed producers and entrepreneurship

Seed production can be beneficial for women as an enterprise, and can at the same time strengthen their key roles in seed systems. In Burkina Faso, women first trained in groundnut seed production in 2015 now make up a community of 540 seed producers, earning up to US$200 in one season. In Malawi, where 27–64 percent of the seed is purchased from local markets and 48 percent of farmers use their own saved seed, 47 percent of members in seed

BOX 3.5 **Empowering women through a participatory plant breeding program**

A gender-responsive PPB program that engaged 12 women farmers from three Syrian villages increased women's self-confidence and visibility as knowledgeable farmers, which in turn increased their decision-making power within the household. Women's participation resulted in their access to and control over improved barley and wheat varieties that responded to the preferences of both women and men and that performed better than locally grown varieties. These improved varieties were sold locally at a premium price. Although traditionally only men farmers engaged in seed marketing, women's access to and control over the new PPB varieties through the program created a novel opportunity for two women to start selling the improved seed and earn a significant income.

Source: Galiè (2013).

producer clubs are women. Women seed producers have indicated profit levels, which are 1.5 times the national average income; they feel more food-secure and have gained a sense of entrepreneurship (ICRISAT 2018). Seed systems can create and expand spaces for women's economic participation (Kandiwa et al. 2018). One way to do this is by supporting women to become community-based seed entrepreneurs with seed operations that start at the local level with one or a few crops and then expand to other areas and more crops. In a similar vein, a PPB program has led to significant empowerment outcomes for women seed producers (Box 3.5).

Most evidence and experience of engaging women as seed producers can be found in the farmer-managed and community-based systems. Social norms related to mobility and access to information appear to influence women's participation in markets as seed sellers. Women are quite invisible in seed production and entrepreneurship and are often referred to as the "daily laborers" in smallholder seed production, particularly in the formal or commercial sector (de Roo and Gildemacher 2016). However, there are a few ongoing efforts to engage women in commercial seed production as producers and entrepreneurs (Ogero et al. 2016). Women are increasingly participating in the formation and management of small seed enterprises (World Bank 2005) that deliver better-quality seed compared with the farmer-managed systems.

BOX 3.6 Women seed producers—opportunities and challenges

In southern Ethiopia, women engaged in seed production benefited more than those engaged in grain production because of their access to the inputs and continuous follow-up by development workers. However, only 30 percent of the seed producers were women and all of them were single (widowed, divorced, or separated). Lack of time affected women's participation in seed and grain production, limiting opportunities for networking or participating in community meetings. Married women's access to information was predominantly through their husbands. Women marketed small quantities, owing to limited networks and mobility, resulting in control over smaller income. However, their engagement in seed and grain production enhanced their capacities, income, and assets, and reduced their vulnerability.

Source: Geleta et al. (2017).

While there are some successful examples, the efforts and related outcomes are constrained by the challenges women entrepreneurs face (Box 3.6).

The weak economic position of women, limited access to production factors, and intra-household dynamics hamper women in their efforts to become entrepreneurs. Farm plot size, land used, soil fertility, ox ownership, access to markets, credit, and extension services had significant effects on outcomes realized by women-headed households[1] from participation in local seed businesses (seed producer cooperatives) in Ethiopia (Mulate et al. 2018). Women seed entrepreneurs face challenges as a result of limited access to finance and seed processing machinery, inability to attract and retain skilled workers, and delayed payments. These are further complicated by societal bias toward women in business (Adam et al. 2018).

Women in households headed by men face several challenges in starting professional seed production. Husbands may not see the advantage of their wives becoming seed producers. Sometimes, men do not support their wives starting a business or an activity for which they have to travel outside the village. The heavy work burden of women owing to domestic and other care

1 An exclusive focus on women-headed households may result in misleading conclusions and recommendation, as this group represents a small proportion of women with constraints and needs that differ from those of the majority of women in households headed by men (Doss and Kieran 2014).

responsibilities compounds this problem. Meanwhile, commercial seed production is capital-intensive and requires more land; however, women's access to capital and assets is often limited (Vice Versa 2017).

Given the gender asymmetries in ownership of assets and access to resources/land and financial capital, women are often unable to invest in the agriculture sector. This trend is reflected in the seed sector, which sees a dominance of male-owned companies and operations. For example, the maize seed sector in East and Southern Africa is male-dominated, with men owning and running most seed companies. In this region, the International Maize and Wheat Improvement Center (CIMMYT) has been building capacity and nurturing commercial women seed producers, with a particular focus on making stress-tolerant maize seed reach smallholder farmers. In this process, CIMMYT supports women in breaking social barriers, contributing to improving household nutrition and livelihoods by providing both jobs and improved seed varieties (Adam et al. 2018).

Promising methods to support women's entrepreneurship include (a) gender-specific laws and strategies; (b) integrated loan and training programs; and (c) alternative credit assessment strategies. Targeting groups and individual seed producers is required (de Roo and Gildemacher 2016). Few groups perform production activities collectively, as individual production is seen to encourage better performance. Farmer group efforts focus on the acquisition of inputs and basic infrastructure, training, and joint marketing. Building organizational and financial management capacity is critical to sustain collective action. Developing an intervention that promotes women seed producers and entrepreneurs should take into account social norms that affect women, be flexible enough to include unpaid care responsibilities, and encourage collaboration among family and community members.

Seed systems as pathways to women's empowerment?

Women and men farmers' roles have been segregated in many seed systems, with little change over the past decades in most countries and for most crops. Social and gender relations and norms in households and communities mediate these roles. In all types of seed systems, women's effective access to quality seed is determined by gender inequities in access to and control over resources (including money, credit, extension services, information, and land). These are also related to limited recognition of women's involvement and knowledge in seed systems, sometimes combined with negative attitudes toward women's leadership and management.

While there is extensive literature on seed systems and how they need to be organized to ensure seed security and resilience for smallholder farmers, the body of research unpacking gender dynamics within these seed systems has only recently started to emerge. Formal seed systems, either governmental or private sector-based, tend to reach out to men—as farmers and household heads—without formally recognizing women as seed users and making no specific efforts to reach them. In farmer-managed systems, across cultural and spatial contexts, women often play central roles as custodians, savers, and managers of seed in the household and within communities (Abdelali-Martini et al. 2008, FANRPAN 2011, Pschorn-Strauss 2016, Paris and Rola-Rubzen 2018). Such farmer-managed systems and community-based seed systems seem more responsive to women's needs and interests—but here also gender inequalities constrain women's access to and benefits of using seeds. Challenges persist such as lack of participation in seed-related decision-making, ensuring good seed quality, and access to novel varieties and related information.

Our review has shown that women's awareness and use of seed of new varieties can be limited as a result of lack of knowledge and access to training (AGRA 2016). Women farmers fail to access good quality seed when they do not control income from crops or lack income from other sources (Mudege et al. 2015). Women may also opt for more affordable but lower-quality seed. To access seed, women tend to rely on social networks and kinship structures based on reciprocity and the "moral economy." Gender-responsive packaging and thoughtful, well-designed subsidies may make seed more accessible to women.

Agricultural extension and advisory services enable farmers and other rural actors to gain knowledge and, importantly, provide information on seeds, inputs, technology, credit, markets, and other value chain/business development services. Such services offer the opportunity to women farmers to strengthen their capacity to improve livelihoods, and health and nutrition outcomes for their families. Knowledge was the most important characteristic conferring status to farmers in Syria (Galiè et al. 2012).

However, extension services often do not reach women farmers. Besides, the mere participation of women in extension services and programs is not sufficient for them to benefit significantly if this occurs in a context where gender inequities and asymmetries persist. Factors such as literacy levels, work burden (balancing of household chores with farm work), and managing limited household budgets all play a part in women's negotiation capacity and ability to attend training sessions.

Taking note of the emerging knowledge and understanding how gender inequalities affect seed system outcomes for women as seed users, the key question is how seed system interventions can advance women's empowerment and gender equality. Some studies demonstrate how this can happen (Galiè et al. 2012, Mudege and Torres 2017). Access to good quality and relevant plant seed can increase productivity, resilience, nutrition, and food security (Mudege and Walsh 2016) and, potentially, enhance benefits and lead to empowerment of women, contributing to gender equality. Access to quality seed alone can increase yield in farmer fields, for example 10–15 percent for maize (Abebe and Alemu 2017) and rice (Haque et al. 2012) and 30–50 percent for potato (Wang 2008). Access to good seed of multiple varieties can be a useful strategy to respond to changing climatic conditions, increasing farmers' resilience (McGuire and Sperling 2016) and enable women to realize their full potential as farmers and seed producers in their own right. In particular, access to and control over seed can change gender relations in a community. Seed sales can enhance women's economic empowerment when they strengthen their decision-making in the household.

Women who were provided access to stress-tolerant rice variety seed and trained in seed production in eastern India and Bangladesh claimed that their social status had improved, and they perceived themselves as farmers and not just housewives (Cueno 2014). They also gained confidence in decision-making, enhanced their knowledge, had marketable surplus, and experienced better status within the household and community. The increased income and control over the use of crops helped them produce quality seed based on their own resources, to be used in the subsequent planting season (ibid.). These are all notable empowerment outcomes of seed availability and access to and use and control of seeds.

The evidence on women's benefit and empowerment outcomes as a result of engagement in seed systems is slightly more tangible when looking at women seed producers. The engagement of women as producers appears to hold high potential to expand economic opportunities. Nonetheless, efforts to promote sustainable and viable women-led small-, medium- and large-scale seed enterprises have to contend with gender inequities in access to resources, particularly capital and market linkages. Gender norms that affect mobility, access to networks for information, and market linkages can influence these initiatives negatively. However, even in societies where social norms are highly

BOX 3.7 Bangladesh—the specialized women's seed network

The Nayakrishi Seed Network in Bangladesh is the active farmers' network of Nayakrishi Andolon, a large agroecological movement in the country. Women farmers are key actors and leaders in the network, covering three major agroecological zones and bringing together 300,000 farmers. The network is made up of seed huts at the village level and community seed wealth centers at district level. This network creates space for women to have their own sphere of knowledge and practice. Among the results has been an increase in the use of local crop varieties, most of which are better adapted to the local agroecological conditions and the reintroduction of (rice) varieties that had disappeared

Source: Ubinig and Nayakrishi Andolon (2015), Ubinig (2018).

constraining, engagement of women as seed producers can provide pathways for empowerment.

In eastern India, for example, women who had no say in decision-making about sourcing seeds were trained in quality seed production and engaged as seed producers. They were very confident of the breadth and depth of knowledge they had acquired, and some started training other women in their communities and beyond. They are now able to discuss at length seed quality parameters in technical terms and do the basic quality testing after production. This has enhanced their self-esteem. They now take part in household decision-making regarding varietal and seed choice (Puskur et al. forthcoming).

Community seedbanks which often engage women actively in seed production and decision-making are important for smallholder farmers, women in particular, to access quality seeds (World Bank 2009, Ubinig and Nayakrishi Andolon 2015, Vernooy et al. 2015, Ubinig 2018). There are some good examples of seed networks managed by women that have created a space for own decision-making and leadership (Box 3.7), going beyond reaching women.

We draw four lessons at this point. First, most seed system interventions by research and development organizations may reach women, and some may benefit them. Second, while gender inequities constrain seed systems from reaching women, gender equality and the recognition and leadership of women's involvement in seed management and production not only contribute to women gaining access to quality seeds but also offer important opportunities for them to obtain benefits and also experience empowerment outcomes.

A third lesson is that it is necessary to consider institutional dimensions of both gender relations and seed systems. Institutional dimensions of gender relations include the extent to which women's involvement in seed management, production, and use is recognized and valued. This also includes gender norms regarding expected and appropriate roles, behavior, voice, and mobility of women. With respect to seed systems, institutional dimensions pertain to seed policies, management, and governance. The evidence about whether and how women's involvement in seed management translates to women's empowerment is limited. Evidence suggests that empowerment of women depends on several social factors beyond the realm of seed systems, and no automatic causal relationship can be established.

Fourth, acknowledging the institutional dimension is critical for approaches that explicitly aim to integrate or mainstream gender in seed systems. This requires a critical analysis of the influence of community norms and practices, customary laws, and formal policies and laws. In some cases, gender mainstreaming approaches are overly technical, ignoring the systemic causes of inequality. Seed systems interventions therefore need to consider gendered power asymmetries and dynamics at play.

Toward gender equality through a seed systems research agenda

Seed systems per se do not directly lead to women's empowerment and gender equality. To achieve this, seed system interventions require specific and targeted objectives, such as the promotion of women's access to resources; women's voice in household and community decision-making; and addressing structural barriers to women's participation. Gender-responsive seed policy is required to ensure gender equality goals are achieved. Here, we propose some elements of a forward-looking research for development agenda that will begin to build the systemic evidence base needed to inform the design and implementation of such gender-responsive seed systems, bolstered by gender-responsive seed policy.

Build evidence on gender dynamics in seed systems. As described, there is little evidence available on gendered aspects of seed systems. There is little sex-disaggregated and much lesser qualitative data to support a deeper gender analysis. Available quantitative data mostly concern women heads of household, who constitute a very small proportion of rural women; most women live in households headed by men. We need better and intentional integration of gender in seed systems studies and also specific efforts to bridge knowledge gaps. Understanding the local social and gender context is critical in designing seed systems that allow equitable access to seeds and that empower women either as users or producers of seed. This should be based on a systematic analysis and understanding of gender roles and the social norms and power relations at play that determine access and use.

Adapt and track gendered seed access indicators. Development of the Access to Seed Index (https://www.accesstoseeds.org/) is a commendable effort, and this a useful tool to gather evidence on access to seed by the private sector. However, the current index does not make a gender-informed analysis of access and does not include gender indicators. It covers crops and varieties that private companies produce (mostly hybrids and cash crops) but not other crops that may be important for women, including vegetatively propagated crops and varieties important for nutrition. What is not measured is not managed. National governments and other research and development agencies could take a cue from this and develop indicators of seed access for a broader range of crops and varieties important for women, and track them.

Analyze gendered impacts of seed systems policies. A very relevant and urgent area of research involves understanding how policies affect seed systems reaching, benefiting, and empowering women. Seed policies developed without a gender lens could have negative impacts on women's ability to access seed. Even though farmer-managed seed systems are the principal source of seed for food crops in both Africa and Asia, national and regional seed policies often do not support them—and may even undermine them. No concerted efforts to date have studied gender dimensions in formal seed policy processes and seed governance, and this is a key knowledge gap to fill.

Flip the question—ask what seed systems can do for women and not just what women can do for seed systems. While unpacking gender dynamics of seed systems, it is important to move beyond the instrumental use of gender analysis to improve seed system outcomes to examine how seed systems can facilitate, support, and spur women's empowerment. Some key questions to address include:

- How should the formal seed sector reinvent itself to be more relevant to women seed users—including understanding and responding to their preferences and needs; designing effective quality seed and information delivery channels; and enhancing capacity of women seed users to be able to use the seed?

- What institutional and marketing innovations and incentives make quality seed affordable to smallholder women seed users?
- What mechanisms and approaches support nurturing women seed entrepreneurs/producers, including gender-responsive financial products, business development services, skills development, and policy incentives and support?
- Do social and behavioral change communication approaches, if integrated into seed system interventions, influence gender norms, attitudes, and behavior that constrain women seed users and producers from obtaining positive benefits from engagement?

Use integrated seed sector development approaches. The integrated seed sector development approach (Louwaars et al. 2013), adopted by the Integrated Seed Sector Development (ISSD) Africa program, provides a practical framework for countries to move toward a more balanced seed sector. The approach proposes to build stronger linkages between formal and farmer-managed systems and create more space for community-based seed initiatives, for example seedbanks, seed cooperatives, and women-led seed enterprises. A better understanding of gender dynamics in integrated seed sector development could strengthen the approach. The theme on Gender and Seed Systems of the ISSD Africa program, in which several CGIAR centers participate, has put this at the core of its agenda (https://issdafrica.org/gender-and-seed-systems/).

There are other opportunities to integrate gender into global seed system development frameworks and initiatives. For example, in 2017, the CGIAR Collaborative Platform for Gender Research (since January 2020 the GENDER[2] Platform) funded five projects on gender dynamics in seed systems. The ISSD Africa program, already mentioned, is another example. However, the gender transformational potential of seed systems remains under-researched and under-profiled. Interventions designed to systematically test the impact of gender-responsive approaches on women's empowerment are very much needed: they represent a contemporary call to arms for seed sector actors and programs. Gender equality and women's empowerment should be the next frontier for seed system development.

2 https://gender.cgiar.org/

Geography of study sites for publications cited in Chapter 3

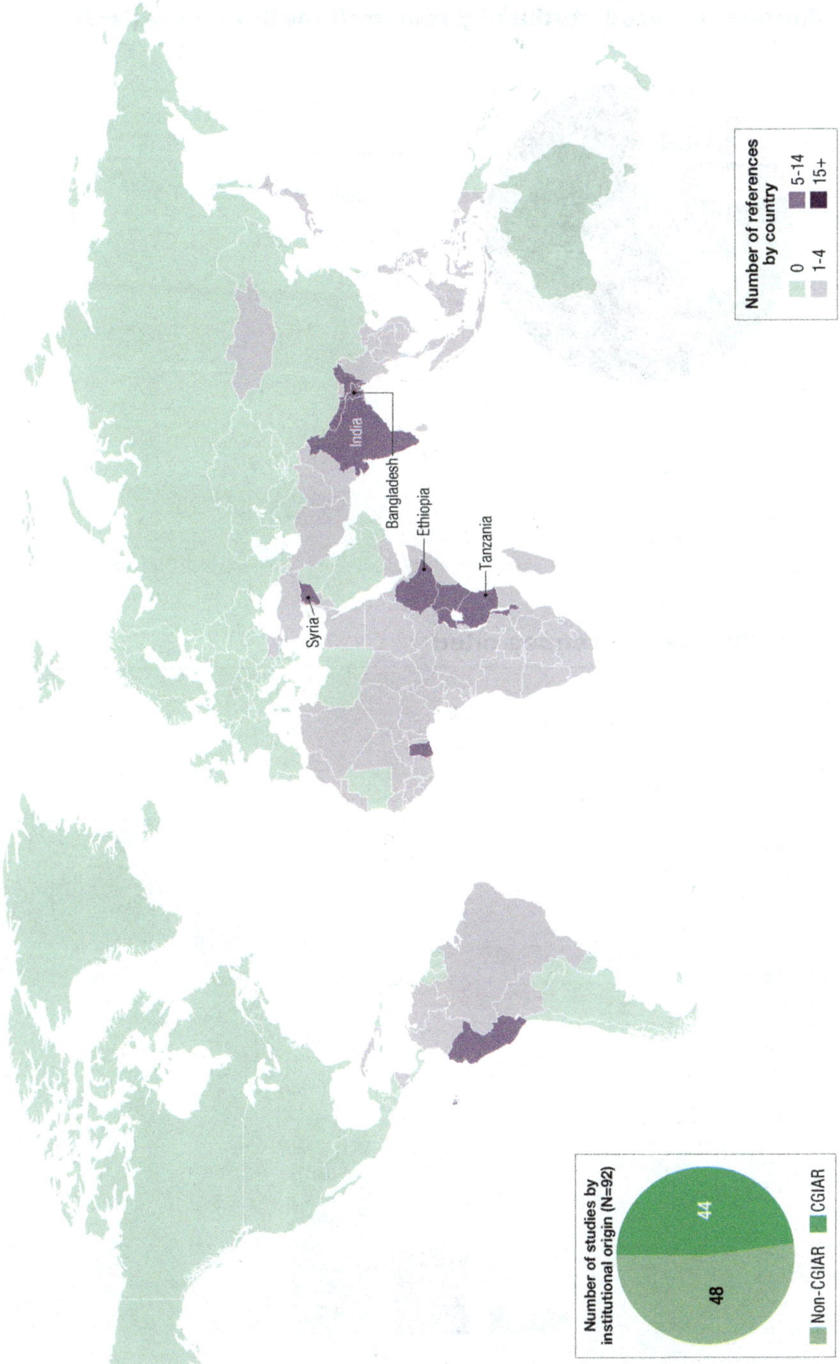

Number of cited studies by research methodology (N=92)

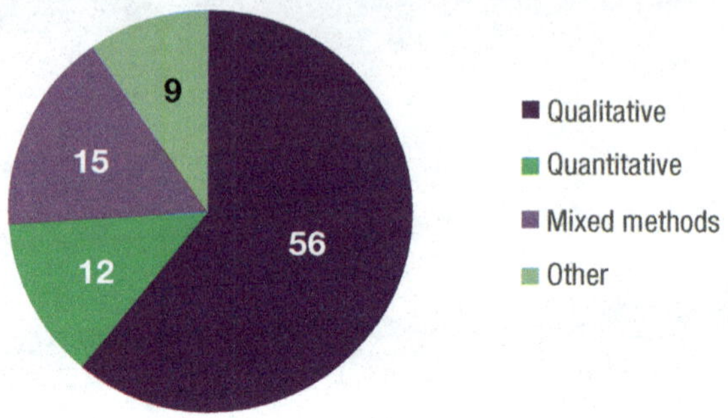

Timeline for references cited

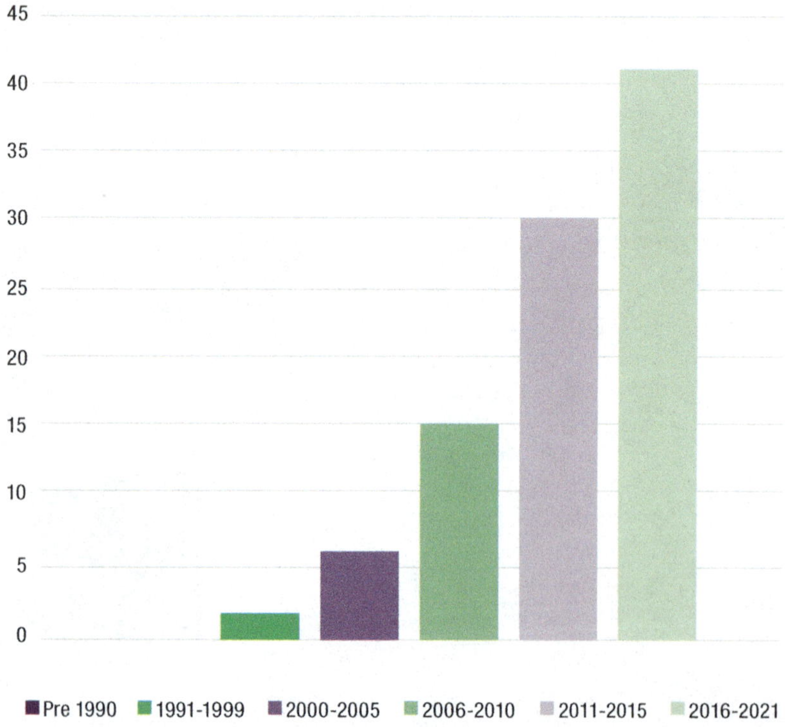

References

Abdelali-Martini, M., A. Amri, M. Ajlouni, R. Assi, Y. Sbieh, and A. Khnifes. 2008. "Gender Dimension in the Conservation and Sustainable Use of Agro-Biodiversity in West Asia." *The Journal of Socio-Economics* 37 (1): 365–383.

Abebe, G., and A. Alemu. 2017. "The Role of Improved Seeds towards Improving Livelihood and Food Security in Ethiopia." *International Journal of Research - Granthaalayah* 5 (2): 338–356.

Access to Seeds Index. 2019. *Access to Seeds Index 2019. Synthesis Report*. Amsterdam: Access to Seeds Foundation.

Adam, R. I., V. Kandiwa, and P. Muindi. 2018. "Gender-Responsive Budgeting Tool for the Promotion of Improved Maize Seed in Africa." Nairobi: CIMMYT.

Agarwal, B. 2018. "Gender Equality, Food Security and the Sustainable Development Goals." *Current Opinion in Environmental Sustainability* 34: 26–32.

AGRA (Alliance for a Green Revolution in Africa). 2016. *Africa Agriculture Status Report 2016: Progress towards Agricultural Transformation in Africa*. Nairobi: AGRA.

Akca, H., M. Sayili, O. Duzdemir, and Y. Serin. 2008. "Information Sources for Farmers and Factors Affecting Seed and Agrochemicals Usage: The Case of Turkey." In *Proceedings of 43rd Croatian and 3rd International Symposium on Agriculture*, 211–214, Opatija.

Aktar, R., A. Chowdhury, A. Zakaria, and C. R. Vogl. 2010. "Seed Information and Communication Networks of Male and Female Farmers: A Micro Level Study in Bangladesh." In *Building Sustainable Rural Futures. The Added Value of Systems Approaches in Times of Change and Uncertainty*, edited by I. Darnhofer and M. Grötzer. 760–769. Vienna: Universität für Bodenkultur.

Almekinders, C., G. Thiele, and D. Daniel. 2007. "Can Cultivars from Participatory Plant Breeding Improve Seed Provision to Small-Scale Farmers?" *Euphytica* 153: 363–372.

Amri, E. 2010. "The Role of Gender in Management and Conservation of Seed Diversity of Crops and Varieties: A Case Study in Bariadi, Tanzania." *American-Eurasian Journal of Sustainable Agriculture* 8 (4): 365–369.

Bates, R. M., A. J. Bicksler, C. Mai, R. Burnette, T. Gill, L. M. Yoder et al. 2011. "Strengthening Informal Indigenous Seed Systems in Southeast Asia." ECHO Asia Impact Center.

Bentley, J. W., J. Andrade-Piedra, P. Demo, B. Dzomeku, K. Jacobsen, E. Kikulwe, and K. Ogero. 2018. "Understanding Root, Tuber, and Banana Seed Systems and Coordination Breakdown: A Multi-Stakeholder Framework." *Journal of Crop Improvement* 32 (5): 599–621.

Bezner Kerr, R. 2013. "Seed Struggles and Food Sovereignty in Northern Malawi." *The Journal of Peasant Studies* 40 (5): 867–897.

Bishaw, Z., P. C. Struik, and A. J. G. Van Gastel. 2012. "Farmers' Seed Sources and Seed Quality: Physical and Physiological Quality." *Journal of Crop Improvement* 26 (5): 655–692.

Bogale, S. A., F. J. Verhees, and H. C. V. Trijp. 2018. "Customer Evaluation of Supply Systems: The Case of Ethiopian Seed Supply Systems." *Journal of African Business* 19 (4): 550–570.

Bravo-Baumann, H. 2000. "Livestock and Gender: A Winning Pair." Working Document. Bern: SADC.

Ceccarelli, S., and S. Grando. 2007. "Decentralized-Participatory Plant Breeding: An Example of Demand Driven Research." *Euphytica* 155: 349–360.

Chant, S., and C. Sweetman. 2012. "Fixing Women or Fixing the World? 'Smart Economics', Efficiency Approaches, and Gender Equality in Development." *Gender & Development* 20 (3): 517–529.

Cueno, A. 2014. "The Outcomes of Ensuring Women's Access to Stress-Tolerant Variety (STV) Seeds and Seed Preservation Training: The Cases of India and Bangladesh." *GRiSP Gender Research Network Workshop Proceedings*, edited by S. Akter and B. Hardy.

De Roo, N., and P. Gildemacher. 2016. *Promoting Sustainable Seed Sector Development*. Amsterdam: Wageningen UR Centre for Development Innovation, KIT, and IFDC.

Doss, C., and C. Kieran. 2014. *Standards for Collecting Sex-Disaggregated Data for Gender Analysis: A Guide for CGIAR Researchers*. CGIAR Gender and Agricultural Research Network.

FANRPAN (Food, Agriculture and Natural Resource Policy Analysis Network). 2011. "Seeds without Borders." Volume XI: Policy Brief 1 (March).

FAO. 2015. *Voluntary Guide for National Seed Policy Formulation*. Rome: FAO.

Galiè, A. 2012. "Equal Access to Seed and Food Security in Syria." *Insights* 82. Brighton: Institute of Development Studies.

Galiè, A. 2013. "Governance of Seed and Food Security through Participatory Plant Breeding: Empirical Evidence and Gender Analysis from Syria." *Natural Resources Forum* 37 (1): 31–42.

Galiè, A., and P. Kantor. 2016. *From Gender Analysis to Transforming Gender Norms: Using Empowerment Pathways to Enhance Gender Equity and Food Security in Tanzania*. London: Routledge.

Galiè, A., J. Jiggins, and P. Struik. 2012. "Women's Identity as Farmers: A Case Study from Ten Households in Syria." *NJAS: Wageningen Journal of Life Sciences* 64/65: 25–33.

Galiè, A., A. Mulema, M. Benard, S. Onzere, and K. Colverson. 2015. "Exploring Gender Perceptions of Resource Ownership and Their Implications for Food Security among Rural Livestock Owners in Tanzania, Ethiopia, and Nicaragua." *Agriculture & Food Security* 4: 2.

Galiè, A., J. Jiggins, P. C. Struik, S. Grando, and S. Ceccarelli. 2017. "Women's Empowerment through Seed Improvement and Seed Governance: Evidence from Participatory Barley Breeding in Pre-War Syria." *NJAS – Wageningen Journal of Life Sciences* 81: 1–8.

Galiè, A., S. Grando, and S. Ceccarelli. 2018. "Gendered Seed: From Variety Selection to Control over Seed." In *State of the Knowledge for Gender in Breeding: Case Studies for Practitioners*, edited by H. A. Tufan, S. Grando, and C. Meola, 109–121. Gender and Breeding Initiative. Lima: CGIAR.

Gebeyehu, S., J. Kangile, and E. Mwakatobe. 2019. "Assessment of Seed Quality along the Rice Seed Value Chain in Tanzania." *Development in Practice* 29 (7): 854–866.

Geleta, E. B., P. Elabor-Idemudia, and C. Henry. 2017. "Scaling-up: Gender Integration and Women's Empowerment in Southern Ethiopia." *Cogent Food & Agriculture* 3 (1): 1415100.

GRAIN. 2000. "Potato: A Fragile Gift from the Andes." *Seedling* (September).

Haque, A., F. A. Elazegui, M. A. Taher Mia, M. M. Kamal, and M. Manjurul Haque. 2012. "Increase in Rice Yield through the Use of Quality Seeds in Bangladesh." *African Journal of Agricultural Research* 7 (26): 3819–3827.

Johnson, N., M. Balagamwala, C. Pinkstaff, S. Theis, R. Meinzen-Dick, and A. Quisumbing. 2018. "How Do Agricultural Development Projects Empower Women? Linking Strategies with Expected Outcomes." *Journal of Gender, Agriculture and Food Security (Agri-Gender)* 3 (2): 1–19.

ICRISAT (International Crops Research Institute for the Semi-Arid Tropics). 2018. "A Thriving Community of Women Seed Producers." ICRASAT Happenings Newsletter, July 26.

Kabeer, N. 1994. *Reversed Realities: Gender Hierarchies in Development Thought*. London: Verso.

Kabeer, N. 2012. "Women's Economic Empowerment and Inclusive Growth: Labour Markets and Enterprise Development." SIG Working Paper 2012/1. International Development Research Centre, London.

Kandiwa, V., R. Adam, K. Lweya, P. Setimela, L. Badstue, and P. Muindi. 2018. *Gender-Responsive Approaches for the Promotion of Improved Maize Seed in Africa*. Mexico City: CIMMYT.

Kassie, G. T., A. Abdulai, W. H. Greene, B. Shiferawd, T. Abate, A. Tarekegne, and C. Sutcliffe. 2017. "Modeling Preference and Willingness to Pay for Drought Tolerance (Dt) in Maize in Rural Zimbabwe." *World Development* 94: 465–477.

Khan, T., A. Kishore, and P.K. Joshi. 2016. "Gender Dimensions on Farmers' Preferences for Direct-Seeded Rice with Drum Seeder in India." IFPRI Discussion Paper 1550, IFPRI, Washington, DC.

Lamontagne-Godwin, J., F. E. Williams, N. Aslam, S. Cardey, P. Dorward, and M. Almas. 2018. "Gender Differences in Use and Preferences of Agricultural Information Sources in Pakistan." *Journal of Agricultural Education and Extension* 24 (5): 419–434.

Louwaars, N. P., W. S. de Boef, and J. Edeme. 2013. "Integrated Seed Sector Development in Africa: A Basis for Seed Policy and Law." *Journal of Crop Improvement* 27 (2): 186–214.

Lukonge, E., R. Gibson, L. Laizer, R. Amour, and D. P. Phillips. 2015. "Delivering New Technologies to the Tanzanian Sweetpotato Crop through Its Informal Seed System." *Agroecology and Sustainable Food Systems* 39 (8): 861–884.

Manfre, C., D. Rubin, A. Allen, G. Summerfield, K. Colverson, and M. Akeredolu. 2013. "Reducing the Gender Gap in Agricultural Extension and Advisory Services: How to Find the Best Fit for Men and Women Farmers." Modernizing Extension and Advisory Services Discussion Paper, USAID, Washington, DC.

Maredia, M., R. Shupp, E. Opoku, F. Mishili, B. Reyes, P. Kusolwa et al. 2019. "Farmer Perception and Valuation of Seed Quality: Evidence from Bean and Cowpea Seed Auctions in Tanzania and Ghana." *Agricultural Economics* 50 (4): 495–507.

McGuire, S., and L. Sperling. 2011. "The Links between Food Security and Seed Security: Facts and Fiction That Guide Response." *Development in Practice* 21 (4–5): 493–508.

McGuire, S., and L. Sperling. 2016. "Seed Systems Smallholder Farmers Use." *Food Security* 8 (1): 179–195.

Mudege, N. N., and S. Torres. 2017. "Gender Mainstreaming in Root Tuber and Banana Crops Seed Systems Interventions: Identification of Lessons Learnt and Gaps." RTB Working Paper, CIP, Lima.

Mudege, N. N., and S. Walsh. 2016. "Gender and Roots Tubers and Bananas Seed Systems: A Literature Review." RTB Working Paper 2016-2, CIP, Lima.

Mudege, N. N., E. Kapalasa, T. Chevo, T. Nyekanyeka, and P. Demo. 2015. "Gender Norms and the Marketing of Seeds and Ware Potatoes in Malawi." *Journal of Gender, Agriculture and Food Security* 1 (2): 18–41.

Mudege, N. N., S. Mayanja, and D. Naziri. 2016a. "Gender Situational Analysis of the Sweet Potato Value Chain in Central and Eastern Uganda and Strategies for Gender Equity in Postharvest Innovations." RTB Challenge Theme Paper. Lima: CIP.

Mudege, N. N., T. Chevo, T. Nyekanyeka, E. Kapalasa, and P. Demo. 2016b. "Gender Norms and Access to Extension Services and Training among Potato Farmers in Dedza and Ntcheu in Malawi." *The Journal of Agricultural Education and Extension* 22 (3): 291–305.

Mudege. N. N., N. Mdege, P. E. Abidin., and S. Bhatasara. 2017. "The Role of Gender Norms in Access to Agricultural Training in Chikwawa and Phalombe Malawi." *Gender, Place, Culture* 24 (12): 1689–1710.

Mudege N. N., O. M. Robert, N. M. Mwanga, T. Chevo, and P. E. Abidin. 2018. "Scaling Up of Sweet Potato Vine Multiplication Technologies in Phalombe and Chikwawa Districts in Malawi: A Gender Analysis." *NJAS - Wageningen Journal of Life Sciences* 85: 1–9.

Mudege, N. N., S. Sarapura, and V. Polar. 2020. "Gender Topics on Potato Research and Development." In *The Potato Crop: Its Agricultural, Nutritional and Social Contribution to Humankind*, edited by O. Ortiz, and H. Campos, 475–506. Dordrecht: Springer.

Mulate, B., B. Eyayu, M. A. Mohd, and J. Dar. 2018. "The Role of Women Headed Households in the Local Seed Business in Ethiopia: The Case of Jabi Tehnan, Bure and Yilmana Densa Districts." *Global Journal of Human-Social Science* 18 (1): 23–35.

Nyantakyi-Frimpong, H. 2019. "Visualizing Politics: A Feminist Political Ecology and Participatory GIS Approach to Understanding Smallholder Farming, Climate Change Vulnerability, and Seedbank Failures in Northern Ghana." *Geoforum* 105: 109–121.

Njuguna-Mungai, E. M. 2009. "An Evaluation of the Effects of Information and Technology Characteristics on Technology Choice and Adoption: The Case of Striga and Stemborer Control Technologies in Maize Production in Western Kenya." PhD Thesis, University of Nairobi.

Njuguna-Mungai, E. M., M. L. Liani, M. Beyene, and C. O. Ojiewo. 2016. "Exploration of Cultural Norms and Practices Influencing Women's Participation in Chickpea Participatory Varietal Selection Training Activities: A Case Study of Ada'a and Ensaro Districts, Ethiopia." *Journal of Gender, Agriculture and Food Security* 1 (3): 40–63.

Njuguna-Mungai, E., R. Puskur, M. McEwan, B. Lenjiso, N. Mudege, R. Pyburn, and P. Gildemacher. Forthcoming. "Seed Information, Seed Sources and Seed Use for Non-Hybrids Crops in Africa and Asia."

Ogero, K., M. McEwan, and N. Pamba. 2016. "Clean Vines for Smallholder Farmers in Tanzania." In *Case Studies of Roots, Tubers and Bananas Seed Systems*, edited by J. L. Andrade-Piedra, J. W. Bentley, C. Almekinders, K. Jacobsen, S. Walsh, and G. Thiele, 80–97. RTB Working Paper 2016-3, CIP, Lima.

Okali, C. 2012. "Gender Analysis: Engaging with Rural Development and Agricultural Policy Processes." Working Paper 26, Future Agricultures Consortium, Brighton.

Paris, T., and M. F. Rola-Rubzen. 2018. *Gender Dimension of Climate Change Research in Agriculture (Case Studies in Southeast Asia).* Wageningen, Netherlands: CGIAR.

Poudel, D., B. Sthapit, and P. Shrestha. 2015. "An Analysis of Social Seed Network and Its Contribution to On-Farm Conservation of Crop Genetic Diversity in Nepal." *International Journal of Biodiversity* 2015: 312621.

Pionetti, C. 2006. *Seed Diversity in the Drylands: Women and Farming in South India*. London: International Institute for Environment and Development.

Pschorn-Strauss, E. 2016. "African Food Sovereignty: Valuing Women and the Seed They Keep." *Keeping Seeds in Peoples' Hands: Right to Food and Nutrition Watch* 32 (1): 49–51.

Puskur, R., R. Ram Mohan, M. McEwan, and N. Mittal. Forthcoming. "Generating Positive Outcomes for Women Seed Producers: What Kind of Engagement Models Work?".

Razavi, S. 2009. "Engendering the Political Economy of Agrarian Change." *Journal of Peasant Studies* 36 (1): 197–226.

Reddy, R. C., V. A. Tonapi, P. G. Bezkorowajnyj, S. S. Navi, and N. Seetharama. 2007. *Seed System Innovations in the Semi-Arid Tropics of Andhra Pradesh.* Monograph. Patancheru: ICRISAT.

Remington, T., J. Maroko, S. Walsh, P. Omanga, and E. Charles. 2002. "Getting off the Seeds-and-Tools Treadmill with CRS Seed Vouchers and Fairs." *Disasters* 26 (4): 316–328.

RTB (CGIAR Research Program on Roots, Tubers and Bananas). 2016. "Multi-Stakeholder Framework for Intervening in RTB Seed Systems." Working Paper 2016-1, RTB, Lima.

Sarapura, S. 2012. "Gender Analysis for the Assessment of Innovation Processes: The Case of Papa Andina in Peru." Module 7: Innovative Activity Profile 3. In *The Agricultural Innovation Systems Sourcebook*, 598–602. Washington, DC: World Bank.

Sen, G., P. Ostlin, and A. George. 2007. "Unequal, Unfair, Ineffective and Inefficient Gender Inequity in Health: Why It Exists and How We Can Change It." Final Report to the WHO Commission on Social Determinants of Health.

Schöley, M., and M. Padmanabhan. 2017. "Formal and Informal Relations to Rice Seed Systems in Kerala, India: Agrobiodiversity as a Gendered Social-Ecological Artifact." *Agriculture and Human Values* 34: 969–982.

Shiferaw, B. A., T. A. Kebede, and L. You. 2008. "Technology Adoption under Seed Access Constraints and the Economic Impacts of Improved Pigeon Pea Varieties in Tanzania." *Agricultural Economics* 39 (3): 309–323.

Sperling, L. 2008. *When Disaster Strikes: A Guide for Assessing Seed Security.* Cali: CIAT.

Sperling, L., and S. Boettiger. 2013. "Impacts of Selling Seed in Small Packs: Evidence from Legume Sales." AgPartnerXChange.

Subedi, A., and R. Vernooy. 2019. "Healthy Food Systems Require Resilient Seed Systems." In *Agrobiodiversity Index Report 2019: Risk and Resilience*, edited by R. Remans, S. Jones, E. Dulloo, C. Viliani, N. Estrada Carmona et al., 127–134. Rome: Bioversity International.

Suma, T. R., and K. Großmann. 2017. "Exclusions in Inclusive Programs: State-Sponsored Sustainable Development Initiatives amongst the Kurichya in Kerala, India." *Agricultural Human Values* 34: 995–1006.

Tatlonghari, G., T. Paris, V. Pede, I. Siliphouthone, and R. Suhaeti. 2012. "Seed and Information Exchange through Social Networks: The Case of Rice Farmers of Indonesia and Lao PDR." *Sociology Mind* 2 (2): 169–176.

Thuo, M., A. A. Bell, B. E. Bravo-Ureta, M. A. Lachaud, D. K. Okello, E. Nasambu Okoko et al. 2014. "Effects of Social Network Factors on Information Acquisition and Adoption of Improved Groundnut Varieties: The Case of Uganda and Kenya." *Agriculture and Human Values* 31: 339–353.

Tripp, R., N. P. Louwaars, W. J. van der Burg, D. S. Virk, and J. R. Witcombe. 1997. "Alternatives for Seed Regulatory Reform: An Analysis of Variety Testing, Variety Regulation and Seed Quality Control." Agricultural Research and Extension Network Paper 69. London: Overseas Development Institute.

Ubinig. 2018. "Bangladesh: The Importance of Farmers' Seed Systems and the Roles of Community Seed Wealth Centers." Dhaka and Rome: Ubinig and Bioversity International.

Ubinig and Nayakrishi Andolon. 2015. "FAO Honoured Nayakrishi Farmer." Ubinig, November 12.

Vernooy, R. 2003. *Seeds That Give. Participatory Plant Breeding.* Ottawa: International Development Research Centre.

Vernooy, R., B. Sthapit, M. A. Dibiloane, N. L. Maluleke, T. Mukoma, and T. Tjikana. 2015. "South Africa: A New Beginning for Community Seed Banks." In *Community Seed Banks: Origins, Evolution and Prospects*, edited by R. Vernooy, P. Shrestha, and B. Sthapit, 254–258. Abingdon: Routledge.

Vice Versa. 2017. "Farmers Gain Access to and Control over Seeds." Food and Business Knowledge Platform, March 12.

Wang, F. 2008. "The Importance of Quality Potato Seed in Increasing Potato Production in Asia and the Pacific Region." In *Proceedings of a Workshop to Commemorate the International Year of the Potato*, edited by M. Papademetriou, 46–53. Bangkok, May 6.

Weltzien, E., and K. von Brocke. 2001. "Seed Systems and Their Potential for Innovation: Conceptual Framework for Analysis." In *Targeted Seed Aid and Seed System Interventions: Strengthening Small-Farmer Seed Systems in East and Central Africa*, edited by L. Sperling, 9–14. Proceedings of Workshop, Kampala, June 21.

World Bank. 2005. *Agricultural Investment Sourcebook.* Washington, DC.

World Bank. 2009. *Gender in Agriculture Sourcebook.* Agriculture and Rural Development Series. Washington, DC.

Yapa, A. 2015. "Sowing the Seeds of Productivity: Weighs the Ideal Mix of Farm Seed Inputs." *LMD* 147.

Chapter 4

PROMISE AND CONTRADICTION: VALUE CHAIN PARTICIPATION AND WOMEN'S EMPOWERMENT

Markus Ihalainen, Sumer Shaikh, Gaudiose Mujawamariya, Sarah Mayanja, Sounkoura Adetonah, Katie Tavenner, and Marlène Elias

With the expansion of agricultural production for the global market, interest among research and development actors in developing more "inclusive" value chains has grown (Stoian et al. 2018a). While mainstream value chain development (VCD) has focused on enhancing the efficiency of processes along the chain, policies increasingly endeavor to address poverty, emphasizing the inclusion of poor and marginalized people in global markets (Bolwig et al. 2010). In recent decades, scholars and development agencies have directed attention toward the "gender gap in agriculture" (Huyer 2016), encompassing, among other things, women's lesser access to market opportunities and to the benefits from cash crops (Pyburn et al. 2015). As a consequence, agricultural development initiatives increasingly aim to support the engagement of women in agricultural value chains (Sachs 2019). However, while promoting the engagement of women in markets and value chains is becoming entrenched in the global agenda for gender equality (Stoian et al. 2018a), systematic evidence demonstrating causal linkages or laying out a clear theory of change between participation and empowerment remains limited (Said-Allsopp and Tallontire 2014, Johnson et al. 2018).

This chapter places empowerment at the core of our analysis, with the explicit aim of understanding *how and under what conditions participation in agricultural value chains can advance or constrain women's empowerment.* Through examining empowerment as a potential outcome of participation, we aim to better understand the nature of empowerment processes in the context of value chains, as well as to discuss the extent to which such processes help catalyze broader structural change. Based on a critical review of the literature, we observe that, while value chain approaches are gaining currency within the global development and gender equality agenda, few studies adopt empowerment as the primary topic of inquiry. Moreover, there is considerable inconsistency in conceptualizing and measuring empowerment in the context of value chains. Evidence linking value chain participation to different dimensions of empowerment is mixed at best. In addition, contextual factors, both within and beyond the value chain, have a strong influence on this link. The following section explains our conceptualization of empowerment and its relation to gender equality in the context of value chains.

The chapter has four sections. The first introduces our operationalization of empowerment and lays out the scope of the chapter. The second reviews the evidence on empowerment outcomes linked to participation in value chains. The third section assesses ways in which VCD efforts advance (or constrain) such outcomes. Here, in order to provide more a nuanced, in-depth insight into the complexity of gender relations, power relations, and empowerment in the context of agricultural value chains, we present two case studies: one on the dairy value chain in Kenya and one on the shea value chain in Burkina Faso. The fourth and concluding section outlines pertinent gaps and identifies priorities for future research on empowerment and value chain participation.

Operationalizing empowerment in value chains

There is considerable discrepancy in the ways in which major development actors frame the relationship between empowerment and value chain participation. For instance, in describing its approach to inclusive market systems development, Oxfam views empowerment *as a necessary precondition*, arguing that "women's empowerment and rights need to be prioritized for more equitable market systems to be achieved" (Kidder et al. 2017, 26). Through a partnership with CARE, Mars claims to empower women through *enhancing opportunities to participate* in the cocoa value chain by providing them with financial services and market access (Mars 2019). Going beyond mere participation, the UN Secretary-General's High-Level Panel on Women's Economic Empowerment stresses the importance of "supporting and enabling women to reach their *full potential at all levels of the value chain*" (Klugman and Tyson 2016, 6 [our emphasis]), while the International Labour Organization stresses *links between empowerment and decent work* (Mayoux and Mackie 2007). CARE International stresses *empowerment as a broader outcome*, viewing global value chains as a potentially "*powerful lever for empowering women*" (Boyle 2016, online [our emphasis]), while at the same time emphasizing women's *empowerment as an instrument to achieve broader societal benefits*, such as food and nutrition security.

The emphasis in this chapter is on empowerment as an outcome of value chain participation. This framing holds empowerment as the dependent variable rather than as a potential ripple effect of other economic or development goals. It goes beyond treating participation or increased income as an "end goal." In defining our scope, we refer to the framework, distinguishing between agricultural interventions aimed at **reaching, benefiting**, or **empowering** women (Johnson et al. 2018), as Chapter 1 introduced. Within the value chain context, gender equality has often been understood in terms of equal participation and benefits (for example FAO 2018, 13). Projects aiming to **empower women**, however, focus on measures including "… outcomes that are inherently empowering (for example women's agency), inherently disempowering (for example gender-based violence, time burden) or indicators of women's position relative to men (for example degree of control over income, participation in joint decision making, gender-asset gap)" (Johnson et al. 2018, 5).

Instead of merely discussing differences in women and men's engagement, this chapter focuses on processes that change "… *gender* relations in order to enhance *women*'s ability to shape their lives" (Riisgaard et al. 2010, 7, emphasis in the original). It therefore includes studies focusing on dynamics of participation and benefits only insofar as they relate to empowering or disempowering processes and outcomes. In addition, while we consider that summarizing the literature on "reach" and "benefit" is beyond the scope of this chapter, it is worth noting that there is a considerable body of research, within and outside CGIAR and across commodity sectors, on identifying and addressing the multiple economic, technological, sociocultural, and political barriers limiting women's opportunities to equitably participate in and benefit from value chains (see reviews by, for example, Njuki et al. 2013, Mutua et al. 2014, Ingram et al. 2016, Gumucio et al. 2018, Kruijssen et al. 2018).

Given the multifaceted nature of empowerment, we avoid prescribing strict operationalizations and indicators. We are interested in understanding if and how participation in different segments or nodes of different agricultural value chains, or different direct benefits (for example income, training, social networks), influence women's agency at individual (for example confidence), household (for example economic decision-making), and collective (for example participation in collective decision-making bodies) levels. We further explore if and how that agency can contribute to achieving changes in economic and social relations within the household (for example reproductive labor burden), value chain (for example job security), or society (for example norms around "women's and men's labor").

Inspired by Naila Kabeer (2017, 651), our understanding of empowerment situates it within structures of inequality, and envisions a theory of change in which empowerment challenges those very structures to enhance gender equality. We view empowerment as an inherently political process and, as such, one that is impossible to disassociate from the structural context within which it is situated. While Kabeer's analysis is multidimensional (that is, political, economic, and social), many contemporary guides to "gender-equitable value chain development" assume structural transformation through individual-level gains across various market-oriented indicators, such as skills, capacities, or participation in collective enterprises (Stoian et al. 2018a, 504). This neoliberal co-optation of empowerment, emphasizing the power of money in allowing women to realize their potential, has produced "... empowerment lite, a version of empowerment pared of any confrontation with the embedded social and power relations that produce societal and material inequities" (Cornwall 2018, 3). We argue that broader socioeconomic, institutional, technological, and environmental processes associated with the expanding reach of agricultural markets do not merely induce changes in women and men's resources and agency but also introduce new structures and power relations. Hence, we assess the various outcomes discussed above in relation to a broader structural context.

We recognize that it is also possible to enhance gender equality, understood in terms of equal rights, responsibilities, and opportunities of women and men (see discussion in Chapter 1) through top-down approaches, including legal reforms (see Chapter 6 on natural resource management, this volume). However, given the prominence of the bottom-up vision linking empowerment gains from value chains to a transformation of unequal structures, both within and beyond the value chain, this chapter has a two-fold objective:

1. To understand the *directionality and magnitude* of effects across individual and relational empowerment indicators (for example women's agency, gender-based violence, time burden, degree of control over income, participation in joint decision-making, gender-asset gap) (Johnson et al. 2018); and,

2. To investigate the *theory of change* across these aspects, and broader structural transformation toward greater gender equality.

Empowerment as an outcome of participation

This section reviews evidence across a number of agricultural value chains and considers the different *forms* of women's participation—for example as wage laborers, smallholder, or self-employed—as well as the *conditions* under which participation and women's empowerment occur.

Within the value chain

Amid the reorganization of agricultural production into global production systems (Carr and Chen 2004), many value chains have witnessed a shift toward more casual employment practices. The expansion of commercial agriculture has, in many cases, opened up opportunities for women to engage in wage and salaried labor, but the situation is complex, fraught with tradeoffs.

Studies on export horticulture (Barrientos 2001, Barrientos et al. 2003), oil palm plantation work (Li 2015), gum processing (Shackleton et al. 2011), and wood furniture production (Nansereko 2010) have found that while women may comprise the majority of the casual workforce, men tend to occupy the permanent positions. Insecure contracts, long working hours, poor pay, and a lack of social protection characterize women's employment (see also Singh 2003, Tallontire et al. 2005). Evidence from the Kenyan horticultural export sector demonstrates how companies hedge against losses incurred by sudden changes in commodity orders by employing women on temporary terms and paying them less than they pay men (Dolan and Sutherland 2002). In some sectors, essentialist perceptions of women's patience (Shackleton et al. 2011), "nimble fingers" (Li 2015), obedience (Singh 2003), and lack of skills and strength (Nansereko 2010, Tallontire et al. 2005) are used to justify the relegation of women into low-skill, low-pay, and often hazardous activities (Dolan 2004, Chen 2009, Grace et al. 2015).

In the Senegalese groundnut sector, flexible work arrangements have allowed women to combine productive work with reproductive responsibilities; however, narratives presenting women's productive work as a "natural continuum of household activities" legitimize insecure and underpaid employment for women (Baglioni 2018, 127). Casual employment opportunities not only reinforce existing gender inequalities but also come with additional negative effects for women. In Kenya's cut flower sector, women identify job insecurity as a major obstacle to reporting sexual harassment (Jacobs et al. 2015). In Indonesia, aggressive land acquisition processes precipitated women's entry into casual labor on oil palm plantations, which eroded their traditional livelihoods and left particularly land-poor households with few alternatives (Li 2015, Elmhirst et al. 2017).

For smallholders and self-employed women, value chains similarly offer opportunities and risks. As smallholder households participate in value chains, there is a risk that male-controlled cash crops will displace female-controlled, low-market value crops—reducing women's access to independent income and/or increasing unpaid labor spent on subsistence production (for example Handschuch and Wollni 2016, Elmhirst et al. 2017). In some instances, women's entry into conventionally male-dominated value chain activities, such as charcoal production in Mozambique, can owe to an absence or transformations of rigid gender norms (Jones et al. 2016). However, in the Malian charcoal and Guatemalan coffee value chains, it was economic distress following increased outmigration of men, rather than more inclusive social norms, that contributed to women's entrance into the male-dominated sectors (Lyon et al. 2010, Djoudi and Brockhaus 2011).

As markets develop for historically female-dominated commodities—for example babassu[1]—there is potential for a shift in gender relations (Gumucio et al. 2018). Men have even taken on domestic responsibilities as a result of women's involvement in babassu production. The creation of "women-centered value chains"—that is, supporting the commercialization of commodities and activities in which women have customarily occupied a central role—can effectively overcome gender-based constraints and enable women's leadership within the value chain (Riisgaard et al. 2010).

However, outcomes are not only positive. Across a number of commodity sectors, including shea (Chen 2017, Kent 2018), bananas (Fisher and Qaim 2012), fisheries (O'Neill et al. 2018), non-timber forest products (Ingram et al. 2014), and small-scale oil palm (Sarku 2016), male involvement and/or appropriation of benefits has increased when the profitability of the value chain has gone up. In Malawi and Tanzania, efforts to promote orange-fleshed sweet potato as a means of alleviating malnutrition created high demand for sweet potato vines. This resulted in wealthier, well-connected men with the means to invest taking over the formerly female market segment of seed production and trade (Sindi et al. 2013, Mudege et al. 2018). Similarly, in Nigeria, the establishment of fish trading centers allowed male aquacultural farmers to bypass women traders and sell directly to consumers (Veliu et al. 2009).

In the Philippines, women in more remunerative nodes, such as trading and marketing, experience a stronger sense of empowerment, as they generally

1 Babassu is an edible oil derived from the South American Babassu palm seeds.

have more autonomy and decision-making power over their work (Malapit et al. 2020). However, existing gender relations and norms often limit women's access to more profitable value chain segments.

For instance, in the Kenyan dairy value chain, men commonly make major financial decisions regarding commercially traded milk whereas women decide over milk that is consumed or traded locally on informal markets (Tavenner and Crane 2018). In the Malaysian aquaculture value chain, women often perform "hidden labor", such as preparing the gear and food for men for their fishing trips (Krujssen et al. 2013). In Nepal's bay leaf value chain, women producers are more prone to exploitation by traders, because they often lack market information and bargaining power (Choudhary et al. 2010). Religious and social beliefs prevent women in the Burkinabe gum value chain from travelling to more lucrative markets and negotiating prices with men (*"Women should not interact with men"; "Women should not travel, women cannot ride bicycles"*) (Shackleton et al. 2011). In northern Mali, women from a lower social class took up charcoal production as a means of adapting to increasing droughts and male outmigration (Djoudi and Brockhaus 2011).[2] They often lacked the political connections to access production permits and were hence relegated to producing informally and trading on less profitable local markets.

Beyond the value chain

Women's participation in agricultural value chains can have empowering and disempowering outcomes beyond the value chain itself. To start, the income women's value chain participation generates can contribute to poverty reduction and well-being improvements (Dolan and Sutherland 2002, Shackleton et al. 2011, Butz 2013, Elias and Arora-Jonsson 2017, Smith et al. 2017, Mulema et al. 2017, Farnworth et al. 2018). In Kenya, many migrant women workers in the cut flower sector send remittances to their families. Despite poor employment, characterized by job insecurity, long hours, and arduous tasks, these earnings make a difference in their lives and those of their families, and this work is seen as a better alternative to unpaid domestic tasks in their ancestral homes (Dolan and Sutherland 2002). In several non-timber forest product value chains, wages for women are very low but the employment opportunities make it possible to avoid risky forms of work (including transactional sex), and the income allows some women to diversify their livelihood portfolio (Shackleton et al. 2011).

2 Interestingly, although similar challenges faced them, women from a higher social class could not take up charcoal production as it conflicted with respectable femininity.

Generally, due to the gendered division of household labor and decision-making, women tend to use a greater share of their income on food whereas men spend more on assets (IFAD 2009, Oduol et al. 2017, Smith et al. 2017). Such pre-existing gender relations may limit the catalytic effect of income in their economic life, as spending is directed away from strengthening their position within the value chain. This is not always the case: in a number of studies women participating in value chains have reported increased human capital, in terms of knowledge and skills for better farm management, other technical aspects of production, and marketing, as well as leadership and negotiation skills (Kanji 2004, Verhart and Pyburn 2010, Shackleton et al. 2011, Quisumbing et al. 2014).

At the household level, women's participation in value chain activities can contribute to increased decision-making power over household finances (Verhart and Pyburn 2010, Said-Allsopp and Tallontire 2014) and more equal distribution of household expenditure (Quisumbing et al. 2013, 2014). For a number of female charcoal producers in Mozambique, financial autonomy is in fact a key motivation for participation (Jones et al. 2016). Increased decision-making power can, in turn, catalyze further livelihood enhancement—for example significantly higher productivity increments on plots where women control outputs (Elson 1995, in Agarwal 2003). In the dairy value chain in Mozambique, decision-making authority has remained with men, who have nevertheless begun to recognize the skills their wives have acquired and increasingly consult them on matters regarding dairy business (Quisumbing et al. 2013).

Other studies—on tea and fish value chains—find no correlation between women's participation and their control over generated income (Loconto 2015, Limuwa and Synnevåg 2018). In Kenya's poultry value chain, men tend to control the accruing benefits even though it is often women and children who rear the chickens (Maina et al. 2014). Control over assets is a key determinant of whether women control income (Quisumbing et al. 2013, 2015, Stoian et al. 2018b). In the Kenyan avocado value chain, sole-male decision-making tends to be more predominant in households producing export-quality avocados than in households producing lower-quality avocados for local markets (Oduol et al. 2017). Lacking control over the production process and outputs, women—particularly in male-headed households—in fact also opt out of participating in more profitable value chains, instead focusing on lower-value markets and commodities over which they enjoy more control (Turner et al. 1997, Oduol et al. 2017, Curry et al. 2019).

At community level, women's engagement in productive activities can support the transformation of norms that limit their mobility, or perceptions of certain activities as unfit for women (Shackleton et al. 2011, Quisumbing et al. 2015, Hunt and Samman 2016). As a consequence of gaining employment and income, women also report increased dignity and pride (Lyon et al. 2010, Said-Allsopp and Tallontire 2014).

In the Indonesian aquaculture sector, however, efforts to strengthen women's economic empowerment did not translate into expanded mobility, social freedoms, or reduced workloads (Sari et al. 2017). Similarly, in the coffee value chain in Mesoamerica, women's increased participation in the formerly male-dominated value chain has not translated into enhanced legal, political, and social status (Lyon et al. 2010). Moreover, in some cases, women—particularly those entering into male-dominated value chains—may lose social respect for engaging in activities portrayed as unsuitable for women (Fisher 2004, Ahmed et al. 2018). In Bangladesh, female entrepreneurs and wage workers are more susceptible to losing social respect than women in smallholder households, as working outside the home is seen as a deviation from social conventions (Ahmed et al. 2018).

Despite some exceptions (see for example Gumucio et al. 2018), women's engagement in economic activities rarely translates into a reduced domestic workload (for example Fröcklin et al. 2013, Quisumbing et al. 2013, 2015). This may force them to limit hours spent on economic activities (Arora et al. 2017). At the same time, value chains allowing for flexibility in labor, such as the acaba, swine, seaweed, and coconut value chains in the Philippines, can be more conducive to women's participation, by allowing women to better combine productive and reproductive work (Malapit et al. 2020). However, in some instances, women's increased incomes and/or reduced time available for reproductive work have led to—sometimes violent—resistance from husbands (Rahman 1999, Elias and Arora-Jonsson 2017). In the Philippines, for instance, the project-level Women's Empowerment in Agriculture Index (pro-WEAI) points to attitudes toward gender-based violence as a key source of disempowerment across all of the above-mentioned value chains (Malapit et al. 2020).

Discussion: empowerment as an outcome of participation?

In sum, the relationship between women's participation in value chains and empowerment is anything but straightforward. Women often obtain multiple and valuable economic and non-economic benefits from their engagement and consider themselves more empowered than before (Said-Allsopp and Tallontire 2014). They gain professional confidence, self-efficacy, and social status, which in turn can help challenge restrictive gender norms that relegate women to subsistence and to the domestic sphere. Income earned through value chain activities can enhance women's household bargaining power and increase their livelihood options. However, women's greater engagement in increasingly commercialized and globalized agricultural production systems is by no means an automatic indication of empowerment.

Our review points to a wide range of both empowering and disempowering features and outcomes, mediated through factors operating at the household and community level and within the value chain. In particular, empowerment gains or losses are associated with the structure and nature of the value chain, including levels of competition, horizontal and vertical power relations, the location of markets, and the terms and conditions under which women participate. Many critical conditions, for example job security, employment quality, and the ability to organize collectively, are identified as prerequisites to women's empowerment *within* global value chains. These sit uneasily with the prevailing trend of increased flexibilization and casualization of the workforce (Said-Allsopp and Tallontire 2014)—a trend that ironically is also a major enabling factor of women's entry into many value chains (for example Loconto 2015).

A second critical factor mediating empowerment gains or losses are gender relations and norms within the household, community, and society. Women's control over resources, decision-making power, and freedom of movement, among others, is in many cases a critical precondition of equitable and empowering participation. Representations of women's labor as low skilled or merely an extension of the domestic sphere may legitimize gender inequalities within value chains. Also, notions of masculinity, constructed around certain commodities or activities, can prompt backlashes against women engaging in "unfeminine" activities or taking on male-coded responsibilities.

Importantly, processes of empowerment and disempowerment can run concurrently, sometimes at different levels. Independent income earned through casual labor may enhance women's standing within the household while the precarious nature of the employment may constrain their ability to improve their working conditions. Also, gains in one dimension of empowerment do not necessarily translate into gains in another dimension—nor does empowerment gained in the economic sphere automatically follow into the household. Earning money may extend women's options while also intensifying their workload and responsibilities without increasing their autonomy (Pearson 2007). Even worse, women may face social repercussions for challenging predominant gender norms.

Our findings not only caution against assuming an automatic, linear relationship between enhancing women's (economic) resources and agency and achieving broader transformational change. They also call on initiatives that aim to enhance women's empowerment through VCD to acknowledge how broader social, political, economic, environmental, and gender inequalities and transformations affect potential empowerment gains from increasingly commercializing value chains. These systems and relationships in which value chains are embedded can both constrain and expand the range of options available for women's (and men's) strategic life choices. This is especially critical as more intensive engagement in value chains can also expose particularly poorer participants to increased competition and risk, while commercial agriculture may displace or erode women's traditional livelihoods (Elmhirst et al. 2017).

Box 4.1 provides an example of applying a gender analysis to ongoing and socially embedded systems of market relations in the Kenyan dairy value chain.

BOX 4.1 Intensifying inequality? Gendered power dynamics and market participation in Kenyan dairy

Private and public sector stakeholders in Kenya widely promote dairy intensification as a means to improve rural livelihoods through increased cash income. However, recent studies suggest that the direct economic benefit of participating in milk markets tends to be skewed toward men. This owes to existing gender inequalities in labor, ownership, and decision-making in dairy production: women tend to be the primary dairy laborers, responsible for the arduous daily tasks of milking, feeding, and caring for cows, yet are marginalized in ownership and decision-making. The question of how women can achieve a commensurate income from the sale of milk is perhaps the most pressing gender issue in Kenyan dairy intensification—one complicated by social norms that ascribe masculinized meaning to cattle, milk, and commercial production in Kenya and that mediate women's and men's engagement with both formal and informal milk markets.

In intensified dairy production, farmers usually milk their cows twice daily. Formal markets collect the first milking (morning milk) from farmers associated with dairy cooperative societies. The members and leaders of these societies are often men, who may inadvertently promote men's control over milk income. Most cooperatives deposit payments with members on a monthly basis to a single account, usually in the name of the "household head". While women can open accounts in their own name, most married women with accounts report the income to their husband and receive an agreed-upon proportion of the total proceeds. In some cases, they turn over all earnings to their husband. Despite their lack of control over proceeds, participation in formal markets can offer tangible benefits for women: cooperatives provide access to loans and veterinary services through "check-off" systems, whereby farmers can purchase dairy inputs and supplies using delivered milk as collateral. By not requiring cash up front, women are able to use the check-off system with minimal spousal negotiations or conflict.

Informal market transactions are those that involve selling milk to neighbors or local shops, or through intermediary salespeople known locally as "hawkers". Women customarily have the discretion to sell informally or consume the second milking of the day (evening milk), with the husband not monitoring the proceeds. In this way, women engaging in informal transactions are often able to exert more control over dairy income. However, the effects of increased milk yields and associated profits under intensification, and the focus on the sale of milk to the formal sector (a space that men customarily dominate), mean women are at risk of losing control over the evening milk sales. Furthermore, although women working as hawkers may be able to circumvent patriarchal constraints to production and marketing, including mobility constraints and issues related to cattle and land ownership, they also face physical security risks by operating outside the law. Hawking is often considered a dangerous and "unfeminine" practice and this therefore endangers their social standing in their family and community.

In Kenya, the process of dairy intensification appears to be generating both potentially empowering and disempowering outcomes for women via their participation in formal and informal milk markets. Greater milk yields lead to higher incomes for evening milk, resulting in heightened risk of intra-household contestation and conflict. While informal market outlets may facilitate the deflection of certain patriarchal norms, the dynamics of intensification accelerate the pushing of the boundaries of "acceptable" femininity, creating opportunities for gendered conflict and exacerbated inequality. The potential benefits of income-generating opportunities available via intensification and formal market participation must be weighed within the contested context of these socially embedded markets.

Value chain development for empowerment

The past few years have witnessed an increased emphasis on "inclusive" value chain efforts, which often encompass the provisioning of effective services and strengthening of links between value chain actors to enhance outcomes for less powerful chain actors (Stoian et al. 2018a). Various forms of VCD efforts can improve value chain outcomes and lead to women's empowerment (Coles and Mitchell 2011). This section reviews empirical evidence on VCD with the explicit aim of understanding if and how such efforts support or constrain women's empowerment. We assess how some of the most prominent VCD trends—namely, process upgrading, horizontal coordination, and vertical coordination—relate to the processes of empowerment and disempowerment discussed in the previous section.

Our review does not capture the full spectrum of value chain upgrading options[3] (for example Gereffi and Kaplinsky 2001, Gereffi and Fernandez-Stark 2011), because our literature search aimed predominantly to solicit studies that provide information on empowerment outcomes. That evidence tends to be restricted mainly to **process upgrading** (including mechanization) as well as various forms of vertical and horizontal coordination (Riisgaard et al. 2010). Further, there is a paucity of literature on gender and chain or **functional upgrading** strategies (Coles and Mitchell 2011). As a point of clarification, while certifications and standards are sometimes described in terms of product upgrading, channel upgrading, or chain governance, we discuss them as a form of vertical coordination. We also recognize that our review does not cover broader efforts aimed at enhancing women's rights, which in turn might enhance empowerment-related outcomes in value chains. Finally, most of the discussed interventions focus predominantly on farmers within the production node. This may be because of the increased emphasis on VCD for poverty reduction and rural development.

---

3 Gereffi and Fernandez-Stark (2011, 13) identify four types of upgrading within the global value chain framework: "process upgrading, which transforms inputs into outputs more efficiently by reorganizing the production system or introducing superior technology; product upgrading, or moving into more sophisticated product lines; functional upgrading, which entails acquiring new functions (or abandoning existing functions) to increase the overall skill content of the activities; and chain or inter-sectoral upgrading, where firms move into new but often related industries." Riisgaard et al. (2010, 14) also include "improved horizontal coordination (the types, systems and levels of cooperation between the same type of value chain actors)... [and]... vertical coordination (the links between actors in different value chain positions)." (Emphasis added.)

Process upgrading

Process upgrading is generally understood to encompass a range of strategies aimed at transforming inputs into outputs in a more efficient manner (Gereffi and Kaplinsky 2001). Empowerment-related outcomes have been discussed predominantly in the context of mechanization and capacity-building. We look at both in this section.

The introduction of various technologies and machinery—**mechanization**—along value chains can contribute toward empowerment through increasing women's assets (Theis et al. 2018) and freeing up women's time from agricultural work (Amare and Endalew 2016). Furthermore, mechanization has the potential to facilitate women's entry into new value chains, as demonstrated in Zambia, where access to protective clothing and modern, more easily accessible beehives has encouraged more women to participate in beekeeping (Shackelton et al. 2011). Yet mechanization can also have the converse effect. In historically female-dominated value chains, new technologies have often displaced female producers and small-scale processors, when machinery replaces their labor (Shackleton et al. 2011, Turner 2014, Sarku 2016).

Moreover, restrictive norms, limited investment capital, and lack of training often limit women's opportunities for technological upgrading (see for example Farnworth 2011, O'Neill and Crona 2017). For instance, in the same Zambian case, the introduction of modern honey presses undermined the role of women in processing because they lacked the skills to operate the presses (Shackleton et al. 2011). In a Mozambican cashew nut processing factory, women were relegated to labor-intensive peeling activities as it was perceived that they could not operate cutting machines (Kanji 2004). In Malawi and Zambia, the introduction of groundnut shelling machinery helped reduce women's drudgery but also resulted in increasing competition from men (Tsusaka et al. 2016). In Benin, the introduction of machinery to process cassava into *gari* displaced manual processing tasks that women usually performed while the processing units created employment opportunities for both women and men (Forsythe et al. 2016, Adégbola et al. 2013). Meanwhile, the wages paid to employees are often minimal, while some tasks—particularly feeding cassava into the mechanical grater—are associated with significant health risks (Andersson et al. 2016).

Across these studies, the gender consequences of mechanization depend on the types and scale of mechanization, rather than on mechanization per se. Mechanization and technological upgrading can facilitate more equitable participation when technologies are culturally appropriate, when they address gender-specific barriers or constraints, and when women have sufficient skills. Technological innovations can help transform restrictive gender divisions of labor through facilitating women's entry into male-dominated value chains (for example Shackleton et al. 2011). However, in many instances, displacing women's traditional labor can have far-reaching consequences where women's traditional labor activities are constitutive of identity (for example Elias 2010 on women shea butter producers). As women in many contexts face considerable financial and social barriers to adopting production-enhancing machinery, there is little evidence linking mechanization to empowerment.

The second set of empowerment-related outcomes is discussed in the context of **capacity-building**. It should be noted that various forms of capacity-building can be embedded in other types of VCD initiatives and hence—perhaps—discussed less frequently as the main independent variable. However, a few of the reviewed studies provide important insights into the relationship between capacity-building and empowerment. While lesser access to trainings and information often hampers women's abilities to apply process upgrading (Riisgaard et al. 2010), capacity-building efforts aimed at skills development can lead to empowerment gains. In Nepal, women who participated in trainings on honey and bay leaf production gained leadership skills and confidence; these in turn allowed them to take on non-traditional gender roles, such as negotiating with traders and buyers or participating in household decision-making (Gurung et al. 2016). In Malaysia, participatory learning tools aimed at exploring market opportunities similarly enhanced women's confidence in negotiating with buyers, by increasing their knowledge of the quality of their products and improving their marketing skills (Faridah Aini et al. 2017).

However, gender-blind capacity-building can exclude women and contribute to disempowerment. For instance, in Kenya's export horticulture sector, men took over representational activities (including trainings, decision-making, sales), leaving women increasingly dependent on their husbands for market information (Velte and Danneberg 2014).

In Malawi, participatory household methodologies—such as the Gender Action Learning System (GALS)—hold potential with regard to enhancing women's participation in household decision-making and challenging rigid gender divisions of labor through involving women and men as well as emphasizing the benefits of "jointness" in household decision-making. However, these outcomes are counteracted if prevailing social norms do not support joint decision-making (Farnworth et al. 2018). An assessment of a number of VCD interventions did not find an automatic positive connection between participation in trainings and changes in household decision-making (Riisgaard et al. 2010).

Value chain coordination

Horizontal coordination—that is, enhanced cooperation between similar types of value chain actors at the same node—is increasingly being promoted to enhance the inclusiveness and equity of agricultural value chains (Turner 2014). Particularly from an empowerment perspective, it has been argued that horizontal coordination increases women's bargaining power vis-à-vis other chain actors (Coles and Mitchell 2011); improves women's access to resources and services (Riisgaard et al. 2010, Coles and Mitchell 2011); enhances women's participation in decision-making at different levels (FAO 2012); and increases women's confidence in demanding change through facilitating a forum for sharing experiences (Said-Allsopp and Tallontire 2014).

In many instances, such gains are mediated through improved **vertical coordination**—that is, improved coordination between actors at different nodes of the value chain—particularly in the context of "women-centered" value chains (Riisgaard et al. 2010). At the same time, horizontal coordination (for example producer organizations) is often a key feature of vertical coordination efforts, such as socially conscious certification schemes and standards (for example Said-Allsopp and Tallontire 2014).

While recognizing the overlaps and synergies between horizontal and vertical coordination efforts, we discuss the related empowerment outcomes in two distinct sections for conceptual clarity.

HORIZONTAL COORDINATION

Despite compelling theories of change that lay out multiple forms of empowerment gains associated with improved horizontal coordination, these connections prove complex. For instance, when cooperatives or producer organizations enforce a "one member per household" rule, they tend to be biased toward male household heads (Oduol et al. 2017, Stoian et al. 2018b, Wijers 2019). The exclusion of women from such groups limits their access to various membership benefits and services, including more remunerative export markets (Oduol et al. 2017), credit (McCarthy and Moon 2018), technologies, and trainings (Fischer and Qaim 2012). In certain certification schemes, women producers struggle to claim premiums disbursed through male-dominated cooperatives (for example Kasente 2012). In Kenyan households that join banana farmer groups through representation by male heads of household, women tend to lose control over banana production and revenue; women's control over outputs can increase as they themselves join the banana farmer groups (Fischer and Qaim 2012).

While cooperatives are often promoted as businesses, they commonly rely heavily on the unpaid organizational labor of members (Lyon et al. 2017). Women's relative time poverty resulting from compounded triple labor burdens can significantly limit their ability to engage meaningfully in various groups (for example Lyon et al. 2017, Gumucio et al. 2018, Stoian et al. 2018b). In Kenya's poultry value chain, men were more likely to participate in producer and marketing groups, whereas women mainly participated in kin-based social groups, such as religious groups. Compared with heterogeneous producer networks, these kin-based, socially homogeneous groups are often associated with "poor business prospects and self-policing in communities that require women to uphold strict moral codes of conduct" (Maina et al. 2014, 3). Male-dominated cooperatives with limited female representation not only limit women's access to benefits but also are less likely to develop policies and programs to address gender inequities (Lyon et al. 2017).

It is possible to enhance women's meaningful participation in collective bodies when interventions require female participation (Mulema et al. 2017) or participation of individuals, rather than households, in the value chain (Fischer and Qaim 2012); when gender quotas are implemented (Agarwal 2010, Woldu et al. 2015); when women occupy leadership positions (Coleman and Mwangi 2013, Bhalla 2016); and when meeting schedules take women's reproductive responsibilities into account (Stoian et al. 2018b). Ethiopian women are more likely to join cooperatives with members from the same community, educated leaders who can read and write, and trainings and input services. Here, women from more educated households are more likely to participate (Woldu et al. 2015). Women may be more comfortable voicing their needs and opinions in all-women groups as opposed to mixed groups. However, mixed groups may have greater access to financial and social resources. Participation in women's groups can be more socially acceptable—but women-only groups may also prompt resentment among men (Riisgaard et al. 2010). In Tanzania, such groups in the mud crab value chain created male resentment, leading to acts of sabotage and strained relationships with male fishers upon whom the women were reliant (Coles and Mitchell 2011).

Strengthening women's collective action, such as that derived from cooperatives and producer groups, can strengthen positive interactions between women's social, human, and financial capital (Mayoux 2012). After training, a women's fish retailer committee in Egypt successfully proposed the establishing of a marketplace, where women subsequently engaged with local government officials and conveyed their needs to them. This resulted in a dialogue on the need for water supply, which the council finally provided (Dickson et al. 2016). Women's increased participation and leadership in Village Savings and Loans Associations in Ethiopia, Malawi, and Tanzania has led to more men paying attention to women's decisions and a gradual shift toward more equal household decision-making. The inclusion of men in gender-awareness trainings was a key success factor (FAO and CARE 2019).

In Burkina Faso, members of the women's union and producer groups in the shea value chain cite not only income generation opportunities but also social benefits: forging friendships through union assemblies, working together and supporting each other's production, and reducing isolation (Elias and Arora-Jonsson 2017). In addition to their improved financial situation, many women expressed particular appreciation of increased opportunities to collaborate with other women (Chen 2017). Similarly, in Kenya's tea sector, women formed various informal groups, such as rotating savings and credit associations, even though the union was viewed as largely ineffective in terms of improving employment quality. In this case, stable employment was identified as a critical prerequisite to enabling women to form associations. Having earned greater confidence in their individual and collective abilities, some groups also used their collective agency to campaign for the election of a woman supervisor. A supportive female leader was a key success factor in women's collective achievements (Said-Allsopp and Tallontire 2014). Elsewhere, women leaders are also found to help enhance women's social networks and their access to information (Loconto 2015, Lyon et al. 2017).

Whereas most of the studies above emphasize the importance of women's collective agency and voice vis-à-vis men and powerful chain actors, they pay little attention to heterogeneity among women and to intersecting social power relations. In the shea value chain in Burkina Faso (Box 4.2), women experience exclusion along socioeconomic status, ethnicity, and age. Women's groups may in themselves be sites of contestation and power struggles, despite providing multiple benefits to their members.

VERTICAL COORDINATION

Vertical coordination efforts that forge stronger inter-nodal relationships can enhance women and men's financial resources and social status (Coles and Mitchell 2011). Such efforts can strengthen women's position within value chains and encourage them to take on new roles (Riisgaard et al. 2010). Many vertical coordination efforts reviewed here are geared toward accessing more

BOX 4.2 Shea in Burkina Faso

In the "shea belt," stretching across the savanna south of the Sahara, women have long collected shea nuts and processed them into an edible butter that has a central role in local diets, trade, and cultures. Internationally, shea butter is better known for its skin healing and moisturizing properties. Growing demand for shea butter in specialty cosmetics and pharmaceutical markets, coupled with global consumer concerns for social justice, gender equality, and environmental sustainability, has led to new prospects for rural shea producers (Elias and Saussey 2013). As one of the few economic opportunities in the hands of women in West African countries such as Burkina Faso, the shea value chain has garnered the interest of private sector, non-governmental, governmental, and multilateral actors with regard to promoting rural development and women's economic empowerment (Laube 2015). This has resulted in efforts to organize associations of women's producer groups; strengthen their capacities to produce to international quality standards; reduce the intense labor requirements of butter processing through technologies; and connect producer associations to certified fair trade and organic markets to enhance incomes.

Shea producer associations have offered innumerable benefits to their members. In central-south Burkina Faso, members of the women-only Fédération Nununa consider that working with other producers has given them much-needed economic, physical, and moral support. The ability to collectively negotiate high-volume contracts with international buyers, rather than selling individually on the local market, has translated into improved prices for their product. Moreover, when orders are placed with the association, women come together to produce shea butter, sharing equipment and using mutual aid arrangements to relieve some of the physical constraints of production.

The association has fostered closer ties and collaboration among women within and, importantly, across villages, as well as with other actors in the shea value chain (for example non-governmental organizations, local and international buyers, other producer associations). Women also feel less isolated: Fédération Nununa members describe how, at production sites, they laugh, talk, and enjoy each other's company instead of working alone in their homestead. As the president of one women's group explains, "When you're alone at home, you have too much time to think of your problems, but when you're surrounded by other people, you feel better because you see that they also suffer, maybe even more than you do, and with their husbands as well" (Léo, February 5, 2007, in Elias 2010). The association strengthens the sense of identity, common culture, and community among producers. In fact, producers consider the enhancement of social capital as a primary benefit of the association, equally if not more important than improved incomes (Elias 2010).

Yet not all shea producers benefit equally from the association, and collectives are not necessarily emancipatory (Arora-Jonsson 2013). In south-central Burkina Faso, benefits from the association are unevenly distributed, and have created generated exclusions based on producers' location of residence, age, ethnicity, or interpersonal relationships with leaders (Elias and Arora-Jonsson 2017). For instance, town-based producers have access to equipment that enables them to produce large quantities of butter and gain lucrative contracts, unlike rural producers (Pouliot and Elias 2013). Spatial divides also create cleavages: some rural members distrust the association's town-based leaders.

Meanwhile, as competition for shea nuts has increased with market demand, older women can no longer compete with young women, who race to common collection areas (such as woodlands) to gather prized shea nuts before dawn. Women's groups are formed along ethnic lines, with autochthonous women capturing the most promising opportunities arising within the association. Moreover, in some areas, access rights to shea nuts have been renegotiated along ethnic lines, resulting in lower production capacities of "strangers" (non-autochthonous women who have migrated into the area), and their partial or total exclusion from more remunerative contracts.

Spaces within the association, and the shea butter value chain, are thus both collaborative and intensely negotiated among socially differentiated producers. As Elias and Arora-Jonsson (2017, 122) argue, "Rather than mere inclusion of women in value chains, this calls for an acknowledgment of power relations throughout the value chains' vertical but also horizontal links to bring about more equitable relations."

profitable markets and linked to certification schemes or standards. Factors enhancing or constraining empowerment in such schemes may be associated more with specific governance aspects than with a specific vertical relation. We therefore discuss empowerment outcomes in a broader manner in relation to various schemes promoting vertical coordination.

Women's limited resources and heavy reproductive burdens can hinder their participation in export-oriented contract farming, in which it can be hard to match the stringent quality and compliance standards (Oduol et al. 2017; see also Chen 2017, Amare et al. 2019). Further, outgrower or contract farming schemes have increased the marginalization of women and contributed to the invisibilization of their labor where it is male household heads who sign contracts despite women's and other family members' significant labor contributions (Dolan 2001, Eaton and Shepherd 2001, Singh 2003, Lyon 2008, Maertens and Swinnen 2012, Li 2015). In Papua New Guinea, women working with their husbands in oil palm used to receive very low remuneration for their labor. A vertical coordination scheme that paid women and men into their separate bank accounts resulted in a significant increase in women's share of the household income and a reduction in economically motivated household conflict and violence (Koczberski 2007).

Fair trade certification has achieved little change in household decision-making dynamics in Peru (Ruben and Fort 2012). The opposite is true in Uganda, where women in certified households enjoy significantly more control over production activities and income than do those in non-certified households. Over time, men in certified households have agreed to joint control over coffee production and revenues (Chiputwa and Qaim 2016). Certification schemes can increase women's labor burden, for instance with schemes oriented at enhancing environmental sustainability that entail adoption of new, often labor-intensive, farming practices (Bolwig and Odeke 2007, Kasente 2012, McArdle and Thomas 2012, Loconto 2015). In certified coffee-producing households in Uganda, post-harvest labor demands increased greatly and women bore the brunt, even though organic certification resulted in increased household income associated with higher yields. As a result, women had less time to spend on their micro-enterprises, hence earned less own income compared with men (Chiputwa and Qaim 2016).

In Ethiopia, women participating in a Rainforest Alliance certification scheme reported gaining social relations and knowledge through interacting with other women farmers during meetings. However, the certification did little to increase their active participation in collective decision-making bodies (Riisgaard et al. 2009). In a woman-only coffee certificate and label developed in Guatemala, Café de Mujer, positive changes occurred in women's access to new knowledge, management of production processes, and participation in capacity-building (Verhart and Pyburn 2010). This scheme required that each farm be managed or owned by women, and salaries be paid directly to the women farmers. In order to comply, a major cooperative adjusted its statutes to enhance women's representation in its governing bodies. For certified women farmers, the most notable change was the sense of dignity and pride as their family and community recognized their work and roles. In a similar coffee scheme in 17 countries, operated by the non-governmental organization Twin, women's membership and decision-making within mixed cooperatives increased. It also provided producers with a premium that is reinvested in gender justice projects, such as developing a gender policy at producer organization level or GALS training at the household level (Jennings et al. 2018).

In Guatemala, certification schemes requiring women producers to be present at sales provided women with access to market channels while altering norms that perceived women handling cash as socially unacceptable (Lyon et al. 2010). In Kenya's horticulture sector, working conditions have improved as a result of the widespread adoption of certification schemes. In addition, the establishment of gender committees—as required by the Kenya Flower Council standard—have helped decrease sexual harassment in the workplace, improve the dissemination of information to workers, and address issues such as credit or family planning that have proved important to women (Said-Allsopp and Tallontire 2014). However, while codes set by the United Kingdom Ethical Trading Initiative address a number of pertinent gender issues, including maternity leave and equal pay, they are generally weak at addressing poor labor standards—thereby excluding women in precarious jobs from the above benefits (Barrientos and Smith 2007).

Discussion: VCD for empowerment?

VCD efforts can contribute to women's empowerment within value chains through enhancing their social, physical, and human capital as well as improving market linkages. This in turn can enhance women's status within the household and community and provide avenues for challenging inequitable gender norms. However, the direction and magnitude of "empowerment effects" turn out to be influenced by the specific features of the interventions in combination with a number of contextual factors. Effects can differ from case to case or even run concurrently. For instance, the introduction of labor-saving technologies may help reduce women's labor burden while increased mechanization also risks disenfranchising women for whom financial, social, or cultural factors constrain access to technologies. Producer organizations and unions can enhance women's social capital as well as their access to credit, labor, and more lucrative markets while the groups themselves may be subject to elite capture and exclusionary practices. Certification schemes and standards may similarly contribute to recognizing women's work and improving market access while male-biased benefit-sharing mechanisms and the financial and labor costs associated with compliance may disproportionately disenfranchise women.

An important conclusion is that most interventions associated with empowering outcomes adopt a deliberate and often exclusive focus on women and gender relations. In practice, this often entails identifying gendered constraints, such as in women's access to technologies and cooperatives or control over income; and deploying deliberate responses to these, such as gender-responsive technologies, more inclusive membership rules or women-only cooperatives, or individual payment schemes. Women-only interventions can be suitable for addressing certain gender-based constraints, such as lack of mobility and access to assets and markets (Riisgaard et al. 2010). However, "even women-only focused interventions require specific gender strategies that seek to counter the special constraints of operating in 'gender conservative' areas" (ibid., 52). Indeed, addressing unequal norms and relations at multiple levels and engaging a broad range of stakeholders—including community leaders, policymakers and employers—in the process are of key importance.

Interventions with a women-only focus risk limiting their aim to enhancing women's agency without addressing underlying power structures, rather than seeking to challenge unequal structures in favor of more equitable and enabling institutions and mindsets (Cornwall and Edwards 2010; see also Chapter 10 on gender transformative approaches, this volume). Without meaningfully engaging both men and women in processes of social change, interventions with a single-minded focus on women risk placing unrealistic expectations on women to transform structures that are rigged against them. They can generate backlash through sparking resentment among excluded men or mask social divisions among women. Approaches meaningfully engaging men can facilitate women's empowerment while enhancing the well-being of the entire household.

Gender-intentional VCD efforts can create spaces for women to come together to access mutual financial, social, and technical support, to collectively market their products for better prices, and to gain increased confidence and social status. Such approaches demonstrate how VCD efforts can facilitate multidimensional empowerment within and beyond the value chain through concertedly building women's social capital while addressing financial and capacity needs. At the same time, our review finds little evidence of enhanced collective agency resulting in women successfully challenging poor working conditions or exploitative labor practices. Instead, changes in labor practices stem mainly from third-party schemes or guidelines, providing companies with economic incentives for putting in place certain policies. While many certification schemes and standards have been found to perform rather poorly on gender issues (Lyon et al. 2010), these findings speak to the importance of embedding gender provisions in such schemes.

At the same time, while collective organization certainly generates multiple benefits for women, this also raises questions about the extent to which one can expect women's collective agency to effectively challenge, resist, or transform some of the stronger power imbalances and exploitative relations that currently characterize many global value chains. Further, as successes in challenging exploitative and gender-insensitive labor practices appear to be contingent on consumer-driven pressure and premised on the viability of a niche market, we need to ask questions about the potential they hold in terms of achieving transformative change at scale in global value chains, which to date "rely on the utilization (some would say exploitation) of cheap labour" (Said-Allsopp and Tallontire 2014, 17). Indeed, analyses and interventions should not decouple women's intrinsic and collective agency—or *power to, with and within* (Said-Allsopp and Tallontire 2014)—from more powerful actors' *power over* women's labor in agricultural value chains, or the market relations, institutional arrangements, regulatory frameworks, and societal norms that permit or incentivize exploitative practices.

Concluding remarks

This chapter has aimed at answering the question: *How and under what conditions can participation in agricultural value chains advance or constrain women's empowerment?* Regarding the directionality and magnitude of empowerment effects, we identify a range of potential economic and non-economic gains. Within the value chain, positive empowerment outcomes pertain especially to women's enhanced positioning, social capital, confidence, skills, and leadership. Beyond the value chain, outcomes include enhanced status, increased agency in household economic decision-making, and—often indirectly and to a lesser extent—the challenging of norms around the gender division of labor.

However, we also observe that women's participation in value chains as well as VCD efforts can have disempowering effects, including loss of control over production processes, loss of social status, exploitative labor conditions, marginalization, and time poverty. These outcomes are often contingent on the nature of pre-existing gender relations and inequalities within the household, community, and value chain. Importantly, processes of empowerment and disempowerment can run concurrently and produce contradicting effects within and beyond the value chain as well as between different dimensions of empowerment. This underscores the need to situate outcomes from value chain participation in a broader conceptualization of empowerment, rather than reducing empowerment to whatever the market can deliver.

While gender-blind VCD efforts can exacerbate disempowering processes, our review also illustrates ways in which gender-responsive efforts can enhance women's empowerment, through:

- Strengthening women's positioning and bargaining power within the value chain;
- Enhancing professional confidence;
- Facilitating spaces for sharing information and building collective agency; and,
- Enhancing household bargaining power through increasing skills, knowledge, and access to independent income.

Our review suggests that it is possible to enhance such outcomes by combining regulatory interventions (for example gender quotas in producer organizations) with services and interventions targeted at addressing gender-specific barriers and building women's social, financial, and human capital. In addition to value chain interventions, efforts aimed at engaging both men and women and fostering jointness and more equitable gender relations can be critical to facilitating empowerment beyond the value chain.

Regarding the relationship between individual empowerment and broader structural transformations toward gender equality, however, we do not find conclusive support that empowerment gains at the individual level allow women to effectively challenge broader inequalities (for example the devaluation of women's labor) that agricultural value chain processes can exploit. Our review hence finds little support for theories of change assuming structural transformation through individual, market-oriented gains, particularly when taking into account 1) the social and political dimensions of empowerment and 2) women's ability to effect broader structural change.

Indeed, it is worth questioning the extent to which we can realistically expect women's incorporation into commercial markets to address broader societal inequalities or power imbalances, particularly when addressing them may run counter to the prevailing market logic or risk a tradeoff with other objectives (for example Said-Allsopp and Tallontire 2014). Overloading agricultural value chains with expectations of enhanced empowerment may water down the concept of empowerment to merely market access and financial decision-making. In order to place empowerment and gender equality at the core, we propose the devotion of more effort to understanding how gender-transformative change happens in different contexts—and then assessing whether, when, and how various forms of value chain engagement can support these processes. Instead of considering women's empowerment as a ripple effect of agricultural commercialization, placing empowerment at the core allows for a much more nuanced and comprehensive understanding of the empowering and disempowering features of value chain engagement and VCD efforts. This constitutes a more honest assessment of potential synergies and contradictions between empowerment and other value chain priorities.

Directions for future research

Future research should seek to fill some of the pertinent gaps in the literature on value chains and women's empowerment. First, while gender dynamics have been studied across a wide range of value chains, evidence remains patchy and biased toward certain geographies, nodes, and specific value chains. For instance, a majority of studies—spanning a number of commodities—have taken place in Africa, possibly as a result of the increasing emphasis on value chains in international development (Chan 2010, Stoian et al. 2018a). Interestingly, studies in Asia tend to focus on forestry and aquaculture whereas coffee is a focus in Latin American studies. At the same time, studies on certification and empowerment in Latin America are likely to focus on coffee (for example Lyon et al 2010, Verhart and Pyburn 2010, Jennings et al. 2018), which may contribute to the overrepresentation in this chapter.

Second, despite recent attention to various processing nodes, the majority of studies assessing empowerment outcomes focus on the production node. Few papers provide any information on traders and retailers, particularly in urban and peri-urban settings. Across commodity sectors, more attention is given to nodes and value chains where women have historically played an active role. Furthermore, despite the complex ways in which households engage with multiple commercial and non-commercial activities (for example Stoian et al. 2012), empowerment outcomes are assessed exclusively in relation to a single value chain. More analysis utilizing integrated household livelihood approaches is critical to understanding the ways in which changes in engagement with commercial value chains affect various economic and non-economic activities and responsibilities, including, for instance, women's time poverty.

Third, empowerment is often measured solely among women, and few studies disaggregate by other social categories or discuss the effects of intersecting power relations. However, accounting both qualitatively and quantitatively for multiple axes of social divisions and power relations can be key to identifying heterogeneous outcomes and understanding multiple, concurrent processes of empowerment and disempowerment. Relatedly, a narrow emphasis on women risks missing ways in which masculinities (and femininities) are intertwined with various economic assets and activities (Stoian et al. 2018a). Resistance at household and/or community level can arise in response to transgressions of conventional gender roles in labor and decision-making. Understanding the relationships between locally constructed notions of masculinity and femininity, labor, and decision-making is thus vital to identifying points of contestation, anticipating social reactions, and uncovering potential entry points for promoting jointness across productive and reproductive domains.

Lastly, and perhaps most critically, while women's empowerment increasingly features as a key objective for VCD interventions, few studies adopt empowerment as the primary topic of inquiry. While many studies provide some information on some dimensions of empowerment at some levels, there is considerable heterogeneity in the ways in which studies operationalize and measure empowerment. In the absence of comparable conceptualizations and methodologies, as well as clear and transparent interrogations of impact pathways linking participation to empowerment, it is difficult to draw definitive conclusions about the role of context factors and the scalability of results.

The literature shows a somewhat dichotomous relationship between studies examining gender relations and empowerment within the value chain and those addressing the impact of participation on household and community dynamics (see also Said-Allsopp and Tallontire 2014). While the former focus on empowerment in terms of labor conditions, collective agency, human capital, etc., the latter assess empowerment as household bargaining power, control over income, and shifting gender roles. In addition, the increasing attention to market-led empowerment indicators such as assets, income decision-making, or professional skills and confidence risks depoliticizing empowerment by removing it from the broader material and sociopolitical context (Cornwall 2018). The following six points are critical in future gender and value chain studies:

1. Enhance conceptual clarity and develop more transparent operationalizations of "empowerment"—particularly in relation to value chains—in order to enable cross-case comparisons;[4]

4 See for example Ahmed et al. (2018) and Malapit et al. (2020) for efforts to amend the WEAI to value chain research. For a broader discussion on the WEAI as a measure for assessing empowerment, see Chapter 9 on measuring women's empowerment, this volume.

2. Develop a clear theory of change laying out various pathways from participation to empowerment across multiple dimensions, spheres, and levels;
3. Build approaches that can link processes within the value chain and processes in the broader systems within which women and men are embedded (Bolwig et al. 2010);
4. Critically assess ways in which economic, institutional, and environmental processes and transformations shaped by the expansion of commercial value chains influence the choices and options available to women and men;
5. Invest in methodologies that can account for intersectionality[5] and engage with notions of masculinity and femininity in the context of value chain-related processes; and,
6. Collect quantitative and qualitative baseline data across multiple dimensions of empowerment in order to make it possible to assess change over time.

In addition to contributing to a more comprehensive and nuanced body of research on value chains and empowerment, these points are critical to deliver research that allows policymakers and practitioners to design more effective and appropriate interventions as well as to monitor impacts along a more refined and empirically grounded theory of change.

5 See for example Sugden et al. (2014), Ravera et al. (2016) and Colfer et al. (2018) for approaches to operationalizing and applying intersectionality in CGIAR (and other agricultural) research.

Geography of study sites for publications cited in Chapter 4

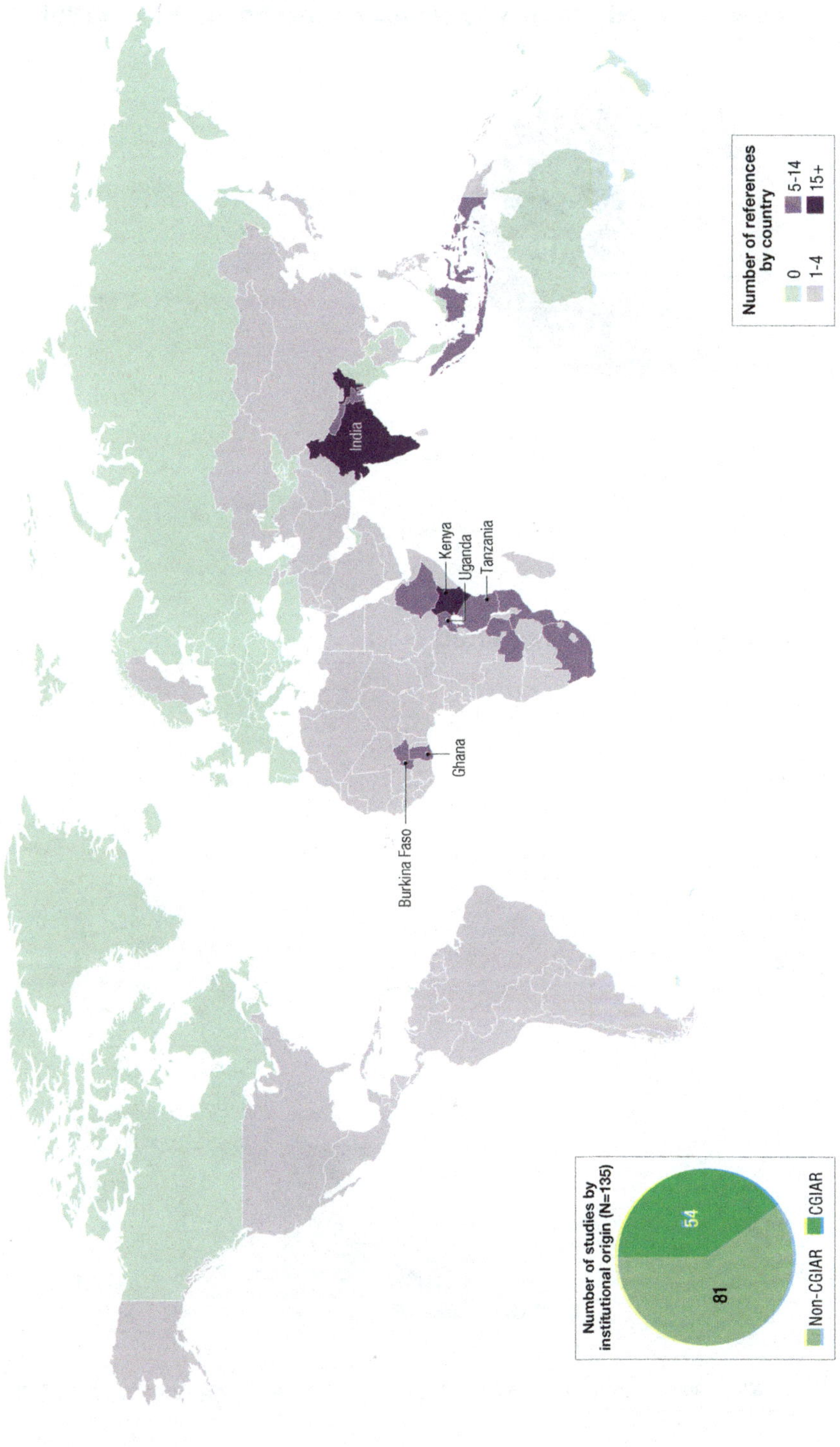

Number of cited studies by research methodology (N=135)

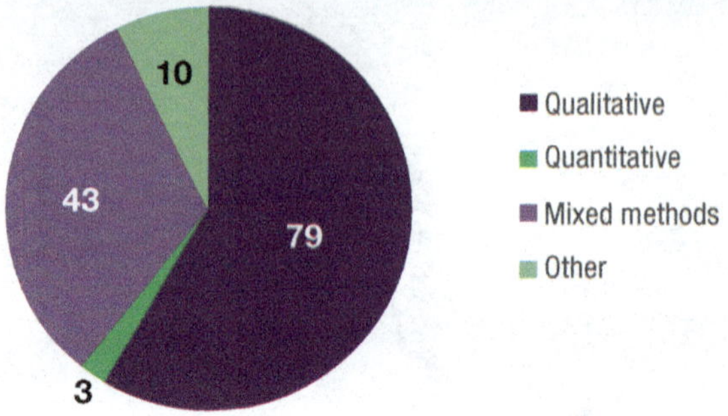

Timeline for references cited

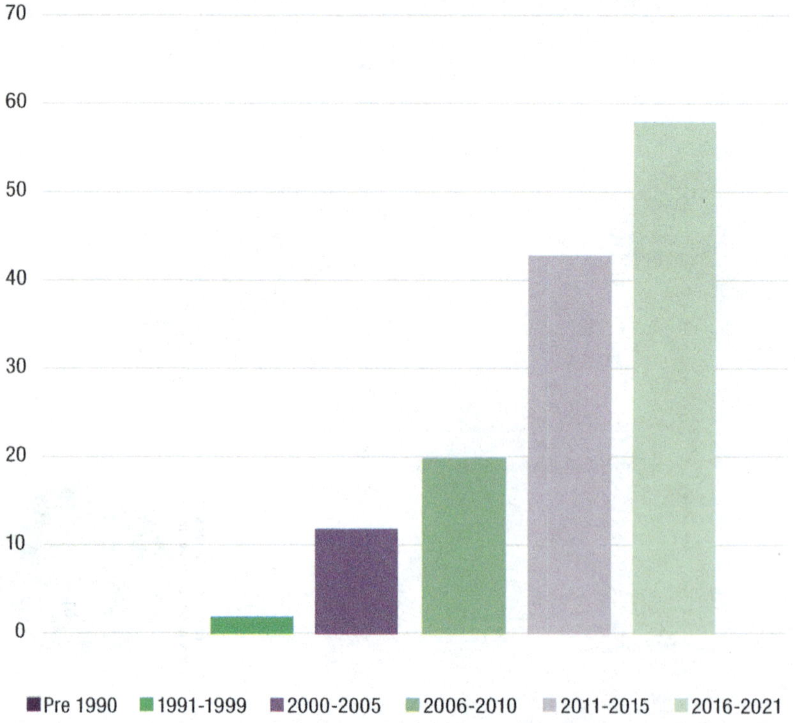

References

Adégbola, Y., R. Yegbemey, I. Djenontin, A. Hibon, K. Hell, G. Thiele, O. Koudande. 2013. "Les marchés du manioc et du 'gari' dans le Sud et le Centre du Bénin: performances et principales contraintes à leur développement." *Cahiers Agricultures* 22 (4): 293–302.

Agarwal, B. 2003. "Gender and Land Rights Revisited: Exploring New Prospects via the State, Family and Market." *Journal of Agrarian Change* 3 (1–2): 184–224.

Agarwal, B. 2010. "Does Women's Proportional Strength Affect Their Participation? Governing Local Forests in South Asia." *World Development* 38 (1): 98–112.

Ahmed, A., H. Malapit, A. Pereira, A. Quisumbing, D. Rubin, J. Ghostlaw, M. Haque, N. Hossain, S. Tauseef. 2018. "Tracking Empowerment along the Value Chain: Testing a Modified WEAI in the Feed the Future Zone of Influence in Bangladesh." Report, IFPRI, Washington, DC.

Amare, D., and W. Endalew. 2016. "Agricultural Mechanization: Assessment of Mechanization Impact Experiences on the Rural Population and the Implications for Ethiopian Smallholders." *Engineering and Applied Sciences* 1 (2): 39–48.

Amare, M., J. Mariara, R. Oostendorp, M. Pradhan. 2019. "The Impact of Smallholder Farmers' Participation in Avocado Export Markets on the Labor Market, Farm Yields, Sales Prices, and Incomes in Kenya." *Land Use Policy* 88: 104–168.

Andersson, K., J. Bergman Lodin, L. Chiwona-Karltun. 2016. "Gender Dynamics in Cassava Leaves Value Chains: The Case of Tanzania." *Journal of Gender, Agriculture and Food Security (Agri-Gender)* 1 (302): 84–109.

Arora, D., J. Arango, S. Burkart, N. Chirinda, J. Twyman. 2017. "Gender [Im]balance in Productive and Reproductive Labor among Livestock Producers in Colombia: Implications for Climate Change Responses." Info Note, February, CCAFS, Copenhagen.

Arora-Jonsson, S. 2013. *Gender, Development and Environmental Governance: Theorizing Connections*. London: Routledge.

Baglioni, E. 2017. "Labour Control and the Labour Question in Global Production Networks: Exploitation and Disciplining in Senegalese Export Horticulture." *Journal of Economic Geography* 18 (1): 111–137.

Barrientos, S. 2001. "Gender, Flexibility and Global Value Chains." *IDS Bulletin* 32 (3): 83–93.

Barrientos, S., and S. Smith. 2007. "Do Workers Benefit from Ethical Trade? Assessing Codes of Labour Practice in Global Production Systems." *Third World Quarterly* 28 (4): 713–729.

Barrientos, S., C. Dolan, A. Tallontire. 2003. "A Gendered Value Chain Approach to Codes of Conduct in African Horticulture." *World Development* 31 (9): 1511–1526.

Bhalla, P. 2016. "Gender Dynamics in Odisha's Forest Rights Act." In *Gender and Forests: Climate Change, Tenure, Value Chains and Emerging Issues*, edited by C. Colter, B. Basnett, M. Elias, 186–205, New York: Routledge.

Bolwig, S., and M. Odeke. 2007. *Household Food Security Effects of Certified Organic Export Production in Tropical Africa—a Gendered Analysis.* Bennekom: EPOPA.

Bolwig, S., S. Ponte, A. du Toit, L. Riisgaard, N. Halberg. 2010. "Integrating Poverty and Environmental Concerns into Value-Chain Analysis: A Conceptual Framework." *Development Policy Review* 28 (2): 173–194.

Boyle, T. 2016. "Three Key Steps to Empowering Women in Global Value Chains." July 11, 2016. Accessed September 25, 2019. https://insights.careinternational.org.uk/development-blog/three-key-steps-to-empowering-women-in-global-value-chains

Butz, R. J. 2013. "Changing Land Management: A Case Study of Charcoal Production among a Group of Pastoral Women in Northern Tanzania." *Energy for Sustainable Development* 17 (2): 138–145.

Carr, M., and M. Chen. 2004. "Globalization, Social Exclusion and Gender." *International Labour Review* 143 (1–2): 129–160.

Chan, M-K. 2010. "Improving Opportunities for Women in Smallholder-Based Supply Chains: Business Case and Practical Guidance for International Food Companies." Discussion Paper, BMGF, Seattle.

Chen, M. 2009. "Informalization of Labour Markets: Is Formalization the Answer?" In *The Gendered Impacts of Liberalization*, edited by S. Razavi, 207–234. Abingdon: Routledge.

Chen, T. 2017. "Impact of the Shea Nut Industry on Women's Empowerment in Burkina Faso: A Multi-Dimensional Study Focusing on the Central, Central-West and Hauts-Bassins Regions." Social Protection and Forestry Working Paper 3, FAO, Rome.

Chiputwa, B., and M. Qaim. 2016. "Sustainability Standards, Gender, and Nutrition among Smallholder Farmers in Uganda." *Journal of Development Studies* 52 (9): 1241–1257.

Choudhary, D., B. Bhattarai, B. Pandit, M. Kollmair. 2010. "Empowering Women through Bay Leaf Value Chain Development. Gender Perspective in Mountain Development: New Challenges and Innovative Approaches." *Sustainable Mountain Development Bulletin* 57.

Coleman, E. A., and E. Mwangi. 2013. "Women's Participation in Forest Management: A Cross-Country Analysis." *Global Environmental Change* 23 (1): 193–205.

Coles, C., and J. Mitchell. 2011. "Gender and Agricultural Value Chains: A Review of Current Knowledge and Practice and Their Policy Implications." ESA Working Paper 11-05, FAO, Rome.

Colfer, C. J. P., Sijapati Basnett, B., M. Ihalainen. (2018). *Making Sense of 'Intersectionality': A Manual for Lovers of People and Forests.* Bogor: CIFOR.

Cornwall, A. 2018. "Beyond 'Empowerment Lite': Women's Empowerment, Neoliberal Development and Global Justice." *cadernos pagu* 52: e185202.

Cornwall, A., and J. Edwards. 2010. "Introduction: Negotiating Empowerment." *IDS Bulletin* 41 (2): 1–9.

Curry, G. N., G. Koczberski, S. M. Inu. 2019. "Women's and Men's Work: The Production and Marketing of Fresh Food and Export Crops in Papua New Guinea." *Oceania* 89 (2): 237–254.

Dickson, M., A. Nasr-Allah, D. Kenawy, M. Fathi, G. El-Naggar, N. Ibrahim. 2016. "Improving Employment and Income Through Development of Egypt's Aquaculture Sector (IEIDEAS) Project." Program Report 2016-14. Penang: WorldFish.

Djoudi, H., and M. Brockhaus. 2011. "Is Adaptation to Climate Change Gender Neutral? Lessons from Communities Dependent on Livestock and Forests in Northern Mali." *International Forestry Review* 13 (2): 123–135.

Dolan, C. 2001. "The Good Wife's Struggle over Resources in the Kenyan Horticulture Sector." *Journal of Development Studies* 37 (3): 39–70.

Dolan, C. S. 2004. "On Farm and Packhouse: Employment at the Bottom of a Global Value Chain." *Rural Sociology* 69 (1): 99–126.

Dolan, C., and K. Sutherland. 2002. "Gender and Employment in the Kenya Horticulture Value Chain." Globalisation and Poverty Discussion Paper 8, Overseas Development Group, University of East Anglia, Norwich.

Eaton, C., and A. Shepherd. 2001. "Contract Farming: Partnerships for Growth." Agricultural Services Bulletin 145, FAO, Rome.

Elias, M. 2010. "Transforming Nature's Subsidy: Global Markets, Burkinabè Women and African Shea Butter." Doctoral dissertation, McGill University.

Elias, M., and S. Arora-Jonsson. 2017. "Negotiating across Difference: Gendered Exclusions and Cooperation in the Shea Value Chain." *Environment and Planning D: Society and Space* 35 (1): 107–125.

Elias, M., and M. Saussey. 2013. "The Gift that Keeps on Giving: Unveiling the Paradoxes of Fair Trade Shea Butter." *Sociologia Ruralis* 53 (2): 158–179.

Elmhirst, R., M. Siscawati, B. Sijapati Basnett, D. Ekowati. 2017. "Gender and Generation in Engagements with Oil Palm in East Kalimantan, Indonesia: Insights from Feminist Political Ecology." *The Journal of Peasant Studies* 44 (6): 1135–1157.

FAO (Food and Agricultural Organization). 2012. "Agricultural Cooperatives and Gender Equality." International Year of Cooperatives, Issue Brief Series, Rome, FAO.

FAO. 2018. *Developing Gender-Sensitive Value Chains— Guidelines for Practitioners*. Rome: FAO.

FAO and CARE. 2019. *Good Practices for Integrating Gender Equality and Women's Empowerment in Climate-Smart Agriculture Programs*. Atlanta, GA: CARE.

Faridah Aini, M., M. Elias, H. Lamers, U. Shariah, P. Brooke, H. Mohd Hafizul. 2017. "Evaluating the Usefulness and Ease of Use of Participatory Tools for Forestry and Livelihoods Research in Sarawak, Malaysia." *Forests, Trees and Livelihoods* 26 (1): 29–46.

Farnworth, C. R. 2011. "Gender-Aware Value Chain Development." Expert Paper prepared for UN Women. London: UN Women.

Farnworth, C. R., P. Kantor, F. Kruijssen, C. Longley, K. E. Colverson. 2015. "Gender Integration in Livestock and Fisheries Value Chains: Emerging Good Practices from Analysis to Action." *International Journal of Agricultural Resources, Governance and Ecology* 11 (3–4): 262–279.

Farnworth, C. R., C. M. Stirling, A. Chinyophiro, A. Namakhoma, R. Morahan. 2018. "Exploring the Potential of Household Methodologies to Strengthen Gender Equality and Improve Smallholder Livelihoods: Research in Malawi in Maize-Based Systems." *Journal of Arid Environments* 149: 53–61.

Fischer, E., and M. Qaim. 2012. "Gender, Agricultural Commercialization, and Collective Action in Kenya." *Food Security* 4 (3): 441–453.

Fisher, M. 2004. "Household Welfare and Forest Dependence in Southern Malawi." *Environment and Development Economics* 9 (2): 135–154.

Forsythe, L., H. Posthumus, A. Martin. 2016. "A Crop of One's Own? Women's Experiences of Cassava Commercialization in Nigeria and Malawi." *Journal of Gender, Agriculture and Food Security* 1 (2): 110–128.

Fröcklin, S., M. de la Torre-Castro, L. Lindström, N. S. Jiddawi. 2013. "Fish Traders as Key Actors in Fisheries: Gender and Adaptive Management." *Ambio* 42 (8): 951–962.

Gereffi, G., and K. Fernandez-Stark. 2011. *Global Value Chain Analysis: A Primer.* Durham, NC: Center on Globalization, Governance & Competitiveness, Duke University.

Gereffi, G., and R. Kaplinsky. 2001. "Introduction: Globalisation, Value Chains and Development." *IDS Bulletin* 32 (3): 1–8.

Grace, D., K. Roesel, E. Kang'ethe, B. Bonfoh, S. Theis. 2015. "Gender Roles and Food Safety in 20 Informal Livestock and Fish Value Chains." Discussion Paper 1489, IFPRI, Washington, DC.

Gumucio, T., M. D. A. Alves, N. Orentlicher, M. C. Hernández Ceballos. 2018. "Analysis of Gender Research on Forest, Tree and Agroforestry Value Chains in Latin America." *Forests, Trees and Livelihoods* 27 (2): 69–85.

Gurung, M. B., U. Partap, D. Choudhary. 2015. "Empowering Mountain Women through Community-based High Value Product Value Chain Promotion in Nepal." *International Journal of Agricultural Resources, Governance and Ecology* 11 (3-4): 330-345.

Handschuch, C., and M. Wollni. 2016. "Traditional Food Crop Marketing in Sub-Saharan Africa: Does Gender Matter?" *Journal of Development Studies* 52 (3): 343–359.

Hunt, A., and E. Samman. 2016. "Women's Economic Empowerment. Navigating Enablers and Constraints." Working Paper 529, ODI, London.

Huyer, S. 2016. "Closing the Gender Gap in Agriculture." *Gender, Technology and Development* 20 (2): 105–116.

IFAD (International Fund for Agricultural Development). 2009. *Gender in Agriculture Sourcebook.* Washington, DC: World Bank.

Ingram, V., J. Schure, J. C. Tieguhong, O. Ndoye, A. Awono, D. M. Iponga. 2014. "Gender Implications of Forest Product Value Chains in the Congo Basin." *Forests, Trees and Livelihoods* 23 (1–2): 67–86.

Ingram, V., M. Haverhals, S. Petersen, M. Elias, B. S. Basnett, S. Phosiso. 2016. "Gender and Forest, Tree and Agroforestry Value Chains: Evidence from the Literature". In *Gender and Forests: Climate Change, Tenure, Value Chains, and Emerging Issues*, edited by C. P. Colfer, B. S. Basnett, M. Elias, 221–242. London: Earthscan/CIFOR.

Jacobs, S., B. Brahic, M. M. Olaiya. 2015. "Sexual Harassment in an East African Agribusiness Supply Chain." *The Economic and Labour Relations Review* 26 (3): 393–410.

Jennings, S., E. Sahan, A. Maitland. 2018. "Fair Value: Case Studies of Business Structures for a More Equitable Distribution of Value in Food Supply Chains." Discussion Paper, Oxfam, Oxford.

Johnson, N., M. Balagamwala, C. Pinkstaff, S. Theis, R. Meinzen-Dick, A. Quisumbing. 2018. "How Do Agricultural Development Projects Empower Women? Linking Strategies with Expected Outcomes." *Journal of Gender, Agriculture and Food Security* 3 (2): 1–19.

Jones, D., C. M. Ryan, J. Fisher. 2016. "Charcoal as a Diversification Strategy: The Flexible Role of Charcoal Production in the Livelihoods of Smallholders in Central Mozambique." *Energy for Sustainable Development* 32: 14–21.

Kabeer, N. 2017. "Economic Pathways to Women's Empowerment and Active Citizenship: What Does the Evidence from Bangladesh Tell Us?" *Journal of Development Studies* 53 (5): 649–663.

Kanji, N. 2004. "Corporate Responsibility and Women's Employment: The Case of Cashew Nuts." *Gender & Development* 12 (2): 82–87.

Kasente, D. 2012. "Fair Trade and Organic Certification in Value Chains: Lessons from a Gender Analysis from Coffee Exporting in Uganda." *Gender & Development* 20 (1): 111–127.

Kent, R. 2018. "'Helping' or 'Appropriating'? Gender Relations in Shea Nut Production in Northern Ghana." *Society & Natural Resources* 31 (3): 367–381.

Kidder, T., S. Romana, C. Canepa, J. Chettleborough, C. Molina. 2017. *Oxfam's Conceptual Framework on Women's Economic Empowerment.* Oxford: Oxfam.

Klugman, J., and L. Tyson. 2016. "Leave No One Behind: A Call to Action for Gender Equality and Women's Economic Empowerment." Report of the UN Secretary-General.

Koczberski, G. 2007. "Loose Fruit Mamas: Creating Incentives for Smallholder Women in Oil Palm Production in Papua New Guinea." *World Development* 35 (7): 1172–1185.

Kruijssen, F., J. Albert, M. Morgan, D. Boso, F. Siota, S. Sibiti, A. Schwarz. 2013. "Livelihoods, Markets, and Gender Roles in Solomon Islands: Case Studies from Western and Isabel Provinces." Working Paper, Honiara, WorldFish.

Kruijssen, F., C. L. McDougall, I. J. van Asseldonk. 2018. "Gender and Aquaculture Value Chains: A Review of Key Issues and Implications for Research." *Aquaculture* 493: 328–337.

Laube, W. 2015. "Global Shea Nut Commodity Chains and Poverty Eradication in Northern Ghana: Myth or Reality?" *UDS International Journal for Development* 2 (1): 128–147.

Li, T. M. 2015. *Social Impacts of Oil Palm in Indonesia: A Gendered Perspective from West Kalimantan.* Bogor: CIFOR.

Limuwa, M. M., and G. Synnevåg. 2018. "Gendered Perspective on the Fish Value Chain, Livelihood Patterns and Coping Strategies under Climate Change-Insights from Malawi's Small-Scale Fisheries." *African Journal of Food, Agriculture, Nutrition and Development* 18 (2): 13521–13540.

Loconto, A. 2015. "Can Certified-Tea Value Chains Deliver Gender Equality in Tanzania?" *Feminist Economics* 21 (3): 191–215.

Lyon, S. 2008. "We Want to Be Equal to Them: Fair-Trade Coffee Certification and Gender Equity within Organizations." *Human Organization* 67 (3): 258–268.

Lyon, S., J. A. Bezaury, T. Mutersbaugh. 2010. "Gender Equity in Fairtrade–Organic Coffee Producer Organizations: Cases from Mesoamerica." *Geoforum* 41 (1): 93–103.

Lyon, S., T. Mutersbaugh, H. Worthen. 2017. "The Triple Burden: The Impact of Time Poverty on Women's Participation in Coffee Producer Organizational Governance in Mexico." *Agriculture and Human Values* 34 (2): 317–331.

Maertens, M., and J. F. Swinnen. 2012. "Gender and Modern Supply Chains in Developing Countries." *Journal of Development Studies* 48 (10): 1412–1430.

Maina, I., E. Waithanji, M. Miruka, E. Mutua, L. Korir, L. Miriti, K. Gituma, P. Wahome. 2014. "Economic Opportunities, Empowerment, and Rights for Indigenous Poultry Farmers." Research Brief 35, ILRI, Nairobi.

Malapit, H., C. Ragasa, M. Martinez, D. Rubin, G. Seymour, A. Quisumbing. 2020. "Empowerment in Agricultural Value Chains: Mixed Methods Evidence from the Philippines." *Journal of Rural Studies* 76: 240–253.

Mars. 2019. "Dove Chocolate Empowers Women in Côte d'Ivoire." Accessed September 20, 2019. https://www.mars.com/news-and-stories/articles/dove-empowers-women

Mayoux, L. 2012. "Gender Mainstreaming in Value Chain Development: Experience with Gender Action Learning System in Uganda." *Enterprise Development and Microfinance* 23 (4): 319–337.

Mayoux, L., and G. Mackie. 2007. *Making the Strongest Links: A Practical Guide to Mainstreaming Gender Analysis in Value Chain Development.* Geneva: ILO.

McArdle, L., and P. Thomas. 2012. "Fair Enough? Women and Fair Trade." *Critical Perspectives on International Business* 8 (4): 277–294.

McCarthy, L., and J. Moon. 2018. "Disrupting the Gender Institution: Consciousness-Raising in the Cocoa Value Chain." *Organization Studies* 39 (9): 1153–1177.

Mudege, N. N., S. Mayanja, T. Muzhingi. 2018. "Women and Men Farmer Perceptions of Economic and Health Benefits of Orange Fleshed Sweet Potato (OFSP) in Phalombe and Chikwawa Districts in Malawi." *Food Security* 10 (4): 1139–1139.

Mulema, A. A., and R. E. Mazur. 2016. "Motivation and Participation in Multi-Stakeholder Innovation Platforms in the Great Lakes Region of Africa." *Community Development Journal* 51 (2): 212–228.

Mulema, A. A., C. R. Farnworth, K. E. Colverson. 2017. "Gender-Based Constraints and Opportunities to Women's Participation in the Small Ruminant Value Chain in Ethiopia: A Community Capitals Analysis." *Community Development* 48 (3): 351–369.

Mutua, E., J. Njuki, E. M. Waithanji. 2014. *Review of Gender and Value Chain Analysis, Development and Evaluation Toolkits.* Nairobi: ILRI.

Nansereko, S. C. 2010. *A Gender Perspective to Value Chain Analysis for Wood-Furniture Industry Upgrading: A Case of Jepara, Central Java–Indonesia.* Bogor: CIFOR.

Njuki, J., E. Waithanji, B. Sakwa, J. Kariuki, E. Mukewa, J. Ngige. 2014. "Can Market-Based Approaches to Technology Development and Dissemination Benefit Women Smallholder Farmers? A Qualitative Assessment of Gender Dynamics in the Ownership, Purchase, and Use of Irrigation Pumps in Kenya and Tanzania." Discussion Paper 01357, IFPRI, Washington, DC.

O'Neill, E. D., and B. Crona. 2017. "Assistance Networks in Seafood Trade–A Means to Assess Benefit Distribution in Small-Scale Fisheries." *Marine Policy* 78: 196–205.

O'Neill, E. D., N. K. Asare, D. W. Aheto. 2018. "Socioeconomic Dynamics of the Ghanaian Tuna Industry: A Value-Chain Approach to Understanding Aspects of Global Fisheries." *African Journal of Marine Science* 40 (3): 303–313.

Oduol, J. B. A., D. Mithöfer, F. Place, E. Kang'ole, J. Olwande, L. Kirimi, M. Mathenge. 2017. "Women's Participation in High Value Agricultural Commodity Chains in Kenya: Strategies for Closing the Gender Gap." *Journal of Rural Studies* 50: 228–239.

Pearson, R. 2007. "Beyond Women Workers: Gendering CSR." *Third World Quarterly* 28 (4): 731–749.

Pouliot, M., and M. Elias. 2013. "To Process or Not to Process? Factors Enabling and Constraining Shea Butter Production and Income in Burkina Faso." *Geoforum* 50: 211–220.

Pyburn, R., G. Audet-Bélanger, S. Dido, G. Quiroga, I. Flink. 2015. "Unleashing Potential: Gender and Youth Inclusive Agri-Food Chains." Working Paper 7, KIT, Amsterdam.

Quisumbing, A. R., S. Roy, J. Njuki, K. Tanvin, E. Waithanji. 2013. "Can Dairy Value-Chain Projects Change Gender Norms in Rural Bangladesh? Impacts on Assets, Gender Norms, and Time Use." Discussion Paper 01311, IFPRI, Washington, DC.

Quisumbing, A. R., D. Rubin, C. Manfre, E. Waithanji, M. van den Bold, D. Olney, R. Meinzen-Dick. 2014. "Closing the Gender Asset Gap: Learning from Value Chain Development in Africa and Asia." Discussion Paper 01321, IFPRI, Washington, DC.

Quisumbing, A. R., D. Rubin, C. Manfre, E. Waithanji, M. van den Bold, D. Olney, N. Johnson, R. Meinzen-Dick. 2015. "Gender, Assets, and Market-Oriented Agriculture: Learning from High-Value Crop and Livestock Projects in Africa and Asia." *Agriculture and Human Values* 32 (4): 705–725.

Rahman, A. 1999. "Micro-Credit Initiatives for Equitable and Sustainable Development: Who Pays?" *World Development* 27 (1): 67–82.

Ravera, F., B. Martín-López, U. Pascual, A. Drucker. 2016. "The Diversity of Gendered Adaptation Strategies to Climate Change of Indian Farmers: A Feminist Intersectional Approach." *Ambio* 45 (3): 335–351.

Riisgaard, L., G. Michuki, P. Gibbon, S. Bolwig, N. Warring, N. Lund Rants. 2009. *The Performance of Voluntary Standard Schemes from the Perspective of Small Producers in East Africa.* Copenhagen: DIIS and Traidcraft.

Riisgaard, L., A. M. Fibla, S. Ponte. 2010. "Gender and Value Chain Development." Evaluation Study. Evaluation Department of the Danish Foreign Ministry, Copenhagen.

Ruben, R., and R. Fort. 2012. "The Impact of Fair Trade Certification for Coffee Farmers in Peru." *World Development* 40 (3): 570–582.

Sachs, C. E., ed. 2019. *Gender, Agriculture and Agrarian Transformations: Changing Relations in Africa, Latin America and Asia.* London: Routledge.

Said-Allsopp, M., and A. Tallontire. 2014. "Pathways to Empowerment? Dynamics of Women's Participation in Global Value Chains." *Journal of Cleaner Production* 107: 114–121.

Sari, I., C. McDougall, S. Rajaratnam, C. Park. 2017. *Women's Empowerment in Aquaculture: Two Case Studies from Indonesia.* Penang: WorldFish.

Sarku, R. 2016. "Analyses of Gender Roles in the Oil Palm Industry in Kwaebibirem District, Ghana." *International Journal of Humanities and Social Sciences* 6 (3): 187–198.

Shackleton, S., F. Paumgarten, H. Kassa, M. Husselman, M. Zida. 2011. "Opportunities for Enhancing Poor Women's Socioeconomic Empowerment in the Value Chains of Three African Non-Timber Forest Products (NTFPs)." *International Forestry Review* 13 (2): 136–151.

Sindi, K., L. Kirimi, J. Low. 2013. "Can Biofortified Orange Fleshed Sweetpotato Make Commercially Viable Products and Help in Combatting Vitamin A Deficiency?" Presented at Fourth International Conference of African Association of Agricultural Economists, Hammamet, September 22–25.

Singh, S. 2003. "Contract Farming in India: Impacts on Women and Child Workers." Gatekeeper Series 111. London: IIED.

Smith, H. E., M. D. Hudson, K. Schreckenberg. 2017. "Livelihood Diversification: The Role of Charcoal Production in Southern Malawi." *Energy for Sustainable Development* 36: 22–36.

Stoian, D., J. Donovan, J. Fisk, M. Muldoon. 2012. "Value Chain Development for Rural Poverty Reduction: A Reality Check and a Warning." *Enterprise Development and Microfinance* 23 (1): 54–60.

Stoian, D., J. Donovan, M. Elias, T. Blare. 2018a. "Fit for Purpose? A Review of Guides for Gender-Equitable Value Chain Development." *Development in Practice 28* (4): 494–509.

Stoian, D., G. Meldrum, H. Lamers, T. Blare, M. Elias, J. Donovan. 2018b. "Exploring a Gender-Responsive Asset-Based Approach to Enhance the Transformative Potential of Value Chain Development in Guatemala, India and Peru." Presented at CGIAR PIM Workshop, Vancouver, BC, July 28, 2018.

Sugden, F., N. Maskey, F. Clement, V. Ramesh, A. Philip, A. Rai. 2014. "Agrarian Stress and Climate Change in the Eastern Gangetic Plains: Gendered Vulnerability in a Stratified Social Formation." *Global Environmental Change* 29: 258–269.

Tallontire, A., C. Dolan, S. Smith, S. Barrientos. 2005. "Reaching the Marginalised? Gender Value Chains and Ethical Trade in African Horticulture." *Development in Practice* 15 (3–4): 559–571.

Tavenner, K., and T. A. Crane. 2018. "Gender Power in Kenyan Dairy: Cows, Commodities, and Commercialization." *Agriculture and Human Values* 35: 701–715.

Tavenner, K., and T. A. Crane. 2019. "Implementing 'Gender Equity' in Livestock Interventions: Caught between Patriarchy and Paternalism." In *Gender, Agriculture and Agrarian Transformations: Changing Relations in Africa, Latin America and Asia*, edited by C. Sachs, 147–161. London: Routledge.

Tavenner, K., M. van Wijk, S. Fraval, J. Hammond, I. Baltenweck, N. Teufel, E. Kihoro, N. de Haan, J. van Etten, J. Steinke, D. Baines, P. Carpena, T. Skirrow, T. Rosenstock, C. Lamanna, M. Ng'endo, S. Chesterman, N. Namoi, L. Manda. 2019. "Intensifying Inequality? Gendered Trends in Commercializing and Diversifying Smallholder Farming Systems in East Africa." *Frontiers in Sustainable Food Systems* 3.

Theis, S., N. Lefore, R. Meinzen-Dick, E. Bryan, E. 2018. "What Happens after Technology Adoption? Gendered Aspects of Small-Scale Irrigation Technologies in Ethiopia, Ghana, and Tanzania." *Agriculture and Human Values* 35 (3): 671–684.

Tsusaka, T. W., A. Orr, H. W. Msere, S. Homann-Kee Tui, P. Maimisa, G. H. Twanje, R. Botha. 2016. "Do Commercialization and Mechanization of a 'Women's Crop' Disempower Women Farmers? Evidence from Zambia and Malawi." Presented at Agricultural and Applied Economics Association Annual Meeting, Boston, MA, July 31–August 2, 2016.

Turner, B. 2014. "Neoliberal Politics of Resource Extraction: Moroccan Argan Oil." *Forum for Development Studies* 41 (2): 207–232.

Turner, T. E., W. M. Kaara, L. S. Brownhill. 1997. "Social Reconstruction in Rural Africa: A Gendered Class Analysis of Women's Resistance to Export Crop Production in Kenya." *Canadian Journal of Development Studies/Revue canadienne d'études du développement* 18 (2): 213–238.

Veliu, A., N. Gessese, C. Ragasa, C. Okali. 2009. "Gender Analysis of Aquaculture Value Chain in Northeast Vietnam and Nigeria." Agriculture and Rural Development Discussion Paper 44, FAO, Rome.

Velte, M., and P. Dannenberg. 2014. "Export Horticulture–Empowering Female Small-Scale Farmers in Kenya?" *DIE ERDE–Journal of the Geographical Society of Berlin* 145 (3): 135–141.

Verhart, N., and R. Pyburn. 2010. "The Rough Road to Gender Equitable Growth: The Case of Café de Mujer Guatemala." *Development* 53 (3): 356–361.

Wijers, G. D. 2019. "Inequality Regimes in Indonesian Dairy Cooperatives: Understanding Institutional Barriers to Gender Equality." *Agriculture and Human Values* 36 (2): 167–181.

Woldu, T., F. Tadesse, M. Waller. 2015. "Women's Participation in Agricultural Cooperatives in Ethiopia." ESSP Working Paper 57, IFPRI, Washington, DC.

Chapter 5

NUTRITION-SENSITIVE AGRICULTURE FOR GENDER EQUALITY

Hazel Malapit, Jessica Heckert, Jessica Scott, Ravula Padmaja, and Agnes Quisumbing

Globally, malnutrition[1] remains unacceptably high, and its burden falls disproportionately on women and girls. The 2018 Global Nutrition Report states that women experience a disproportionate burden of some forms of malnutrition: one in three women of reproductive age has anemia; women have a higher prevalence of obesity than men—yet millions of women are underweight (Development Initiatives 2018). Women and adolescent girls have greater nutritional needs. For example, young women's iron requirements are higher at puberty, and caloric and micronutrient needs are higher during pregnancy and lactation. Poor nutritional status for women and girls also has direct intergenerational consequences via pregnancy and childbirth outcomes (Victora et al. 2008, Black et al. 2013a). The first 1,000 days (start of pregnancy until the child's second birthday) represent a critical window during which poor nutrition leads to irreversible deficits in children's development, the ability to learn, and productivity and health in adulthood (ibid.).

Many empirical studies and programmatic approaches have focused on the instrumental value of leveraging women's empowerment for improved nutrition outcomes; nutrition-sensitive agriculture programs (NSAPs) have been one such area (Malapit 2019). (See recent systematic reviews on gender in NSAPs by Newton et al. 2018 and Ruel et al. 2018.) This chapter flips the framing of how women's empowerment and gender equity can lead to improved nutrition outcomes in NSAPs by examining how investments in NSAPs contribute to women's empowerment and gender equality. Empowerment is the process by which women acquire the ability to make strategic life choices in areas where they were previously denied that ability (Kabeer 1999).

1 Malnutrition refers to forms of undernutrition and overnutrition. Commonly assessed forms of undernutrition include stunting (low height for age), wasting (low weight for height), and anemia (low blood hemoglobin concentration). Overnutrition includes overweight and obesity.

In this chapter, we operationalize empowerment as the ability to make important decisions, especially regarding production, livelihoods, food provision, and other areas that are critical for the interventions implemented in NSAPs, as well as shifting gender attitudes and norms toward a more equitable division of labor and resources. We begin by describing the background on NSAPs in relation to gender equality and women's empowerment. The next section reviews four NSAPs to highlight what we can learn about the potential of using NSAPs to achieve women's empowerment and gender equality. We then discuss the four cases, drawing out lessons. In the final section, we outline a forward-looking research agenda for this theme.

Nutrition-sensitive agriculture programs vis-à-vis gender equality and women's empowerment

Despite increasing global commitments, it is clear that scaling up even the most effective nutrition-specific interventions and programs will not achieve global nutrition targets (Black et al. 2013b, Ruel and Alderman 2013). Doing so requires large-scale nutrition-sensitive programs in sectors that address the underlying determinants of nutrition, alongside nutrition-specific interventions (Ruel and Alderman 2013). Nutrition-sensitive interventions cannot simply address food security or income with the expectation that these will improve nutrition. Rather, nutrition-sensitive programs and interventions must include nutrition-specific goals and actions targeted to populations with nutritional deficits (see Box 5.1). Nutrition-specific goals and actions must address the immediate determinants of nutrition, such as diets and nutrient intakes, feeding and care practices, and infectious diseases. An agricultural program or intervention that promotes production of a nutritious food, without nutrition-specific goals and actions for increasing consumption, cannot be considered nutrition-sensitive.

Given agriculture's role in the production, availability, and consumption of food, and in providing livelihoods and incomes in rural areas, the need for investments in NSAPs is undeniable. To enhance nutrition outcomes, it is necessary to redirect agriculture from merely producing large quantities of food toward producing nutritious foods and ensuring they are available to, and consumed by, those who need the key nutrients (Willett et al. 2019).

NSAPs must also recognize that factors outside agriculture determine health and nutrition status, and that men and women face differential health needs and risks that vary across contexts and the lifecycle (A4NH 2016).

BOX 5.1 Definitions of nutrition-specific and nutrition-sensitive interventions or programs

Nutrition-specific interventions or programs are those that address the *immediate determinants* of fetal and child nutrition and development: adequate food and nutrient intake, feeding, caregiving and parenting practices, and low burden of infectious diseases.

Nutrition-sensitive interventions or programs are those that address the *underlying determinants* of fetal and child nutrition and development: food security; adequate caregiving resources at the maternal, household, and community levels; and access to health services and a safe and hygienic environment—and incorporate specific nutrition goals and actions.

Source: Ruel and Alderman (2013).

Poor households do not have the same access to the right kinds, qualities, and quantities of food, and the household dynamics that dictate the intrahousehold distribution of food may place some individuals in poor households at greater risk of poor diets and undernutrition. For example, recent country-specific studies including macro and micronutrient assessment found inequitable shares of food and nutrients among household members (Wibowo et al. 2015, Harris-Fry et al. 2018).

Agriculture can contribute to nutritional outcomes through six general pathways (Ruel and Alderman 2013) (see Box 5.2). NSAPs justifiably focus on women, given their prominent role in three of the six pathways (Pathways 4–6). However, gender matters for all the pathways, because gender differences in roles, preferences, and power mediate each step in the series of actions leading to nutrition and health outcomes. Different genders and social groups may also receive differential benefits and risks associated with these pathways owing to their specific health needs and sources of resilience that vary across contexts and the lifecycle (A4NH 2016, Harris and Nisbett 2018).

NSAPs could potentially improve women's empowerment and gender equality in two ways. First, if NSAPs succeed in improving the nutritional status of women and girls, the immediate consequences are improvements in well-being and the narrowing of existing gender gaps in nutritional status. If sustainable, these impacts lay the foundation for better health and nutrition

BOX 5.2 **Agriculture–nutrition pathways**

Pathway 1: Agriculture is a source of food, both for households directly engaged in **production** and for the food system as a whole.

Pathway 2: The sale of agricultural products is a source of **income**, which can be used to purchase nutritious food, and goods and services that support good health.

Pathway 3: Agricultural markets determine food **prices**, which influences the relative cost of buying and selling nutritious foods.

Pathway 4: Women's empowerment and social status influence how resources are allocated within the household.

Pathway 5: Women's participation in agriculture influences their **time use,** which can have positive or negative consequences for their ability to provide care for children.

Pathway 6: Women's participation in agriculture may also have positive or negative consequences for **women's own health and nutrition**, through exposure to health hazards related to agricultural activities, and the balance of energy expenditure and consumption.

Source: Ruel and Alderman (2013).

outcomes not only for themselves but also for their future children. For women, improved nutritional status may unlock the empowerment process, whereby the benefits of improved health can enable them to take advantage of opportunities and exercise agency. Whether these changes result in long-term improvements in gender equality is still unknown. Longitudinal studies with sufficiently long follow-up periods have traced the benefit of good nutrition in early childhood over the life course, but these studies are few (for example Maluccio et al. 2009, Hoddinott et al. 2013). Additionally, nutrition impact studies typically measure nutritional status (for example height and weight) only for the target populations of interest (usually women and children within a specific age range), and rarely document how gender gaps in nutritional status have changed as a result of the intervention.

Second, many NSAPs recognize that gender dynamics and women's empowerment play an important role in achieving nutrition impacts. By targeting pregnant women and mothers of young children, these programs have capitalized on the instrumental role of women's empowerment and leveraged it for enhancing nutrition and health outcomes. NSAPs may adopt specific strategies and actions that shift gender norms and promote women's empowerment as a pathway for achieving their nutrition goals. While initially motivated by instrumental objectives, in practice this presents an opportunity to directly empower women as part of the intervention. Such impacts are measured alongside other outcomes that the NSAPs are seeking to improve within the program's timeframe.

Evidence from selected case studies

This section presents evidence from four types of NSAPs that used different approaches in addressing gender dynamics in design and implementation. We purposely selected these cases from among technologies that CGIAR and other agricultural research centers have developed, as well as those implemented by partners and evaluated by a CGIAR research center. We selected only nutrition-sensitive programs—that is those that incorporate nutrition-specific goals and actions targeted at nutritionally vulnerable populations (see Box 5.1). We therefore excluded CGIAR programs that have nutrition objectives (for example reducing malnutrition, increasing dietary diversity) but that do not have nutrition-specific goals and actions (for example nutrition training), as they do not meet this definition of "nutrition-sensitive." We also selected cases based on the strength of the evaluation design, including particularly those with peer-reviewed publications on their impacts and those that provided more systematic information and analyses on how the program had addressed gender dynamics. Finally, cases selected show a range of approaches to gender programming in a diverse set of interventions and contexts, to highlight what we can learn from different approaches to NSAPs.

The cases feature programs that introduced biofortified orange-fleshed sweet potato (OSP) in Uganda, improved vegetables and fish technologies in Bangladesh, a dairy value chain project in northern Senegal, and homestead food production (HFP) and home gardening in Bangladesh and Burkina Faso. All four cases address the production pathway (Pathway 1), two of the cases address the income pathway (Pathway 2), and three address the women's empowerment pathway (Pathway 4). The last case also reports findings on the program's negative influence on time use (Pathway 5). At the minimum, most NSAPs aim to both reach and benefit women, as reflected in the four cases featured; in three of the selected cases, the programs also included activities to empower women, whether implicitly or explicitly. In reviewing the evidence, we acknowledge that not all of these NSAPs were designed to empower women or to achieve gender equality, and thus not all evaluations included indicators for these outcomes. Nevertheless, the evaluations as well as organizations' reflections on implementation can provide important lessons on how NSAPs can promote women's empowerment and gender equality. Table 5.1 presents a summary of the approaches used in the four cases.

Disseminating orange-fleshed sweet potato to women's farmer groups in Uganda[2]

Biofortification, the process of breeding staple crops rich in essential micronutrients, is a feasible and cost-effective means of delivering micronutrients to populations with limited access to diverse diets and other micronutrient interventions (Bouis et al. 2011, Bouis and Saltzman 2017). Biofortification has been shown to be effective for vitamin A-rich OSP in Mozambique and Uganda, where randomized controlled trials (RCTs) document impacts on vitamin A intake among mothers and young children in both countries and on child vitamin A status in Uganda (Hotz et al. 2012a, 2012b). Additional analyses also show large reductions in the prevalence and duration of diarrhea among young children (Jones and de Brauw 2015), supporting the well-known role of vitamin A in protecting immunity (Ruel et al. 2018).

In Uganda, the HarvestPlus Reaching End User (REU) project introduced OSP to approximately 10,000 farm households between 2007 and 2009, by providing free OSP planting material (vines) and complementary training. The intervention included a one-time distribution of OSP vines to project households, provision of extension services to men and women who were members of project farmer groups on OSP production and marketing, and provision of nutritional knowledge training on vitamin A deficiency to women in the same households (female farmer group member or female spouse of the male farmer group member). The impact evaluation sample included 84 farmer groups from 3 districts where white and yellow sweet potato were commonly grown and consumed.

The REU project was designed to increase production of OSP and increase vitamin A availability in the household (Pathway 1), by promoting its cultivation alongside complementary nutrition training intended to encourage the consumption of OSP by mothers and young children among beneficiary households. As typical in nutrition projects, this NSAP aimed to reach and benefit mothers with young children who resided in the project

2 This section draws heavily on Gilligan et al. (2020).

TABLE 5.1 Summary of case study approaches

	Case 1 *Disseminating OSP to women's farmer groups in Uganda*	**Case 2** *Improved vegetable and polyculture fish management technologies in Bangladesh*	**Case 3.3** *Dairy value chain project in northern Senegal*	**Case 3.4** *HFP programs and their evolution*
Intervention	• Distribution of OSP vines, provision of extension services (primarily to men) • Nutritional knowledge training (primarily to women)	• Training and credit for homestead production of vegetables (targeted to women group members) • Training to some group members and credit to all adopters of group-operated polyculture fishponds (targeted to women's groups) • Training to all and credit to poor adopters of individually operated polyculture fishponds (primarily to men)	• Provision of micronutrient-forti-fied yogurt as incentive to milk producers who met supply quota (implicitly targeted to women) • BCC training provided to all house-holds (primarily to women)	• Provision of inputs and training for homestead production of nutrient-rich food (targeted to women) • BCC training (targeted to women)
Pathways influenced	Pathway 1 (production)	Pathway 1 (production) Pathway 2 (income) In the vegetables and group fishponds only: Pathway 4 (women's empowerment)	Pathway 1 (production) Pathway 4 (women's empowerment)	Pathway 1 (production) Pathway 2 (income) Pathway 4 (women's empowerment) Pathway 5 (time use)
Aim to Reach, Benefit, or Empower women?	Reach, Benefit	Reach, Benefit, Empower (vegetables and group-based fishponds only)	Reach, Benefit, Empower	Reach, Benefit, Empower
Institution originating technology	HarvestPlus	• World Vegetable Center • WorldFish	n/a	• Helen Keller International
Implementation partners	• Volunteer Efforts for Development Concerns and Farming for Food • Development Pro-gram-Eastern Uganda	• Vegetables: Gono Kallayan Trust • Group fishponds: Banchte Shekha • Individual fishponds: Mymensingh Aquaculture Extension Project and Department of Fisheries	• La Laiterie du Berger • Cellule de Lutte Contre la Mal-nutrition	• Helen Keller International • Local partner in Burkina Faso: Association d'Appui et de Promo-tion Rurale du Gulmu
Evaluation partner	IFPRI	IFPRI	IFPRI	IFPRI
Key references	Hotz et al. (2012a), Gilligan et al. (forthcoming)	Kumar and Quisumbing (2011), Quisumbing and Kumar (2011)	Le Port et al. (2017), Bernard, et al. (2019)	Quisumbing et al. (2015), Olney et al. (2016)

Note: BCC = behavior change communication, HFP = homestead food production, IFPRI = International Food Policy Research Institute, OSP = orange-fleshed sweet potato.

households. It targeted mothers as beneficiaries of the nutrition improvements, and they received training on the nutrition content of the biofortified crop, but there was no explicit strategy to involve men in the nutrition training. While men take a lead role in crop choice decisions in the Uganda project areas, women also play an active role in crop selection, particularly for food crops for household consumption, and they often supply labor on household farms.

Gilligan et al. (2020) find that the probability of adoption of OSP is highest on parcels where there is joint (rather than female-exclusive) control over productive decisions, and where women take the lead in deciding which crops are grown. On the other hand, the probability of adopting OSP is lowest on parcels exclusively controlled by men. These results suggest women play an important role in the decision to adopt OSP but that they often make this decision jointly with their husbands. Thus, the strategy of targeting only women with nutritional training may be missing an opportunity to create awareness of the benefits of OSP among men, and recognition that engaging with both men and women may be the best strategy for promoting adoption.

The REU project in Uganda is an example of a NSAP designed to influence OSP production and consumption, but it did not explicitly aim to shift gender norms or empower women in OSP crop choice or consumption. Rather, it viewed gender norms around crop choice and consumption decisions as given and was designed to accommodate these norms. Women were reached in the nutritional training but less so in the extension services on OSP. While the REU project achieved positive nutrition impacts, providing benefits to women through improved vitamin A intake, the impact evaluation was not designed to assess whether it had increased women's ability to exercise greater agency over OSP production and consumption decisions, or whether the changes in dietary intake of vitamin A had closed gender gaps in micronutrient deficiencies. Nutrition information was collected only from women and children, and even the impact results on children in Uganda are not reported separately for girls and boys (Hotz et al. 2012a).

Improved vegetable and polyculture fish management technologies in Bangladesh[3]

In Bangladesh, as in other countries in South Asia, a pro-male bias in food distribution within the household is closely linked to women's micronutrient deficiencies (Kumar and Quisumbing 2011). Boys in this region are also favored

3 This section draws heavily on Kumar and Quisumbing (2011) and Quisumbing and Kumar (2011).

in the distribution of non-food health inputs such as healthcare (Haddad et al. 1996). Thus, many of the food-based interventions that government and civil society organizations in Bangladesh have undertaken have targeted women. Examples include programs that promoted improved vegetable and polyculture fish production technologies that were evaluated for their long-term impacts on household and individual well-being using a quasi-experimental approach.

The agricultural technology interventions included a vegetable intervention targeted to women's group members who grew vegetables on small plots on or near the household compound, and a polyculture fish technology intervention implemented using two alternative dissemination strategies. In one site, the implementing non-governmental organization arranged long-term leases of small ponds managed by groups of 5–20 women who received credit and training in polyculture fish production methods. Some groups also took advantage of a food-for-work program to excavate ponds. In the other fishpond site, a government extension program required beneficiary households to already own or manage a pond or to share pond ownership with other households.

The group fishponds intervention provided training to some members of each adopter group and credit to all group members. The individual fishponds intervention provided training to all adopters but credit only to relatively poorer households. The individual fishponds training and credit was supposed to reach both men and women, but it reached more men than women, and the training tended to reinforce existing gender norms about women's tasks and mobility in public spaces.

All three intervention modalities aimed to increase household well-being, as measured by consumption expenditures, assets, income, and calorie availability; diet, as measured by intakes of calories, protein, iron, and vitamin A consumed by children and adults; and nutritional status, as measured by the concentration of blood hemoglobin and anthropometric measures (height and weight) converted into standardized height-for-age and weight-for-height indicators.

While all three intervention modalities aimed to work through the production and income pathways (Pathways 1 and 2, respectively) by providing training and credit so that beneficiaries could adopt the improved agricultural technologies, the vegetable and group fishponds interventions worked through women's groups and provided them with resources, such as credit and access to a fishpond site. These group-based interventions offered opportunities for the women beneficiaries to manage and make important decisions on different aspects of vegetable and fish production (Pathway 4).

Thus, while all three intervention modalities aimed to reach and benefit women, only the group-based interventions had the potential to contribute to women's empowerment.

The largest monetary returns to early adoption at the household level were in the individually operated fishponds sites, while early adopters of the vegetable technologies experienced insignificant impacts on household-level monetary returns, in large part because the diffusion of the vegetable technologies beyond the original treatment area may have eroded any short-term gains the early adopters enjoyed. Nevertheless, early adopters of improved vegetables achieved sustained improvements in nutritional status, especially for women and girls. The proportion of stunted girls decreased differentially by 28 percentage points while the proportion of thin boys decreased differentially by 43 percentage points. Women's body mass index (BMI) increased as desired, although men's BMI decreased—an unintended effect in this undernourished population. These findings suggest that working through women's groups and disseminating vitamin A- and iron-rich vegetables that are consumed by women had a positive net impact on the nutritional status of women and children, especially girls.

Early adoption of group fishponds had mixed impacts on children's long-term nutritional status, and did not significantly affect men's or women's nutritional status. Among early adopters of the individual fishponds, consumption expenditures and calorie availability at the household level improved and the proportion of women with low hemoglobin levels decreased. However, impacts on long-term indicators of nutritional status for girls were not sustained. Unlike at the vegetables site, BMIs of women in the individual fishponds site did not improve, perhaps because the increase in nutrient intake did not compensate for women's increased work effort. Because the individual fishponds approach did not effectively target women for disseminating technology or nutrition knowledge, the intervention did not enable women to invest the family's resources toward their own nutrition or girls' nutritional status in the long run.

These NSAPs promoting improved vegetable and fish technologies show that group-based dissemination strategies targeted to women are an effective strategy to reach women and can benefit them by improving the nutritional status of women and girls. The group-based approach may also help women build social capital and accumulate other physical assets. Social and physical assets are important resources that support the empowerment process, alongside agency and achievements, which in this case are improvements in nutritional status (Kabeer 1999). A mixed-methods study by Hallman et al. (2007) and qualitative work by Naved (2000) have explored the role of group-based approaches in building social capital and enhancing women's decision-making.

In addition, although the evaluation was not designed to assess empowerment impacts (for example using decision-making indicators), a related paper assesses the impact of these new technologies on men's and women's asset accumulation (Quisumbing and Kumar 2011). The paper finds that women's assets increase more, relative to men's, when technologies are disseminated through women's groups, suggesting that implementation modalities are important in determining the gendered impact of new technologies. Social capital, as embodied through women's groups, not only serves as a substitute for physical assets in the short run but also helps build up women's asset portfolios in the long run. However, even if women's assets increased more than men's in the group vegetable site, men's asset stocks were still much greater than women's. This finding suggests that closing the gender asset gap may require more concerted and deliberate programming to increase women's control of assets and reduce gender asset inequality. This intervention did not include any programming to increase men's involvement in health and nutrition, so it is difficult to compare the results with the counterfactual approach involving men. This continues to be a gap in the design and implementation of NSAPs.

Dairy value chain project in northern Senegal[4]

A nutrition-sensitive value chain integrates nutrition objectives and nutrition-specific interventions along the supply chain while maintaining the product's economic value and taking into account the nutritional needs of multiple actors, including consumers (Le Port et al. 2017). Nutrition-sensitive value chain approaches have the potential to deliver nutrition-specific interventions, such as micronutrient supplements or micronutrient-fortified food products, at scale cost-effectively. Can such approaches also promote women's empowerment and gender equality?

A pilot study in a remote area in northern Senegal tested whether it was possible to use a dairy value chain to distribute a micronutrient-fortified yogurt (MNFY) to improve hemoglobin and reduce anemia among women and children in semi-nomadic milk supplier households (Le Port et al. 2017, Bernard et al. 2019). A local dairy factory (La Laiterie du Berger) produced the MNFY using milk collected from its suppliers. The intervention used

4 This section draws heavily on Le Port et al. (2017) and Bernard et al. (2019).

a novel nutrition-sensitive contract design, whereby milk producers who supplied a minimum amount of cow milk delivered five days a week received daily MNFY the following week and were instructed to give it to their children aged 24–59 months (Le Port et al. 2017). The MNFY was delivered to milk collection points mainly accessed by women. Thus, the scheme implicitly targeted women as the main recipients of the incentive, whereas men are typically the main recipients of cash payments (Bernard et al. 2019).

Additionally, the study conducted a behavior change communication (BCC) strategy with all target households. This strategy focused on optimal infant and young child nutrition, the importance of micronutrients, and the role that diverse diet and fortified food can play in delivering them. Although the BCC training was open to all, it was mostly women who attended the sessions.

The intervention was designed to influence the dairy production and women's empowerment pathways (Pathways 1 and 4, respectively), recognizing that, among the semi-nomadic pastoralists (the Pulaar or Fulani) who dominate milk production in this region, gender norms around milk production are clearly established at a young age, with women in charge of milk production and men in charge of herd management. The MNFY incentive was also viewed as valuable to women, given their role as primary caregivers. It therefore functioned as a valued resource transferred to women dairy farmers directly, benefiting them by increasing the returns to their individual efforts in dairy production. The intervention was also designed around the milk production domain, which women already controlled (men were in charge of contracts and traveling to the central processing plant to receive payment). Because women provided the labor in household dairy production, they could adjust production to fulfill the contract and receive the incentive at milk collection points, which were in areas they could access easily.

Evaluation of the impact of the nutrition-based incentive on children's nutritional status and milk production using a cluster RCT shows that the dairy value chain intervention was effective, increasing the regularity of milk deliveries, although these impacts were limited to the dry season and to those contracts headed by a woman (Bernard et al. 2019). It also improved hemoglobin in children 24–59 months of age at baseline, after one year of intervention, with a statistically significant impact on boys but not girls.

The incentive also increased women's decision-making: the overall female decision-making index increased by 0.53 points, from an average of 4.48 in the control group. It led to significant increases in women being the main decision-makers with respect to veterinary, vaccination, and insemination services, and cow migration decisions.[5] For female-controlled contracts, the incentive also led to women being significantly more likely to be the main decision-maker on domains of selling milk and cow migration.

Overall, the intervention successfully reached, benefited, and empowered women. It effectively reached women by selecting milk collection points that women accessed and by focusing on aspects of milk production that women already controlled. Women benefited through increased returns on their labor, provided in the form of an in-kind transfer that they valued (MNFY). Women were also empowered, as improvements around their ability to make decisions regarding milk production show. These results highlight the potential of using nutrition-sensitive value chain interventions to empower women.

It is also notable that men had a limited response to the incentive. The intervention implicitly targeted women but there was no explicit strategy to exclude men. It is possible that, because the MNFY was distributed at collection points where women more often went, men were less likely to directly receive the incentive. Men may have been less aware of the nutritional benefits of the MNFY that the BCC, which reached more women than men, reinforced. As in the OSP case, this supports the recommendation to include men in BCC strategies to emphasize the shared responsibility of both men and women in enhancing the nutritional status of all household members.

Homestead food production programs and their evolution[6]

HFP programs, which focus on the production of nutrient-rich foods on homestead plots, represent a notable type of NSAP. Helen Keller International (HKI) originally piloted the HFP model in Bangladesh in the 1980s; it expanded and adapted the program for Cambodia, Nepal, and the Philippines in the late 1990s and recently adapted it for West Africa. These programs typically target women in smallholder agricultural households and train them to cultivate kitchen gardens and often to raise poultry or fish. Training is conducted on Village Model Farms (VMFs) and field staff train women to cultivate on their own homesteads. Production of both crops and animal source foods is intended to benefit households via home consumption (Pathway 1) and through the sale of surplus produce for increased incomes

5 Decision-making was assessed over cattle purchases and sales, inputs provided to cows, milk sales and use of money from cows, home consumption of milk, and timing of seasonal migration; responses were also aggregated into an index (Bernard et al. 2019).

6 This section draws heavily on Quisumbing et al. (2015), Olney et al. (2016), and personal correspondence with Stella Nordhagen of Helen Keller International (April 10, 2018).

(Pathway 2). HFP programs are nutrition-sensitive; in addition to cultivating nutrient-rich foods, they include a BCC strategy that teaches beneficiaries optimal nutrition, health, and hygiene practices and provides them with skills to negotiate in favor of these behaviors in their household.

Evaluations of these programs have consistently found positive impacts on the diets of women and households (Girard et al. 2012), and on the nutritional status of children and women in Burkina Faso and Cambodia, among others (Olney et al. 2009, 2016).

Early HKI programming did not deliberately aim to empower women. The original Bangladesh model did not initially challenge gender norms or patriarchal power structures (Hillenbrand 2010). All-male field staff conducted agricultural training while all-female staff delivered nutrition education. The main selection criteria for the VMF owners were possession of a suitable and sizeable land plot and prior experience in farming, which favored men. Inadvertently or deliberately, men were not held responsible for the nutritional side of food production, reinforcing existing gender norms. Agricultural technology transfer in this model reinforced the stereotypes that men are capable of "farming" whereas women are suited for "gardening" and food preparation.

HKI's programs have since evolved to increase attention to empowering women and transforming gender dynamics. Feedback from field managers and beneficiaries indicated positive changes in women's lives related to program participation (Hillenbrand 2010). Earlier evaluations, although not always optimally designed, also found evidence of increased influence on household decisions (Bushamuka et al. 2005, Iannotti et al. 2009). Subsequently, programming in Bangladesh was modified to address gender concerns more directly, by eliminating land size as a criterion for choosing VMF owners, having women's groups themselves choose the Village Farmer Leader (VFL), using group-based marketing, employing new tools to describe and build women's own capacities and needs, and creating opportunities at all levels for staff training and reflection on gender concerns. Many of these changes have been included in the design of HKI's current HFP programs, including the Enhanced Homestead Food Production Program (E-HFP) in Burkina Faso.

However, women taking on the role of VFL faced heavier workloads (Pathway 5), because, despite their added responsibilities in farming, attending meetings, and other program-related activities, their household workloads did not diminish (Hillenbrand 2010). Beneficiaries often cited workloads and time as constraints, which may have not only limited uptake but also discouraged women from growing labor-intensive crops and selling products in the market (Kjeldsberg et al. 2018).

Specific adaptations made for Burkina Faso included training women as VFLs, cultivating model farms on land designated by the village for that purpose, and providing drip irrigation on the VMF. As is the case in much of West Africa, the Burkina Faso site faces more severe water constraints than do HKI's Asian sites, and the process evaluation identified potential program adaptations related to irrigation. Even if both men and women benefit from adaptations addressing water scarcity, women may gain greater benefits, because they are typically responsible for water collection. In addition, increasing space available at VMFs tends to benefit women more, as they typically do not have land of their own that has a reliable water supply (Olney et al. 2013).

The E-HFP program in Burkina Faso reduced wasting and diarrhea and increased hemoglobin among children aged 3–12 months old at the start of the 2-year program, although no significant impacts were found on stunting or underweight prevalence (Olney et al. 2015). The program also reduced underweight among mothers and increased their say in decision-making, notably in areas relating to purchasing decisions and healthcare decisions, as well as overall empowerment (Olney et al. 2016). These improvements in women's say in the domains of spousal communication and decisions on purchasing, healthcare, and family planning contributed to the program's impact on reducing wasting, with the largest share attributable to spousal communication, although improvements in these same domains of women's empowerment did not contribute to the increase in hemoglobin (Heckert et al. 2019).

There were also positive impacts on women's ownership and control of assets, as well as an increased perception among community members of women's ability to manage agricultural land (van den Bold et al. 2015). The change in gender norms on women's landownership is notable, given that the program did not distribute land to women but taught them how to manage their home gardens.

Increased evidence of program impacts on women's decision-making, as well as other aspects of women's empowerment, has led to greater integration of empowerment objectives in HKI's programming over time (Haselow et al. 2016). The current phase of the E-HFP program serves as a platform for *Nurturing Connections©*, a gender-transformative curriculum that aims to change gender norms, attitudes, behaviors, and institutions that underlie or reinforce gendered inequalities, through dialogues with husbands, community leaders, and community members in general (Nordhagen et al. 2017). This approach focuses on enhancing women's self-esteem and self-efficacy and also aims to mitigate the excessive demands that such programs place on women's time.

The *Nurturing Connections*© curriculum has been applied in a number of projects in the CGIAR. For example, WorldFish in collaboration with the Cereal Systems Initiative for South Asia-Bangladesh (CSISA-BD) applied *Nurturing Connections*© activities focused on the intrahousehold distribution of food with the nutrition-sensitive small-scale aquaculture program (Farnworth et al. 2015). Through practical family activities focused on distributing food within the household, women reported a change in husbands' attitudes toward women's practice of eating last and leaving a small amount of food on their plates. Until this exercise, men had not noticed this practice. The men acknowledged that women worked hard all day and should eat equally, and committed to paying more attention to what their wives were eating in the future. *Nurturing Connections*© was also the basis for the gender-sensitization curriculum in the Agriculture, Nutrition, and Gender Linkages (ANGeL) project in Bangladesh (Ahmed et al. 2018), a cluster RCT whose treatment arms included agricultural training, nutrition BCC, and gender-sensitization trainings to husbands and wives together.

HKI continues to use the *Nurturing Connections*© curriculum in new projects. An RCT in Cambodia has three arms: a treatment arm with a gender-transformative HFP model using the *Nurturing Connections*© curriculum, gender messaging within technical trainings, and women elected as homestead producers; a second arm in which households decide who will be the main producer and participant in trainings that are gender-blind; and a control arm (A4NH 2019). HKI's willingness to learn from implementation and to design and test modifications to improve its programming to enhance empowerment impacts illustrates how programming can be more intentional in tracking progress on gender outcomes. Such gender-intentional, well-evaluated NSAPs help build the evidence base to show how these types of programs can empower women and transform gender dynamics.

What have we learned?

Our review of selected NSAPs indicates that, while such programs can contribute to women's empowerment and gender equality, this does not happen automatically. First, there are multiple pathways linking agriculture and nutrition, and, although gender dynamics influence all of these pathways, the design and implementation of NSAPs do not always consider the role of gender dynamics. They are more likely to pay attention to women but not always in relation to men and other influential persons within their households and communities. When NSAPs do pay attention to gender relations, they may design programs to accommodate existing gender norms rather than to transform them. That is to say, most often these programs are gender accommodative (that is, they operate within existing gender norms and dynamics but do not attempt to change them) rather than gender transformative (that is, they aim to address gender inequalities by addressing harmful norms and dynamics).

For example, although both the HarvestPlus REU project that disseminated OSP and the Senegal dairy value chain intervention recognized women's productive roles as farmers growing OSP and as dairy producers in the fortified yogurt intervention, respectively, these NSAPs targeted nutrition trainings to women and not men, reinforcing existing gender stereotypes. The fact that OSP adoption was highest in plots that were jointly owned (but where women had a greater role in deciding what to grow) signals a missed opportunity to educate men on the benefits of growing OSP for their families. In Senegal, the limited impact of the incentive on men may have been an unintended consequence of targeting women, even if men were not intentionally excluded. While there is evidence of gender-accommodative approaches increasing women's empowerment, gender-transformative approaches may lead to larger and sustained increases in women's empowerment (see also Chapter 10, this volume).

Second, even when the design of NSAPs considers gender relations, they may seek only to "reach" (for example including them in nutrition education activities) or "benefit" women (for example leading to improved diets), not to "empower" them. This falls short of providing women opportunities to empower themselves and make strategic life decisions (Johnson et al. 2018). Without explicit empowerment objectives, they may not have specific strategies or actions that can empower women. This is well illustrated in the case of the agricultural technologies in Bangladesh: the group-based strategies had more sustained long-term impacts in reducing the gender asset gap (even if the gap was not completely closed) and in improving the nutritional status of women and girls compared with the program that targeted the household as a whole (but ended up reaching men by default). Qualitative work undertaken in conjunction with the agricultural technology study attested to the increased decision-making power of women who were involved in group-based programs (Naved 2000, Hallman et al. 2007). NSAPs that were more intentional in addressing gender barriers had more transformative impacts, such as the E-HFP program in Burkina Faso, which reduced the gender asset gap, empowered women, and changed local norms around women's land ownership.

Third, even if NSAPs have empowerment objectives, they may fail to attain them because of lack of intentionality and limited consideration of the range of domains that could be affected. NSAPs may be more likely to attempt to increase women's ability to make decisions in domains related to health and nutrition (OSP; HKI Burkina) or in productive domains where women traditionally make decisions (milk production in Senegal; vegetables in home gardens) but have only recently begun addressing nutrition messages to men or improving spousal communication. NSAPs, particularly those dealing with agricultural production, may increase women's control of assets, and perceptions that women are able to manage land (HKI Burkina), yet none of the NSAPs reviewed had attempted to address labor burdens or workloads. Engaging men in recognizing women's roles in agricultural production may help improve decision-making around those domains but involving men in sharing in domestic work and childrearing is a relatively unexplored area that may both shift gender norms around caregiving and reduce women's workload.

Finally, even if NSAPs aim to empower women, their monitoring, evaluation, and learning (MEL) frameworks may not include measurable indicators of empowerment, and thus will not be able to ascertain whether they have empowered women (Santoso et al. 2019). We are unable to ascertain the empowerment impacts of the agricultural technology interventions in Bangladesh or of disseminating OSP through farmer groups in the REU project because these interventions did not have empowerment as a stated objective, and thus no efforts were made to monitor progress toward it.

Moreover, when the earlier NSAPs were implemented, no internationally validated measure of women's empowerment existed to monitor empowerment impacts; as such, projects that assessed such impacts did so with an ad hoc set of indicators, making comparisons difficult. Later projects, more aware of the importance of empowering women, deliberately collected empowerment indicators so as to assess whether they had empowered women, and used internationally validated indicators. Nevertheless, across NSAPs, there is often no uniformity in the empowerment indicators collected, suggesting that future research that articulates what empowerment survey modules could or should look like would be valuable to guide understandings of the complexities of capturing gender equality and empowerment.[7]

In sum, based on the case studies presented here, NSAPs that have the most potential to empower women and to shift underlying barriers that perpetuate gender inequalities are those that share the following features. First, they are intentional about their objective to empower women and to transform gender relations. Second, they recognize potential tradeoffs between women's different roles and the unintended consequences of participation in NSAPs, such as increased work burden. Third, they employ evidence-based strategies that work to empower women, and continue to learn from their experience and adapt, as illustrated by the experience of HKI. These strategies include women's group-based programming, gender-sensitization activities that reach men and communities, building women's technical capacity, and addressing exclusionary gender norms. Fourth, they make deliberate efforts to monitor progress toward women's empowerment and gender equality by using the appropriate indicators and methodologies.

Building a research agenda that supports the next generation of gender-transformative NSAPs

Nutrition-sensitive agriculture holds promise as a vehicle to achieve women's empowerment and gender equality. However, it is vital to address programming and evidence gaps if we are to transform the harmful norms and dynamics that perpetuate gender inequalities. Addressing these gaps begins with recognizing both the comprehensiveness and the blind spots of measures in the area of nutrition and NSAPs, and designing research to close these gaps.

First, for research to shape the next generation of gender-transformative NSAPs, project designers and researchers should work together to ensure that strategies aim to empower women and **assess impacts on women's empowerment as an outcome in its own right, not merely as an instrument for achieving nutrition outcomes.** Significant work is now being conducted under the Gender Assets and Agriculture Project Phase 2 (GAAP2), working with a portfolio of nutrition- and gender-sensitive agricultural development projects, to fulfill the need for improved women's empowerment measures.[8]

---

7 See also Chapter 9, this volume.

8 Descriptions of the projects in the GAAP2 portfolio are available at http://gaap.ifpri.info/portfolio/.

GAAP2 has developed the project-level Women's Empowerment in Agriculture Index (WEAI) (pro-WEAI) (Malapit et al. 2019), adapted from the WEAI, a survey-based tool that measures inclusion and empowerment in the agriculture sector (Alkire et al. 2013). Pro-WEAI has been designed to diagnose disempowerment and assess empowerment impacts in mixed-methods impact assessments of agricultural development projects (Malapit et al. 2019). To complement pro-WEAI's significant focus on agricultural production, GAAP2 is also developing an optional add-on pro-WEAI health and nutrition module for nutrition-sensitive agriculture projects, which measures women's ability to make decisions in the area of health and nutrition (Heckert et al. 2020).

An ongoing project under the Innovative Methods and Metrics for Agriculture and Nutrition Actions (IMMANA) portfolio is developing the Women's Empowerment in Nutrition Index (WENI), designed to measure women's nutritional empowerment in relation to their own nutrition (Narayanan et al. 2019). Its development is based on qualitative and quantitative work in two sites in India. The instrument focuses on measuring knowledge, resources, and agency in the areas of food, health, and fertility. The WENI has expanded the health and nutrition relevant domains in which empowerment is being measured.

Second, research can contribute to understanding how NSAPs can **bring men on board in efforts to transform gender norms.** The majority of nutrition-related interventions and research focused on women's empowerment and gender equality has targeted women as potential agents, and been oriented toward improving the well-being of women and children. Men, the other half of the gender equation, have been understudied in nutrition research, especially the roles they can play in improving nutritional outcomes for women and children and in gender-transformative change. Engaging men may help improve the gender-equitable allocation of food produced or ensure women have a say in controlling the income from it. The shift in the allocation of food and all types of resources may occur through increased communication and more equitable negotiation about the use of these resources.

Engaging men may also help relieve women's time burdens, which are often high as a result of productive and domestic labor requirements. Many nutrition interventions involve trainings that increase women's time commitments, and, for those with young children, potentially increase women's labor in fulfilling the caregiving tasks that these programs encourage (Johnston et al. 2018). Evaluations of such approaches should continue measuring workloads via time use measures. Moreover, mothers and fathers, through their respective social roles, may offer different benefits to their children. Given the potential for male involvement to affect these pathways, understanding how male involvement in nutrition-focused interventions can be channeled toward increasing gender equality and the empowerment of women is an important area for future research.

Third, research on NSAPs should pay attention to the **unintended consequences of women's involvement,** including increased work burdens and the possibility of backlash from men. The gendered division of labor is deeply rooted in beliefs that women are primarily responsible for managing the household and caring for children; women's work burdens at home tended to remain constant despite the additional time required for farming or livelihood activities. Challenging these norms by not performing assigned tasks may result in intimate partner violence (IPV): women and men smallholder pastoralists in Tanzania reported that "physical punishment awaited the wife in case of problems" related to tasks assigned by men (Galiè et al. 2019, 130). Thus, women felt they had limited capacity to gain more control over their own time and inadequate time to engage in revenue-generating activities of their choice (ibid.).

Gender-sensitization trainings are well suited to addressing the household division of labor and resources and shifting the distribution to promote joint household goals on income and nutrition. Because individuals often make tradeoffs in allocating their time, it will be important to measure not only time use but also time-use agency, or the ability to decide how women allocate their time. Complementing agricultural interventions with gender sensitization is a promising approach for transforming entrenched beliefs and reducing adverse outcomes. The *Nurturing Connections©* curriculum, for example, includes messages that encourage men to share domestic responsibilities with women. Moreover, BCC may have beneficial consequences for reducing IPV. A recent study shows that intensive nutrition BCC was critical for sustained reductions in IPV, where women who received both transfers and BCC experienced 26 percent less physical violence post-program compared with women who only received transfers or who were in the control group (Roy et al. 2018).

Intensifying women's engagement in agriculture may also require greater work effort. Women's increased energy expenditures, if uncompensated by adequate food intake, could negatively affect their own health and nutrition. For example, a recent study finds that time spent in agricultural cropping work is negatively associated with BMI for non-overweight individuals, suggesting that gains in nutritional status from increased income and food availability could be offset by an increase in work effort associated with agricultural work (Komatsu et al. 2019). However, impact evaluations tend to focus on the gains in consumption following interventions, and rarely look at impacts on energy expenditures, which are more difficult to measure. Recent advances in wearable devices such as accelerometers are now expanding the ways that NSAPs can analyze other potential impacts. For example, researchers at the International Crops Research Institute for the Semi-Arid Tropics (ICRISAT), in collaboration with the University of Reading, the University of East Anglia, and the National Institute of Rural Development and Panchayati Raj (NIRDPR), are developing methodologies for using these sensor-based technologies to create reliable energy expenditure profiles associated with agricultural and livelihood activities in rural agricultural settings in developing countries.

Fourth, there is a need to collect **data on nutrition and health outcomes from all household members,** not just the target group of the intervention. Many NSAPs focus on maternal and child nutrition but do not necessarily examine the diets and nutrition and health status of other household members, making it impossible to assess whether the intrahousehold distribution of food has shifted. To assess whether the NSAP is contributing to gender equality, we need to collect data on the diets and nutritional status of other household members, including adolescent girls and boys. The 2020 Global Nutrition Report (Development Initiatives 2020) also calls for these data, to investigate and address drivers of nutrition inequalities.

To be effective, NSAPs must recognize that gender relations and intrahousehold dynamics can either impede or facilitate the achievement of their nutrition goals in complex ways. Some NSAPs limit their programs to reaching or benefiting women while accommodating rather than transforming existing gender relations and structures. These restrictive gender relations and structures are not confined to the household. To make the next generation of NSAPs gender transformative will require a different approach to NSAP programming that takes into account the multi-level nature of structures that need to be transformed: from intrahousehold relations between individual men and women to structures that limit women's rights within the community and beyond. It will also require a different approach to MEL that goes beyond the individual and household, and may point to a research agenda that examines how individuals and households are embedded within their communities and societies.

Finally, the lessons from the review of case studies are not specific to NSAPs: lessons learned from evaluations of other agricultural development projects also point toward the importance of gender-transformative approaches that explicitly aim to "empower" women. We expect that impact evaluations of ongoing NSAPs will generate more evidence on how they can not only meet their nutrition objectives but also empower women and promote gender equality. The next generation of research that emerges from these impact evaluations will help sharpen our focus on how NSAPs can be a pathway toward women's empowerment and gender equality.

Geography of study sites for publications cited in Chapter 5

Number of cited studies by research methodology (N=57)

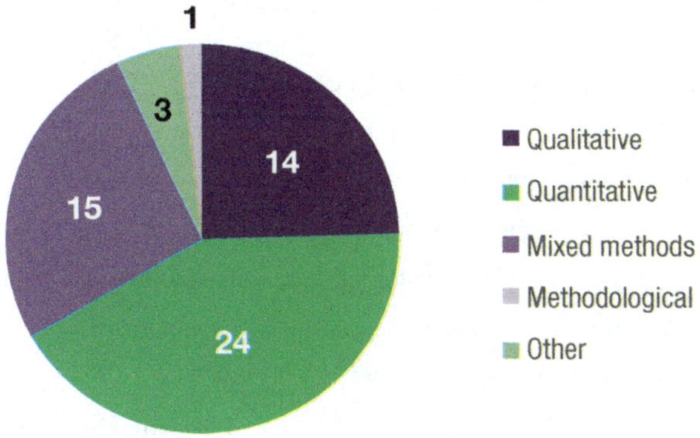

Timeline for references cited

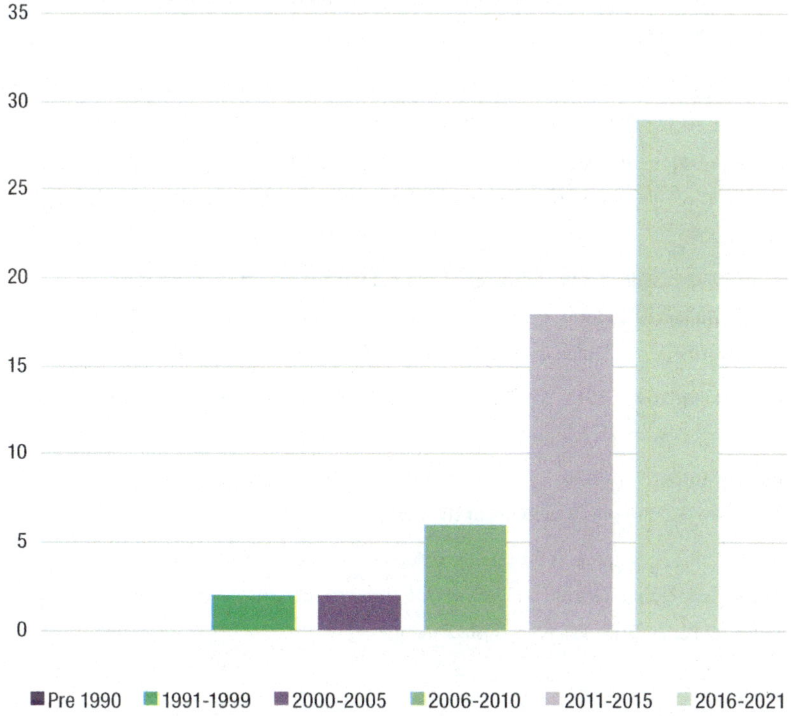

References

A4NH (Agriculture for Nutrition and Health). 2016. *Gender Strategy for Phase II.* Washington, DC: A4NH.

A4NH. 2019. "In Conversation: Discussing Pro-WEAI and Efforts towards Supporting Rural Women." October 15. https://a4nh.cgiar.org/2019/10/15/in-conversation-discussing-pro-weai-and-efforts-towards-supporting-rural-women/

Ahmed, A. U., J. Hoddinott, P. Menon, A. Quisumbing, S. Roy, M. Younus. 2018. *Agriculture, Nutrition, and Gender Linkages (ANGeL) Evaluation Results.* Dhaka: IFPRI.

Alkire, S., R. Meinzen-Dick, A. Peterman, A. Quisumbing, G. Seymour, A. Vaz. 2013. "The Women's Empowerment in Agriculture Index." *World Development* 52: 71–91.

Bernard, T., M. Hidrobo, A. Le Port, R. Rawat. 2019. "Nutrition-Based Incentives in Dairy Contract Farming in Northern Senegal." *American Journal of Agricultural Economics* 101 (2): 404–435.

Black, R. E., C. G. Victora, S. P. Walker, Z. A. Bhutta, P. Christian, M. de Onis, M. Ezzati, S. et al. 2013a. "Maternal and Child Undernutrition and Overweight in Low-Income and Middle-Income Countries." *The Lancet* 382 (9890): 427–451.

Black, R. E., H. Alderman, Z. A. Bhutta, S. Gillespie, L. Haddad, S. Horton, A. Lartey et al. 2013b. "Maternal and Child Nutrition: Building Momentum for Impact." *The Lancet* 382 (9890): 372–375.

Bouis, H. E., and A. Saltzman. 2017. "Improving Nutrition through Biofortification: A Review of Evidence from HarvestPlus, 2003 through 2016." *Global Food Security* 12: 49–58.

Bouis, H. E., C. Hotz, B. McClafferty, J. V. Meenakshi, W. H. Pfeiffer. 2011. "Biofortification: A New Tool to Reduce Micronutrient Malnutrition." *Food and Nutrition Bulletin* 32 (S1): 31S–42S.

Bushamuka, V. N., S. de Pee, A. Talukder, L. Kiess, D. Panagides, A. Taher, M. Bloem. 2005. "Impact of a Homestead Gardening Program on Household Food Security and Empowerment of Women in Bangladesh." *Food and Nutrition Bulletin* 26 (1): 17–25.

Development Initiatives. 2018. *2018 Global Nutrition Report: Shining a Light to Spur Action on Nutrition.* Bristol: Development Initiatives.

Development Initiatives. 2020. *2020 Global Nutrition Report: Action on Equity to End Malnutrition.* Bristol: Development Initiatives.

Farnworth, C. R., N. Sultana, P. Kantor, A. Choudhury. 2015. "Gender Integration in Aquaculture Research and Technology Adoption Processes: Lessons Learned in Bangladesh." Working Paper 2015–17, WorldFish, Penang.

Galiè, A., N. Teufel, A. Girard, I. Baltenweck, P. Dominguez-Salas, M. Price, R. Jones et al. 2019. "Women's Empowerment, Food Security and Nutrition of Pastoral Communities in Tanzania." *Global Food Security* 23: 125–134.

Gilligan, D. O., N. Kumar, S. McNiven, J. V. Meenakshi, A. R. Quisumbing. 2020. "Bargaining Power, Decision Making, and Biofortification: The Role of Gender in Adoption of Orange Sweet Potato in Uganda." *Food Policy*. 95: 101909.

Girard, A. W., J. L. Self, C. McAuliffe, O. Olude. 2012. "The Effects of Household Food Production Strategies on the Health and Nutrition Outcomes of Women and Young Children: A Systematic Review." *Paediatric and Perinatal Epidemiology* 26 (S1): 205–222.

Haddad, L. J., C. Peña, C. Nishida, A. R. Quisumbing, A. Slack. 1996. "Food Security and Nutrition Implications of Intrahousehold Bias: A Review of Literature." FCND Discussion Paper 19, IFPRI, Washington, DC.

Hallman, K., D. Lewis, S. Begum. 2007. "Assessing the Impact of Vegetable and Fishpond Technologies on Poverty in Rural Bangladesh." In *Agricultural Research, Livelihoods, and Poverty: Studies of Economic and Social Impacts in Six Countries*, edited by M. Adato, and R. Meinzen-Dick, 103–148. Baltimore, MD: Johns Hopkins University Press.

Harris, J., and N. Nisbett. 2018. "Equity in Social and Development Studies Research: Insights for Nutrition." *Advancing Equity, Equality and Non-Discrimination in Food Systems: Pathways to Reform, UNSCN News* 43: 57–63.

Harris-Fry, H. A., P. Paudel, N. Shrestha, T. Harrisson, B. Beard, S. Jha, B. Shrestha et al. 2018. "Status and Determinants of Intra-Household Food Allocation in Rural Nepal." *European Journal of Clinical Nutrition* 72: 1524–1536.

Haselow, N. J., A. Stormer, A. Pries. 2016. "Evidence-Based Evolution of an Integrated Nutrition-Focused Agriculture Approach to Address the Underlying Determinants of Stunting." *Maternal and Child Nutrition* 12 (S1): 155–168.

Heckert, J., D. K. Olney, M. T. Ruel. 2019. "Is Women's Empowerment a Pathway to Improving Child Nutrition Outcomes in a Nutrition-Sensitive Agriculture Program? Evidence from a Randomized Controlled Trial in Burkina Faso." *Social Science & Medicine* 233: 93–102.

Heckert, J., E. Martinez, G. Seymour, A. Pereira, H. Malapit, S. Roy et al. 2020. *Development and Validation of a Health and Nutrition Module for the Project-Level Women's Empowerment in Agriculture Index (pro-WEAI)*. Washington, DC: IFPRI.

Hillenbrand, E. 2010. "Transforming Gender in Homestead Food Production." *Gender and Development* 18 (3): 411–425.

Hoddinott, J., J. Behrman, J. Maluccio, P. Melgar, A. R. Quisumbing, M. Ramirez-Zea, A. Stein. 2013. "Adult Consequences of Growth Failure in Early Childhood." *American Journal of Clinical Nutrition* 98 (5): 1170–1178.

Hotz, C., C. Loechl, A. Lubowa, J. Tumwine, G. Ndeezi, A. Nandutu Masawi, R. Baingana et al. 2012a. "Introduction of β-Carotene–Rich Orange Sweet Potato in Rural Uganda Resulted in Increased Vitamin A Intakes among Children and Women and Improved Vitamin A Status among Children." *The Journal of Nutrition* 142 (10): 1871–1880.

Hotz, C., C. Loechl, A. de Brauw, D. Eozenou, M. Gilligan, M. Moursi, B. Munhaua et al. 2012b. "A Large-Scale Intervention to Introduce Orange Sweet Potato in Rural Mozambique Increases Vitamin A Intakes among Children and Women." *British Journal of Nutrition* 108 (1): 163–176.

Iannotti, L., K. Cunningham, M. Ruel. 2009. "Improving Diet Quality and Micronutrient Nutrition: Homestead Food Production in Bangladesh." Discussion Paper 928, IFPRI, Washington, DC.

Johnson, N., M. Balagamwala, C. Pinkstaff, S. Theis, R. Meinzen-Dick, A. Quisumbing. 2018. "How Do Agricultural Development Projects Empower Women? Linking Strategies with Expected Outcomes." *Journal of Gender, Agriculture and Food Security* 3 (2): 1–19.

Johnston, D., S. Stevano, H. Malapit, E. Hull, S. Kadiyala. 2018. "Review: Time Use as an Explanation for the Agri-Nutrition Disconnect? Evidence from Rural Areas in Low and Middle-Income Countries." *Food Policy* 76: 8-18.

Jones, K. M., and A. de Brauw. 2015. "Using Agriculture to Improve Child Health: Promoting Orange Sweet Potatoes Reduces Diarrhea." *World Development* 74: 15–24.

Kabeer, N. 1999. "Resources, Agency, Achievements: Reflections on the Measurement of Women's Empowerment." *Development and Change* 30 (3): 435–464.

Kjeldsberg, C., N. Shrestha, M. Patel, D. Davis, G. Mundy, K. Cunningham. 2018. "Nutrition-Sensitive Agricultural Interventions and Gender Dynamics: A Qualitative Study in Nepal." *Maternal and Child Nutrition* 14 (3): e51923.

Komatsu, H., H. Malapit, M. Balagamwala. 2019. "Gender Effects of Agricultural Cropping Work and Nutrition Status in Tanzania." *PLOS ONE* 14 (9): e0222090.

Kumar, N., and A. R. Quisumbing. 2011. "Access, Adoption, and Diffusion: Understanding the Long-Term Impacts of Improved Vegetable and Fish Technologies in Bangladesh." *Journal of Development Effectiveness* 3 (2): 193–219.

Le Port, A., T. Bernard, M. Hidrobo, O. Birba, R. Rawat, M. Ruel. 2017. "Delivery of Iron-Fortified Yoghurt, through a Dairy Value Chain Program, Increases Hemoglobin Concentration among Children 24 to 59 Months Old in Northern Senegal: A Cluster-Randomized Control Trial." *PLOS ONE* 12 (2): e0172198.

Malapit, H. 2019. "Women in Agriculture and the Implications for Nutrition." In *Agriculture for Improved Nutrition: Seizing the Momentum*, edited by S. Fan, 58–67. Wallingford: CABI.

Malapit, H., A. Quisumbing, R. Meinzen-Dick, G. Seymour, E. Martinez, J. Heckert, D. Rubin et al. 2019. "Development of the Project-Level Women's Empowerment in Agriculture Index (pro-WEAI)." *World Development* 122: 675–692.

Maluccio, J. A., J. Hoddinott, J. R. Behrman, R. Martorell, A. R. Quisumbing, A. Stein. 2009. "The Impact of Improving Nutrition during Early Childhood on Education among Guatemalan Adults." *Economic Journal* 119 (537): 734–763.

Narayanan, S., E. Lentz, M. Fontana, A. De, B. Kulkarni. 2019. "Developing the Women's Empowerment in Nutrition Index in Two States of India." *Food Policy* 89: 101780.

Naved, R. T. 2000. "Intrahousehold Impact of the Transfer of Modern Agricultural Technology: A Gender Perspective." FCND Discussion Paper 85, IFPRI, Washington, DC.

Newton, J., N. Verhart, A. Bake. 2018. *Enhancing the Effectiveness of Agriculture-to-Nutrition Pathways: Key Insights from a Gender Analysis of Impact Evaluation Design.* Amsterdam: Food and Business Knowledge Platform.

Nordhagen, S., C. Bastardes Tort, A. Kes, L. Winograd. 2017. "Nurturing Connections? Evaluating the Impact of a Women's Empowerment Curriculum in Cote d'Ivoire." Working Paper, ICRW, Washington, DC.

Olney, D. K., A. Talukder, L. L. Iannotti, M. Ruel, V. Quinn. 2009. "Assessing Impact and Impact Pathways of a Homestead Food Production Program on Household and Child Nutrition in Cambodia." *Food and Nutrition Bulletin* 30 (4): 355–369.

Olney, D. K., J. Behrman, E. Iruhiriye, M. van den Bold, A. Pedehombga. 2013. *Enhanced Homestead Food Production Program in Burkina Faso: Process Evaluation.* Washington, DC: IFPRI.

Olney, D. K., A. Pedehombga, M. Ruel, A. Dillon. 2015. "A 2-Year Integrated Agriculture and Nutrition and Health Behavior Change Communication Program Targeted to Women in Burkina Faso Reduces Anemia, Wasting, and Diarrhea in Children 3–12.9 Months of Age at Baseline: A Cluster-Randomized Controlled Trial." *The Journal of Nutrition* 145 (6): 1317–1324.

Olney, D. K., L. Bliznashka, A. Pedehombga, A. Dillon, M. Ruel, J. Heckert. 2016. "A 2-Year Integrated Agriculture and Nutrition Program Targeted to Mothers of Young Children in Burkina Faso Reduces Underweight among Mothers and Increases Their Empowerment." *The Journal of Nutrition* 146 (5): 1109–1117.

Quisumbing, A. R., and N. Kumar. 2011. "Does Social Capital Build Women's Assets? The Long-Term Impacts of Group-Based and Individual Dissemination of Agricultural Technology in Bangladesh." *Journal of Development Effectiveness* 3 (2): 220–242.

Quisumbing, A. R., D. Rubin, C. Manfre, E. Waithanji, M. van den Bold, D. Olney, N. Johnson et al. 2015. "Gender, Assets, and Market-Oriented Agriculture: Learning from High-Value Crop and Livestock Projects in Africa and Asia." *Agriculture and Human Values* 32 (4): 705–725.

Roy, S., M. Hidrobo, J. Hoddinott, A. Ahmed. 2018. "Transfers, Behavior Change Communication, and Intimate Partner Violence: Postprogram Evidence from Rural Bangladesh." *The Review of Economics and Statistics* 101 (5): 865–877.

Ruel, M. T., and H. Alderman. 2013. "Nutrition-Sensitive Interventions and Programmes: How Can They Help to Accelerate Progress in Improving Maternal and Child Nutrition?" *The Lancet* 382 (9891): 536–551.

Ruel, M. T., A. R. Quisumbing, M. Balagamwala. 2018. "Nutrition-Sensitive Agriculture: What Have We Learned So Far?" *Global Food Security* 17: 128–153.

Santoso, M. V., R. B. Kerr, J. Hoddinott, P. Garigipati, S. Olmos, S. Young. 2019. "Role of Women's Empowerment in Child Nutrition Outcomes: A Systematic Review." *Advances in Nutrition: An International Review Journal* 10 (6): 1128–1151.

Van den Bold, M., A. Dillon, D. Olney, M. Ouedraogo, A. Pedehombga, A. Quisumbing. 2015. "Can Integrated Agriculture-Nutrition Programmes Change Gender Norms on Land and Asset Ownership? Evidence from Burkina Faso." *Journal of Development Studies* 51 (9): 1155–1174.

Victora, C. G., L. Adair, C. Fall, P. Hallal, R. Martorell, L. Richter, S. Sachdev et al. 2008. "Maternal and Child Undernutrition: Consequences for Adult Health and Human Capital." *The Lancet* 371 (9609): 340–357.

Wibowo, Y., B. Sutrisna, H. Hardinsyah, R. Djuwita, M. Korib, A. Syafiq, A. Tilden et al. 2015. "Relationship between Intra-Household Food Distribution and Coexistence of Dual Forms of Malnutrition." *Nutrition Research and Practice* 9 (2): 174–179.

Willett, W., J. Rockström, B. Loken, M. Springmann, T. Lang, S. Vermeulen, T. Garnett, D. et al. 2019. "Food in the Anthropocene: The EAT–Lancet Commission on Healthy Diets from Sustainable Food Systems." *The Lancet* 393 (10170): 447–492.

Chapter 6

A GENDER–NATURAL RESOURCES TANGO: WATER, LAND, AND FOREST RESEARCH

Deepa Joshi, Iliana Monterroso, Bryce Gallant, Kokila Perera, and Valentina Peveri

Gender relations shape identities, norms, rules, and responsibilities for women and men, and mediate access to, use, and management of water resources, as well as ownership, tenure, and user rights to land and forests (and related infrastructure, services, technologies innovations and interventions). Natural resource management (NRM) interventions thus have important implications for women's labor, time, decision-making, and transformational gains.

In the current context of fluid economic and political changes, together with changes to climate, the gendered dynamics of natural resource use, allocation, and management are also evolving. There are rapid shifts in livelihoods, mobility, and migration for women and men, as well as differing vulnerabilities and capacities for resilience in climate change processes and emergencies. Gender in its intersections with class, race, religion, ethnicity, age, disability, and other dimensions of difference determines who gains and who loses "in the rapid restructuring of economies, ecologies, cultures and polities from global to local levels" (Rocheleau et al. 1996, 3). These intersecting inequalities point to the complex and dynamic character of spaces of assumed common interest, such as "the community" and "households," as well as to the plurality in interests, needs, vulnerabilities, and agency of diverse groups of women and men (Elmhirst 2015).

The upcoming decade of 2021–2030 is dedicated globally to restoring the ecosystem: "to scale up the restoration of degraded and destroyed ecosystems as a means to fight the climate crisis and enhance food security, water supply and biodiversity."[1] The Sustainable Development Goals (SDGs) indicate a broad policy consensus among development actors that the ecological resilience of the planet is not disassociated from people's well-being.

1 https://www.decadeonrestoration.org/

This narrative is mirrored in several other international agreements and conventions, for example the United Nations Declaration on the Rights of Indigenous Peoples, the International Labour Organization Convention Concerning Indigenous and Tribal Peoples in Independent Countries and more recently the Voluntary Guidelines on the Governance of Tenure of Lands, Fisheries, and Forests.

These declarations and guidelines all draw some attention to equality and inclusion—and call on nation states to ensure more equal access to natural resources. Since the first United Nations Summit dedicated to the environment (1972), which marked a turning point in the development of international environmental policies,[2] there has also been progress in articulating the links between gender and who engages, benefits, or is excluded from processes of natural resource governance and management. While this is a hard-won gain, we discuss below the enormity of tasks that still remain in ensuring inclusive natural resource governance.

This chapter tackles the question: *How has NRM research for development (R4D) contributed to gender equality*? In addressing this, the converse question inevitably surfaces: What impacts has gender theory and discourse had on natural resource management? However, the former is the main focus—drawing on an extensive review of natural resource-related research.

The first section looks at how and why gender came to matter in the management and governance of natural resources. It offers an analytical positioning for the chapter and explores the confluence of several bodies of work that have informed the gendered dynamics of natural resources, from political ecology analyses of nature–society intersections, to the political economy of environmental agendas and mandates, to feminist analyses of environmental change processes. Together, these insights help explain the connection between the two questions posed above.

The following three sections each provide an in-depth analysis of how innovative thinking and action on natural resources—forests, water, and land, in particular—was crucial vis-à-vis impact on gender equality. These three natural resources have shared as well as distinct characterizing features. Forests are geographically and also administratively contained, and land is a fixed asset, with easy-to-define dimensions of ownership. Water, on the other hand, is fluid and dynamic because it is inherently mobile and transitory (Meinzen-Dick et al. 1997, 1307). Analyzing the gendered nature of the

2 See https://www.un.org/ga/search/view_doc.asp?symbol=A/CONF.48/14/REV.1

management, governance, and rights to these natural resources makes for an interesting comparative analysis.

Approaches to NRM and governance and recognition of agriculture–environment intersections have grown in promising directions, providing opportunities to address gender equality and inclusion. However, we need to critically reflect: Are we doing enough? Are complex and intersectional gendered inequalities in rapidly changing social, political, economic, and ecological contexts understood and addressed? This is a pressing concern, especially given the peripheral mention of gender in the upcoming Decade on Ecosystem Restoration, which aims to "prevent, halt and reverse the degradation of ecosystems worldwide" (Elias et al. forthcoming). Gender and the environment (or natural resources) still tend to be treated as distinct and parallel agendas across R4D institutions, with simplistic understandings of "nature" as an "economic resource" and "gender" as being mostly "about women." This explains why gender power imbalances persist in the economics and politics of NRM.

In the conclusion of this chapter, we critique progress on gender equality in relation to natural resources, and highlight challenges for transformative change, as well as the potential for forward-looking research agendas.

How gender came to matter in the management and governance of natural resources

Overlapping processes laid the foundation for unpacking gender dimensions of natural resources, bringing attention to "gender and other forms of social difference as relational, dynamic" and making links "between environment, human and nonhuman others across scales and spaces" (Clement et al. 2019a, 5). Breakthroughs included seminal research on plural (von Benda-Beckmann 1979, von Benda-Beckmann et al. 1997, 1998), customary usufructuary rights (Fortmann and Bruce 1988, Fortmann 1990) to land, water, and other natural resources; and polycentric governance, collective action, and management of the "commons" (Ostrom 1990–2007). These trajectories are discussed at length elsewhere. Here, we discuss the cross-fertilization of these ideas with the CGIAR System-wide Program on Collective Action and Property Rights (CAPRi) (see Meinzen-Dick 2017, Meinzen-Dick et al. 1997).

Another strategic milestone was the critical review of irrigation design, infrastructure, management, and outcomes in South Asia. This body of work helped shape water policy discourse from "management" to "governance" of water, consequently demonstrating the gendered nature of access to, use of, and decision-making in irrigation systems and services (van Koppen and

Mahmud 1995, Zwarteveen 1997, Meinzen-Dick and Zwarteveen 1998). Similar socio-political interventions helped change the focus from forests as commercial timber harvesting sites, to forests as environmental resources to be conserved as collectively owned and used natural resources (Agrawal and Ostrom 2001, Otsuka and Place 2001, Colfer and Capistrano 2005, Larson et al. 2010). Attention to the voice and agency of marginalized communities and rural women in forest management significantly highlighted the gendered dynamics of forest access and use (Jackson 1995, Leach et al. 1999, Leach 2007, Mwangi and Dohrn 2008, Mwangi et al. 2009, Mwangi et al. 2011, Colfer 2011).

Work is emerging around intersectional inequalities in the case of forests and of water (Locke et al. 2017, Clement and Karki 2018, Colfer et al. 2018, van Koppen 2018, Clement et al. 2019a, Elias et al. forthcoming). This research, influenced by political ecology scholarship, highlights environmental politics: how environmental changes and challenges are not mere by-products of biophysical changes to the ecosystem but rather outcomes of economic, political, and social interests and mandates (Haraway 1991). The distinctions between the terms "environment," "nature," and "natural resources," and how we use them, have long been recognized as neither casual nor without implications (Harvey 1993). In other words, in the framing of NRM, "values entered [and significantly impacted] processes of scientific reasoning" (Lapniewska 2016, 143).

After the 1992 United Nations Earth Summit, natural resources were defined as "natural assets" (raw materials) that can be subject to economic production or consumption.[3] The intent to *manage* nature (land, water, forests) as a *resource* with ascribed economic values and implications is an outcome of deep-rooted economic and political agendas (Harvey 1993). This narrative, an outcome of a "partial perspective," is precisely what Haraway (1988) said needed to be critically reviewed as the "Science Question in Feminism." Policies, strategies and interventions to manage water (as well as land and forests) have historically been guided primarily by managerial and economic perspectives (see Mosse 1997, 2002, 2008). It is only relatively recently that nature–society interrelations—that is to say, how ecosystem functioning is an outcome of multiple uses, knowledge/s, and social relations between diverse groups of people—have questioned natural resource mandates, innovations, investments, and technologies.

3 See https://stats.oecd.org/glossary/detail.asp?ID=1740

A focus on gender equality and inclusion helps connect environment and natural resources, as well as agriculture and food security agendas. CGIAR—the largest agricultural innovation network globally and framed as "food commodity-centric"—started more than two decades ago to unpack agriculture–environment intersections and reverse environmental degradation. This shift to addressing the agriculture–environment interface was critical in understanding how addressing poverty and vulnerability required looking beyond agriculture to understand diverse rural livelihoods that rely on a wider subset of natural resources.

Recent reforms in CGIAR have been shaped by growing critiques of a narrow focus on *developing* land and water resources to *intensify* production of certain commodity crops (Rockström et al. 2017). The links between poverty, hunger, and intensification have long been questioned outside CGIAR but not necessarily with a focus on gender (Sen 1980, Chappell 2018). It is in this context that we draw attention to CGIAR research analyzing women's empowerment through sectoral lenses: agriculture and women's empowerment (IFPRI 2012), women and irrigation management (IWMI 2017), or women–forest relations (CIFOR 2016).

In sum, diverse bodies of work across multiple disciplines and diverse trajectories enabled a shift in focus within CGIAR agriculture–natural resources. These change processes brought women's agency and empowerment—more than simply engagement and participation—into the frame. In analyzing these trajectories, we point to how innovations geared toward improving the functionality of natural resource interventions have contributed to broader goals of gender equality and inclusivity—despite this not always being deliberate. These trajectories have been influenced by feminist analyses of masculinities, patriarchy, and exclusions in natural resource policies and interventions.

This chapter does not aim to delve into the scope and breadth of these feminist analyses of NRM and governance processes; however, to very broadly set the context here, ecofeminists put "women" on the NRM agenda and pointed out that women's inherent wisdom and commitment to nature had been overlooked in the patriarchal and neoliberal design of natural resource appropriation and accumulation. Engineered by male-dominated institutions and mandates, such an approach to natural resources was identified as detrimental to the well-being of both women and nature (Mies and Shiva 1993). Feminist political ecologists have critiqued the singular focus on women—pointing out that gender was but one construct of difference, divide, and inequality in the politics and economics of natural resource access, use, and governance (Agarwal 1995, Rocheleau et al. 1996, Jackson 1993), and that tying nature to women was often detrimental to them (Leach 2007). Feminist researchers argued that the focus should be on reversing structural inequalities and not just on "fixing women" (Mies 1986, Jackson 1993, Kabeer 1999, Leach 2007).

Recent echoes of this argument by mainstream institutions (WEF 2020) are promising but we should not overlook the differences between academic, activist, and R4D scholarship on the topic of natural resources and gender. In other words, the synergies have not always been deliberate and this is precisely why a depoliticization of gender–power dimensions of commons management persists (Clement et al. 2019a). The point we make here, is that, without the essential cross-fertilization of thinking between natural resource interventions and feminist analyses, there would have been little progress on gender and natural resources. Gender researchers within the CGIAR system have represented a conduit, facilitating these critical intersections. However, much remains to be done.

From management to governance

The first shift in natural resource thinking we highlighted above concerns Ostrom's work on environmental governance. Post-World War II, structural adjustment interventions promoted nationalization and a corresponding state accumulation of natural resources across the global South (Bromley and Cernea 1989). In this context, Ostrom powerfully argued that water bodies, forests, and pastoral grounds were essentially common pool resources and thereby best governed and managed by plural institutional arrangements of and by local communities (Ostrom et al. 1994). Unsurprisingly, these ideas met considerable resistance, as collective property and actions or collectives themselves were perceived as obstacles to efficient development of natural resources (de Soto 1986, 2001).

Ostrom's work, which demonstrated how the flow and benefits derived through plural politico-legal arrangements and collective action improved efficient and equitable management of, access to, use of, and control over these resources was useful in influencing NRM policy and practice. The importance of plural rights and norms, asset endowments, and politico-legal arrangements of natural resource governance made a strong business case for going beyond technocratic approaches to natural resources.[4] This entailed

---

4 While there was feminist critique of technocratic and narrowly econometric NRM interventions, Ostrom's work focused more on alternatives: the "why and how to" of plural decision-making and control of natural resources, so as to enable commons, communing, and collectives for NRM (see Clement et al. 2019a). For a review of environmental governance, see Lemos and Agrawal (2006); for a review of water governance see Roth et al. (2015); for a review of forest governance see Arts (2014).

a pivotal discursive from state management to more shared governance of natural resources (McCulloch et al. 1998, Lemos and Agrawal 2006).

Ostrom's Institutional Analysis and Development Framework provided strategic entry points for CGIAR researchers, working under the broader umbrella of CAPRi, to examine the gendered dynamics of natural resources, including collective institutional arrangements of NRM (Knox and Meinzen-Dick 2001, Meinzen-Dick et al. 2011, 2014, Doss and Meinzen-Dick 2015). Ostrom's work had far-reaching outcomes—as water, forests and large areas of land in the global South were (and are still) managed under plural management and politico-legal arrangements (Zwarteveen and Meinzen-Dick 2001, Mwangi et al. 2011, Wily 2011). Most prominent of these was the setting-up of community-based natural resource user groups and associations, which expanded the scope to consider gender in decision-making spaces (Lemos and Agrawal 2006).

Through the various overlaps discussed above, gender is today an important variable in the structure and functions of NRM collectives; and "good governance" of natural resources has come to be associated with principles of inclusion, participation, transparency, and accountability—as opposed to the earlier focus on effectiveness and efficiency. These shifts in inclusive natural resource governance have also been widely acknowledged as essential to achieving co-determining social, economic, and environmental outcomes (UNDP 1997, FAO 2012, Davis et al. 2013, Arts 2014).

However, while Ostrom's work transformed the dynamics of natural resources from technocratic management interventions to more "polycentric governance, collective action and commons management... creating legitimate space and authority for grassroots structures to self-govern the commons" it blurred the heterogeneity and inequality that characterize "commons," "collectives," and "communities" (Clement et al. 2019a, 2). This, too, is changing slowly but surely as we write, with increasing attention to rights, recognition, power relationships, and norms that shape gender inequalities (Badstue et al. 2018). The focus on gender within CGIAR itself today concerns not only who does what in terms of roles and responsibilities at the household and community levels but equally how gender dynamics are at play in natural resource institutional arrangements and policy directives, as well as natural resource investments and innovations (CGIAR GENDER Platform 2020). The latter especially are key strategic gains.

From governance to the plurality of rights

The second discursive shift we discussed above related to how rights to natural resources are defined by institutional arrangements—who are rights-holders, the scope of rights, and the types of responsibilities and benefits one may obtain from resources (Agrawal and Ostrom 2001, Schlager and Ostrom 1992). Rights to land, forest, and water resources are plural and diverse—and determined by informal more than the formal rules and norms that authorize the exercise of these rights (Schlager and Ostrom 1992, Meinzen-Dick et al. 1997). Rules are institutional arrangements that sustain claims and legitimize rights at different levels. For instance, the nation-state can establish formal rules through legal regulations on land, water, and forests while communities or resource user associations may have local rules concerning who gets to use which resources, and how (Zwarteveen and Meinzen-Dick 2001, Meinzen-Dick and Mwangi 2009). These rules, however, may or may not be recognized by statutory or customary law, and different sets of regulations often overlap, and can even be contradictory. In other words, what is the rule is not always the outcome, in practice.

A groundbreaking body of work showed how plural and often co-existing politico-legal frameworks and arrangements shape social differentiation; it opened new windows to analyzing gendered disparities in relation to natural resources (von Benda Beckman and von Benda Beckman 2009; von Benda Beckman et al. 2006). Research along these lines showed how nature–society relations are continuously subject to negotiation and contestation, often marginalizing certain groups of resource users in diverse local contexts (Perreault 2014, Roth et al. 2015). Power struggles, conflicts over resources, and exclusions from access to and use of natural resources are essentially about the recognition of rights—and issues related to agency or voice often have an impact on this.

Legal pluralism, which explains the relevance of why and how of the co-existence of multiple legal arrangements in legitimizing claims over resources, proved crucial to understanding the social dynamics of NRM (Meinzen-Dick and Pradhan, 2002). CAPRi work on natural resource research showed how women's rights or their exclusions from rights to resources are entangled with their ability to participate in, and their agency to inform, natural resource decision-making (see Were et al. 2008). This work continues to analyze, inform, and monitor inclusivity in natural resource interventions with particular attention to women's individual and collective agency.

Do rights to natural resources contribute to gender equality and women's empowerment?

There is a great deal of discussion and differing opinions on whether and to what extent clear, secure rights to natural resources are key in addressing poverty and food insecurity (Agrawal 1994, Larson et al. 2010, Sunderland et al. 2014, Meinzen-Dick 2017, Bose et al. 2017). On the one hand, nuanced studies on natural resource governance and institutional arrangements show that understanding people's choices, their ability to benefit and decide on how to use these, and the outcomes derived from these resources is as key to understanding and achieving women's empowerment as is securing formal rights (Kabeer 1999). This argument is supported by analyses that show that rights to natural resources are not dependent only on formal recognition but also embedded in social relations that legitimate claims over resources (von Benda-Beckman and von Benda-Beckman 2000). Our intent here is not to further analyze these arguments but to simply state that the focus on natural resource governance, institutional arrangements for NRM, and rights to natural resources have all been central to analyzing the socio-political dynamics of natural resources, and thereby to drawing attention to gendered inequalities, as well as intersectional vulnerabilities—nested in kinship, community, and other social relationships (Li 1998, Kabeer 2005, 2017).

The issues discussed above, coupled with feminist analyses of natural resource policies, institutions, and outcomes, significantly influence the current transformative agenda of "fixing the system" and not just attempting to "add in women, and stir." This shifts the focus from "only" monitoring the extent to which women benefit from natural resource access, to critically analyzing issues of power, politics, and difference, including participation and representation in natural resource governance institutions, policies, and narratives at scale. This was precisely the feminist agenda for change in development policy and practice (Mies 1986).

The focus on natural resource governance, institutions, and rights has thus been a game-changer not just for women but equally for diverse marginalized groups. In the next sections, we discuss in more detail how these conceptual shifts took root in forest, land, and water policies, strategies, and interventions. In doing so, we discuss how institutional arrangements unfolded at different levels, from policy decisions to institutional arrangements of planning, implementation, and practice at the community and household levels—and how all of these were in turn informed, as well as reiterated, by gendered norms, behaviors, opportunities, challenges, choices, and redefining values.

Forest governance and the recognition of rights: contributions to gender equality

The diverse value of services and goods that forest ecosystems provide to both humans and non-humans, as well as the relational value of these ecosystems to local communities, were largely invisible in early programs on forest management (Scott 1998). Management approaches that focused on the economic returns from timber commercialization were promoted by colonial administrations, and even facilitated by scientific forestry institutions. Additionally, agricultural intensification facilitated by the mechanization and modernization of agriculture resulted in increasing deforestation (Angelsen and Kaimowitz 1999). These approaches led to rapid and widespread loss and degradation of forests[5] in the global South.

In the late 1990s, there was an overwhelming call for alternative solutions, driven by two key arguments. First was the need for a sustainable forest management paradigm that emphasized not commercialization but rather conservation of forest resources (Sayer and Palmer 1994, Noble and Dirzo 1997). Second, there was a strong call to involve local people in collaborative governance of forests—enabling different perspectives, plural rights, and shared roles and responsibilities of diverse stakeholders to define forest management (Colfer et al. 2005, Sunderlin et al. 2005, Porter-Bolland et al. 2011, Arts 2014, Seymour et al. 2014). Today, the vast majority of the world's forests are under state *custodian* ownership, with overlapping customary user rights and tenure regimes. This has been a significant change, even though it needs to be acknowledged that these two legal systems are far from harmonized in most cases (Meinzen-Dick et al. 1997, Agrawal et al. 2014).

In the case of forest governance, much more so than for land and water, grassroots initiatives that led to community collectives were significantly impacted by feminist movements. The women-led Kenyan Green Belt movement in sub-Saharan Africa saw the award of a Nobel Peace prize for its proponent, Wangari Maathai. Similarly, in India, the famous Chipko ("to get stuck to") movement, where local communities protested and prevented state-led commercial logging by *hugging* trees, is said to have been essentially led by women—that is, the ecofeminist discourse of women nurturing nature[6]

5 We use the term "forests" following the definition provided by Sunderlin (2005, 1386) including "all kinds of forests, ranging from relatively untouched "natural" ones to those with high levels of intervention and management. 'Natural' forests are the focus of most conservation concern, though highly managed forests can also be an important source of biodiversity.

6 There are contested opinions around this claim of predominantly female leadership in this movement (Guha 2000).

(Shiva 1988). However, as we discussed above, there are feminist critiques of this narrative.

The consideration of gender in forest management was significantly influenced by shifts toward collaborative governance of forests, which called for the recognition of the plural rights of forest-dependent communities, including indigenous and customary groups (Agrawal and Ostrom 2000, Larson et al. 2010, Agrawal 2014). As discussions around forest governance policy and practice began to translate to interventions that favored decentralization of authority over large forest areas—and shifting responsibilities from central to local governments—several questions needed answering. Who should manage forest lands and resources? Who should be involved in which activities? Who should have the right to govern forests and who should set the rules for governance (Ribot and Larson 2005, Ribot et al. 2006, García-Ferández et al. 2008)?

An analysis of 290 forest user communities in Kenya, Uganda, Bolivia and Mexico highlighted the importance of involving forest-dependent, often marginalized, communities in technical and policy discussions, and called for interventions based on the participation of resource users (Colfer 2011). Several other analyses showed that enabling spaces created for local communities—initiatives that built the skills of both men and women in adopting new technologies, monitoring practices, managing conflict, and enhancing cooperation—were more likely to contribute to sustainable practices and more effective management of forests (Mai et al. 2011, Mwangi et al. 2011, Sun et al. 2011, Seymour et al. 2014, Notess et al. 2018).

The importance of clear tenure rights to forests is increasingly an important precondition in the implementation of currently popular interventions related to Payment for Environmental Services (van Noordwijk and Leimona 2010, Blundo-Canto et al. 2018) and Reducing Emissions from Deforestation and Forest Degradation, or REDD+ (see Duchelle et al. 2017, Sunderlin et al. 2014, 2018). Recent analyses of these interventions both within and outside CGIAR highlight the need to critically review potential impacts of such interventions on diversely unequal local communities, including negative implications for local people's livelihoods and strategies, institutions, and socio-cultural systems (Elias et al. forthcoming). These analyses draw attention to gender power issues in relation to unequal benefit-sharing, food insecurity, introduction of new powerful stakeholders, illegal land acquisition, unfair free prior and informed consent, and the introduction of monoculture plantations (Bayrak and Marafa 2016).

In local communities, gender, social status, and membership are significant determinants of who can benefit from acquired forest rights and influence perception around rules, tenure security, and livelihood outcomes (Colfer 2011, Larson et al. 2019a). Having a voice in the management of forest or other common pool resources increases women's recognition in their community (Colfer et al. 2015), although the converse may also be the case—that is, women with more recognition in the community are more likely to have a voice in the management of common pool resources (Meinzen-Dick et al. 2019; Balasubramanya et al. 2019). Additionally, securing tenure rights of forest resources for women can provide security in cases of loss of rights to privately owned assets (land) through death of or separation from their spouse (Quisumbing and Otsuka 2001a, Doss and Meinzen-Dick 2018); enhance their engagement in public processes of negotiation and thereby their self-determination (Larson 2010, Larson et al. 2015); and improve their agency in collective rule-making processes.

Attention to how reforms are being implemented provides not only the opportunity to address inequalities in resource access and participation in decision-making but also insights on gender equality in general. For instance, in Burkina Faso, forest regulations that prevent grazing in forests and customary rules and regulations around land have resulted in serious constraints for women, minorities, and migrant groups (Coulibaly-Lingani et al. 2009). Alternatively, there are many examples of how formalizing community rights to forests has allowed women to be recognized in communal by-laws, usually by establishing mechanisms for their participation in collective decisions around forest resources (Larson et al. 2019b).

For instance, Uganda's Forest Policy (2001) is explicit about increasing tenure security for women, encouraging their active participation in decision-making, resource management, and benefit-sharing. It also goes a step further in initiatives to promote changes in attitudes and organizational cultures in order to break down gender barriers (Banana et al. 2012). Similarly, in Peru, the National Forest Law and the Law of Subnational Governments adopts equity and social inclusion as important principles—although the guidelines to monitor these changes are missing (Larson et al. 2019b).

Social forestry initiatives have thus provided an opportunity to review how collective action in forest management has paved the way to open the institutional spaces for women to engage in forest decision-making processes (Agrawal and Ostrom 2001, Colfer and Capistrano 2005, Colfer et al. 2005, Ribot and Larson 2005, Coleman and Mwangi 2013). The combined outcomes of ecofeminist discourses and grassroots-led forest management interventions have resulted in better understanding and addressing intersecting inequalities.

In sum, rethinking forests as communal, shared spaces has been instrumental in reshaping gender equitable rights to forests, and, in some cases, broader gender equality gains for women. Rights have proved important preconditions for effective management and inclusive governance—that is, collective action and institutional arrangements, processes of rule-making, provision and allocation of resources, monitoring, enforcement of compliance, and decision-making arrangements at scale.

Yet caution is needed in integrating gender into these studies and interventions. Gender stereotyping—that men are the public face for decisions relating to forests management—is pervasive and entrenched among official, non-governmental, and private actors (Nightingale 2011, Elmhirst et al. 2017). This is also evident in the way extension services prioritize men and address limitations. It calls for reviewing the way training, capacity-building, and extension services are being organized to ensure different needs are being addressed at the local level (Nightingale 2006). Similarly, there has long been feminist critique of positioning women as formidable environmental stewards (Jackson 1995, Leach 2007) and a call for more nuanced analyses of women's relations with forests (Gururani 2002).

From water management to water governance: the outcomes for gender equality and women's empowerment

In 2000, the Global Water Partnership referred to the world water crisis as an issue of governance. This did not imply that the availability of accessible water was not an issue or that the technical and financial aspects of service delivery were unimportant. Rather, it emphasized that, the distribution and allocation of water and related services reflected distribution and allocation of power in society. Therefore, addressing water problems required paying attention to issues of power, politics, and inequality (UNDP and SIWI 2005, 3). This shift in focus from water *development or management* to water *governance* significantly paved the way for looking at issues of gender inequality and empowerment in relation to water.

Here, we discuss how CGIAR research informed the shift to water governance from the planning and implementing of water development interventions informed (only) by economic or engineering perspectives. Looking more critically at the complexity of water–society interrelations at scale, in other words, "Who gets what water, when and how, and who has (what kinds of) rights to water and related services, and their benefits" (UNDP and SIWI 2005, 3) has contributed to furthering gender inequality. We also look at how scholarship on legal pluralism has helped raise attention to gender inequalities.

Water's legal pluralism

Research on the legal pluralism of water surfaced the incoherence between *what is said*—that is, outlined in policies, formal laws, and institutional approaches vis-à-vis *what actually happens in practice*—how diverse local communities accessed, managed, and governed water through pluralistically informal ways. This scholarship helped address the ambiguity on the rights to water in state-led irrigation interventions, and challenged the narrative that managing irrigation systems effectively and ensuring agricultural productivity required intervention by engineers to "modernize" water development—its capture, transport, allocation, and delivery to farmers (Roth et al. 2015). The framework for interdisciplinary, legal anthropological approaches made the "legal pluralism" around everyday water access, use, and management visible. In time, it became very evident that a co-existence and interaction between multiple legal orders such as state, customary, religious, project, and local laws were what determined and influenced claims to water rights and the use of water locally (Meinzen-Dick and Pradhan 2002).

In parallel, research at the International Water Management Institute (IWMI) demonstrated different dimensions of the gendered dynamics of water's legal plurality. Research in Bangladesh showed how groups of poor, landless women managed water boreholes to sell water to other water users (van Koppen and Mahmud 1995). While the source of water might determine its accessibility, ownership does not disable water use or its asset value. Those driven by sheer need and poverty, like poor landless women in Bangladesh, had found ways to access and use common pool water resources. Similarly, research in Nepal pointed out that the exclusion of women from irrigation water user associations was not necessarily negative for the women (Zwarteveen and Neupane 1995). The excluded women actually informally accessed (as "free-riders") the irrigation water, which they used for multiple purposes, without having to abide by the financial obligations of association membership (ibid).

In sum, water's fluidity and legal plurality make negotiations on access, use and control dynamic, as well as spatially and temporally contextual (Bruns and Meinzen-Dick 2000). The growing recognition that water access and use is somewhat disassociated from ascribed formal rights not only helped reshape water management and governance in development policy and practice but also set the stage to focus on gender. More recent research on the plurality of interconnected land and water rights in sub-Saharan Africa has helped define approaches that reconcile customary law and formal water regulations in new tools for equitable water allocation (van Koppen et al. 2017; van Koppen and Schreiner, 2019). This work proposes combining customary rights with formal permits to prioritize water access and use in South Africa and Zimbabwe to ensure water to vulnerable groups.

Engendering water governance

Even if inherently fluid, water has also been managed and governed historically and traditionally through centralized approaches (Joshi 2015). Colonial governments appropriated traditional and autonomous governance structures for water resources, for centralized management and control to meet economic agendas across the global South (Agarwal and Narain 1997, Shiva 1988, van Koppen and Schreiner 2018). These changes—especially evident in relation to water for irrigation and urban, industrial use—have persisted post-colonization, and are exacerbated by neocolonial liberal agendas of growth and development involving powerful outside actors and forces (Verzijl et al. 2017). Water policy reforms in the 1980s transferred management from government agencies to community-based governance initiatives, and widely promoted the creation of water users' associations as alternatives to centralized management.

Research led by IWMI in the 1990s assessed the viability and functionality of these shifts, and showed how these interventions were shaped by policies that did not "explicitly consider the possibility that women are water users." They also proved to be based on assumptions that "all users are equally able to pay for water" and consequently, "impact studies" did not assess the links between inclusivity and functionality (Zwarteveen 1998, 301). Such assumptions were challenged by the earlier mentioned studies paying attention to women's irrigation work, and the exclusion of women from irrigation management led to women becoming 'free-riders'—practical realities which, among other things, affected irrigation management performance (Zwarteveen and Neupane 1995). Further research pointed out that, while women might informally access and use (irrigation) water, their lack of formal rights and the mediation of their access to water through relationships with male rights-holders—husbands, fathers, or other male relatives—could reinforce structural gendered inequalities (van Koppen 1990). More importantly, initiatives to involve women in water management did not address complex intersectional inequalities (Joshi 2011).

Research conducted by IWMI in Sri Lanka in the 1990s also showed the mismatch between interlinked domestic/productive water needs of rural communities and sectorally planned and designed water services (Bakker et al. 1999). Because women's roles and responsibilities vis-à-vis water span domestic and productive sectors, the mismatch has implications especially for women's health, nutrition, and gendered social relations (van Koppen and Hussain 2007, van Koppen and Smits 2010, van Koppen et al. 2017, Mitra and Rao 2019). The alternative, Multiple Use Water Services (MUS), is now an established policy intervention and strategy for water resource planning in several countries and inspires further innovations. For example, in remote, rural Nepal, MUS-informed micro-hydropower projects meet electricity, irrigation, and domestic water supply needs of local communities, targeting lower-caste Dalit households and poor, marginalized women (Shah 2016).

CGIAR researchers have analyzed how policy shifts and financial investments in irrigation system rehabilitation and decentralization in Africa, Asia, and Latin America to find that, "Rights to irrigation land and water were rarely vested in poor men, and even less in poor women" (van Koppen 1998, 361). This work on irrigation and gender is continuing in interesting ways in evolving socio-political and socioeconomic landscapes (Balasubramanya 2019).

Unresolved water–gender–poverty links have impacts on health, nutrition, and food security for women and marginalized communities and households (Hussain and Giordano 2003). For the poor, including women, to be counted in and gain from these interventions, it was necessary to include poverty and gendered barriers to land and water rights early on (van Koppen 1998, 361). However, this was easier said than done. "Entrenched masculinities" in the water sector are a key reason for persistent gender water inequalities, and lack of attention to these issues in water planning, implementation, innovations, and investments. That water management organizations in most countries are almost entirely staffed by men is an outcome of deep-rooted social inequalities, and is the reason water infrastructure and technology are not geared to address gendered patterns of water roles, rights, and responsibilities (Zwarteveen 1994, 2008). Gender equality outcomes are not achieved merely by focusing on women, without addressing how increasing their participation in water management, implementation practices, institutional arrangements, and policy- and decision-making is nested in cultures of masculinity (Shrestha et al. 2019).

CGIAR-led research that critically engaged with the dynamics of water access, use, and control, especially in relation to irrigation, has had strategic implications for gender equality. The focus on multiple-use water resources has not only exposed the limitations of sectoral water interventions but also, and more importantly, the fact that these limitations have by far the greatest impacts on the poorest and most marginalized of women. On another note, the focus on institutional arrangements beyond the community—looking into masculine structures and cultures of organizations implementing irrigation projects—has led to an emerging body of work on masculinities in relation to water. Going forward, water is key to achieving the SDGs of sustainable agriculture intensification and ecosystem restoration. The work described in this section challenges interventions to seriously reconsider narrow, sectoral technocratic framings and perspectives, calling for more nuanced understandings of the complexities of gender-power inequalities informed by feminist perspectives (Joshi et al. 2018).

Land and gender—a contested terrain

Unlike forests and water, which are essentially common pool resources, land is a fixed asset and can be classified as public, private, common, and communal property (Bromley 1992). The formalization of private land rights makes land less likely to be a common pool resource. However, as we discuss below, this is not always the case.

Nonetheless, much of the discussion on land in terms of agriculture is in terms of formal ownership. In this section we look critically at narratives and interventions related to women's formal ownership of land—identifying that these interventions do not simplistically lead to equitable outcomes, increased empowerment, or improved agency. This is especially so because gender and the intersection of other social identities—for example race and ethnicity—determine entrenched inequalities, which are deeply nested in institutional structures and cultures and therefore governance systems (Joshi et al. 2018, van Koppen et al. 2017).

Historically, restructuring ownership of and control over land, together with agricultural production, was a key driver of the colonial agenda. In India, for example, colonialism set in place land reforms, new revenue systems, and processes of taxation that irreversibly altered its agrarian economy and society—in the process also creating different types of disparities along ethnicity, caste, and religious lines (Baviskar 2005, Datar 2017). However, in India and elsewhere, land dispossessions through accumulation and appropriation of "common lands"—land not demarcated as private—have been widespread and raised challenging questions. The colonial appropriation of land resulted in unequal rights to and use of land in both the global North and the global South, and led to large-scale displacement and marginalization of local people (Frankema 2010).

North America and sub-Saharan Africa in particular saw wide-scale dispossession of customary, traditional rights to land and the forced displacement of indigenous communities. In areas where feudal and patriarchal histories had already established disparities in ownership and control of land, colonialism further entrenched exclusionary systems. In Africa, the severity of land scarcity is linked to the colonial appropriation of land, promotion of commercial agriculture, and urban sprawl (Whitehead and Tsikata 2003). All these impacts were also distinctly gendered. While this is not discussed in great detail, there is evidence that colonialism resulted in irreversible changes to more equitable traditional practices and systems of inheritance that recognized women's land ownership and use (Akinola 2018).

Land tenure thus refers to larger bundles of highly dynamic land rights, including rights not only to land but also to trees, irrigated lands, water, and woodlands.[7] CGIAR research in the early 2000s on the evolution of land tenure institutions, in Ghana, Indonesia, Uganda, Nepal, Viet Nam, Japan, and Malawi, provided evidence on factors that enabled an understanding of the impacts that changes in land tenure institutions have on NRM (Otsuka and Place 2001). CGIAR research has also paid attention to the gendered impacts of changes in customary land systems. In Indonesia and Ghana, land inheritance systems evolved from matrilineal systems to systems in which both daughters and sons inherited (Quisumbing and Otsuka 2001a).

These studies show that gender-land interrelations are complex and contextual. In Indonesia, women's ability to improve their incomes was impacted by their educational levels, regardless of their inheritance of land (Quisumbing and Otsuka 2001a). Nonetheless, these insights do not dilute concerns—that, regardless of the contextual nature and meaning of land ownership, globally women not only own disproportionately less land in comparison with men (15:85 percent) but also own smaller and less productive pieces of land. Women are also reported to be less able to capitalize on other gains from the ownership of this fixed and vital asset; for example, they have disproportionately less access to agriculture extension services, institutions, credit, and value chains.[8]

In 1994, Bina Agarwal's research in India observed the lack of access to, ownership of, and control over property as the most critical influence on the

7 http://www.fao.org/3/y5744e/y5744e0a.htm

8 https://wle.cgiar.org/content/gender-and-agriculture-infographic

gender gap, along with how it affects women's ability to improve well-being, social status, and empowerment. This work was key in shaping development interventions to secure women's rights to land, through both individual and joint land titling initiatives. Later analyses of statutory and customary land tenure systems demonstrated that it was not only women in Asia who were disadvantaged in their access to and control of land but also women in Africa and Latin America (Lastarria-Cornhiel 1997, Deere et al. 2012, Kieran et al. 2015).

A broad consensus has since then emerged that strengthening women's property rights over land and resources is important for both poverty reduction and equitable growth (Kieran et al. 2015). To address the need for accurate and reliable statistics to monitor these rights, gender researchers from the Policy, Institutions, and Markets (PIM) program developed an analytical framework to assess landownership, in order to expand the statistical content of the Gender and Land Rights Database. They used five indicators: distribution of agricultural holders by sex; agricultural owners by sex; incidence of male and female agricultural landowners; distribution of agricultural land area owned by sex; and distribution of agricultural land value owned by sex.

The application of this framework in different ecological and socio-political contexts is beginning to show large and complex gender gaps in landownership across countries (Doss et al. 2015). While there is wide variation across countries and regions in women and men's ownership of land, the value of land owned by women is disproportionately lower than that owned by men or that owned jointly (ibid.). In Bangladesh, Tajikistan, Viet Nam, and Timor-Leste, gender gaps in land ownership exist, and they vary, especially across the diversity of land tenure systems (Kieran et al. 2015, 2017).

In general, the ownership of cultivated land, including irrigated land, evolving toward more individualized, mainly private forms of property, has had different gendered impacts. A growing body of work is showing that increasing women's ownership of land is unlikely to narrow and reverse the gender gap or deliver empowerment of women (Jackson 2003). First, with women less likely to be listed on ownership documents, and more likely to hold fewer land titles in cases where joint ownership is promoted, a simplistic focus on "title" to land misses much of the reality regarding land tenure, access, and use (Doss et al. 2013, 77). Second, transferring ownership of land to women on its own does not increase productivity if other structural constraints, such as access to and use of other inputs, technology, and credit, are not addressed (Quisumbing et al. 2001b). Landless poor women, who rely on agricultural labor as a means of livelihood, are more likely to benefit from improved wage labor and work conditions as opposed to land ownership per se, especially because a small parcel of land by itself is adequate neither for subsistence nor for productive agriculture (Whitehead and Kabeer 2001).

The allocation of separate land titles for women has also been found to be problematic in patriarchal contexts, as this can result in a loss of social capital for women (Rao 2010). A case study from Kenya showed how formalization of customary rights, through individual titling, resulted in new forms of exclusion, because plural, multiple claims to different types of rights were reduced to singular rights (Meinzen-Dick and Mwangi 2009).

Many other questions remain unanswered around uncultivated lands or "resources" that are dangerously ambiguous—sometimes under communal property regimes as "commons" but more often as land (with resources) to be developed. These were lands that were historically appropriated and colonized to form plantations of tea, coffee, cotton, sugarcane, etc. (Ely 1918). This trend of appropriation of ambiguously co-owned lands continues and is precariously linked to a so-called "development" of natural resources, or what is known as "carbon colonialism" (Lyons and Westoby 2014). This is precisely why Ahlers and Zwarteveen (2009, 409) question the agenda for "individualization and privatization of resource rights as offering possibilities for confronting gender inequalities" vis-à-vis "challenging the individualization, marketization and consumer/client focus of the neo-liberal paradigm."

Recent research in Africa explores the challenges of implementing reforms to ensure gender equality in land governance, including access to services (Ghebru 2019). This work highlights the need for analyses that consider how gender, in conjunction with age, ethnicity, religion, and other factors, affect both individual and joint land ownership, as well as how these intersecting social categories, in turn, influence access to government services, relate to empowerment, and are linked to domestic violence. The pro-WEAI tool (see Chapter 9, this volume) is an attempt to capture some of these dimensions (Malapit et al. 2019). This tool includes 12 indicators that measure three types of agency: intrinsic agency (power within), instrumental agency (power to), and collective agency (power with) in relation to a wide subset of resources (including land). As such, it captures important dimensions of the diversified assets and livelihoods of women, either individually or collectively.

However, standardizations in measures of empowerment can be challenging (see Chapter 9, this volume). Data from Bangladesh, Nepal, and Tajikistan, using WEAI illustrate further questions that need to be considered in relation to gender and land (Clement et. al. 2019b). For example, how can dynamic and evolving changes be assessed? How to assess local meanings and values of gendered norms, roles, and identities? How to analyze structural barriers at scale that keep both marginalized women and men unequal and food-insecure (ibid.)?

In sum, "the land question" in relation to gender, while strategic to women's empowerment, is complicated, and calls for nuanced and transformative analytical frameworks that go well beyond looking at what happens within households and communities (Jackson 2003). In other words, these analyses will need to "articulate with wider political-economic structures and historical dynamics [as well as how these are] characterized by new ways of capitalist expansion" into natural resource regimes (Ahlers and Zwarteveen 2009, 409).

Conclusion: collectives, commons, rights—what next?

CGIAR-led research on NRM has been instrumental in demonstrating the limitations of managerial approaches that see local communities as being composed of rational individuals who are driven by economic necessities and compulsions; they focus on income or livelihoods. Research that distinguishes *management* from *governance* of natural resources has helped capture the plurality of rights and the lived experience of diverse local communities. It underlined the fact that NRM landscapes are multifunctional spaces that cannot easily be compartmentalized into binary categories such as public/private, rich/poor, biophysical/social, material/intangible, human/non-human, or masculine/feminine. Today, it is no longer possible, at least within CGIAR, to conceive natural resource initiatives without attention to gender equality: a significant achievement.

We now stand at a pivotal time in development history where there is increasing consensus to "fix the system" rather than "fix women" (WEF 2020). It is now well acknowledged that securing access rights for women and calling on their participation in NRM does not automatically translate into improved agency and material, political, and social gains to women—that is, to women's empowerment. Research on forest tenure reforms, water user associations, and land reforms shows that, while changes in laws and provisions may provide the basis for more equitable access, use, and management, they do not always guarantee the ability to exercise these rights. Research increasingly shows that natural resource policies and reforms are nested in colonial legal systems and in institutional structures and cultures and driven by neoliberal agendas (Ahlers and Zwarteveen 2009; Joshi et al. 2018, van Koppen and Schriener 2018, Elias et. al. forthcoming).

While it is important to continue to analyze gender and land ownership, tenure, and outcomes (Doss et. al. 2015) and, in general, the efficacy of natural resource interventions and investments (Banana et al. 2012, Colfer et al. 2018, Shrestha et al. 2019), the writing is on the wall: we need to go well beyond popular women–environment narratives. Rights and access to, and control over, natural resources have the greatest impact on the poorest and most marginalized women and men: simply "adding in women and stirring" will not achieve the SDG of reaching the furthest behind (Harding 1995). As we move ahead with much more political agendas of transformative change, it is important to acknowledge that we need to push for approaches that will tackle root causes and the systemic and structural barriers to gender inequality (Hegde et al. 2017, World Fish 2018, Elias et al. forthcoming).

Moving forward and pushing the boundaries

Addressing inequalities across scale, incorporating intersectional approaches, and addressing systemic barriers to gender inequality are all integral to pushing the boundaries toward a next generation of gender and natural resource research. Natural resource governance must speak to and address interconnected and structural dynamics of gender inequality in rapidly changing social, political, and environmental contexts. Nuance as to these complexities needs to continue to inform the study of property rights and collective action, and of formal and informal networks, including social arrangements. When the ground reality is complex, solutions can hardly afford to be simplistic.

This requires a more conscious and deliberate synergy between natural resource R4D agendas and feminist approaches. While instrumental in sharpening the focus on the meanings of gender equality and inclusion, feminist approaches have, until recently, exerted influence only from outside of the R4D arena. Our analysis shows that research findings, data, frameworks, and/or guidelines alone do not easily make a dent in the entrenched cultures, practices, and values of policy and practice related to NRM and agricultural R4D. Masculinities persist not only in social relations but equally in institutions at scale, and in the very definition of what constitutes science (Haraway 1988). And, intersectional inequalities are scalar and deeply entrenched too (Joshi 2011)

The challenge of tackling change in the structures and cultures of organizations engaging in R4D remains, alongside resistance to feminist approaches. That said, several CGIAR centers and research programs now focus on analyzing ways to "transform" complex, dynamic, and resilient social and gender norms at scale (see Chapter 10, this volume). This is the start to synergizing natural resource R4D agendas and feminist calls for "re-politicizing" the power structure and political order of change (Batliwala and Dhanraj 2004) in "integrating power and politics in the analysis of the commons" (Clement et al. 2019a, 1). These shifts are aligned with the "transformative" implications of the SDG agenda, and with calls to "fix the system" rather than just "fixing women" by the drivers of the systems themselves (WEF 2020).

It is promising that recently published work and work in progress within CGIAR explains why the process of "integrating" gender by way of statistical and technocratic solutions that "tackle only the symptoms" of inequality is not enough (Arora-Jonsson and Basnett 2018, Elias et al. forthcoming). Ensuring a focus on gender equality is now well embedded in natural resource agendas. This may work to prevent "gender evaporation," whereby gender priorities are lost in the articulation of wider development goals and sectoral interventions, and between the formulation of a promising policy and its implementation. However, without intersectional analyses, gender policies are still likely to be diluted. To address systemic constraints, we must continuously rethink our framing of gender equality and empowerment, and avoid instrumentalist interpretations. This requires mediating the focus from being just on women and binary framings of inequalities between women and men.

To conclude, natural resource governance is inherently political. Most contemporary interventions—like Payment for Environmental Services and REDD+—tend to commoditize nature, blur complex social differences and disparities in overtly simplistic narratives of "local communities," and reduce multiple, plural rights and access to natural resources through convenient project framings of "rational, technical" institutional arrangements (Rodríguez de Francisco et al. 2013). Feminist scholarship and research asserts that no substantive progress can be claimed unless gender, power, and inclusion are synergistically and systematically incorporated in the design and implementation of natural resource programs and reforms (Sweetman and Ezpeleta 2017, 363). We note with optimism that, given the long and entwined history of gender and natural research within CGIAR organizations, the network is in a good position to rethink politically and strategically how to embrace feminist agendas and "fix the system." First and foremost, this needs to begin by looking inward at our own institutions, research programs, and agendas.

Geography of study sites for publications cited in Chapter 6

Number of cited studies by research methodology (N=185)

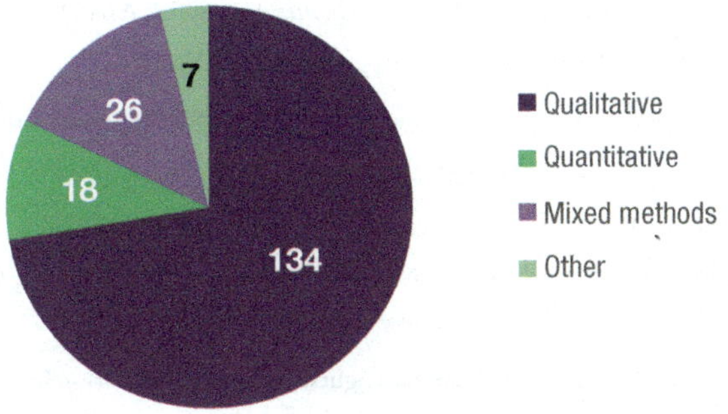

Timeline for references cited

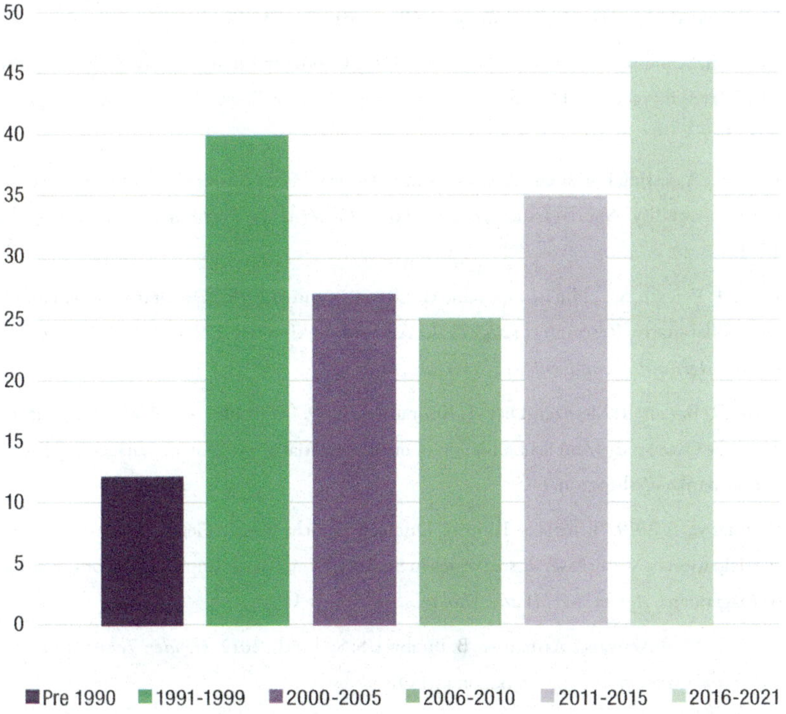

References

Agarwal, B. 1994. *A Field of One's Own: Gender and Land Rights in South Asia.* South Asian Studies 58. Cambridge: Cambridge University Press.

Agarwal, B. 2014. "Food Sovereignty, Food Security and Democratic Choice: Critical Contradictions, Difficult Conciliations." *The Journal of Peasant Studies* 41 (6): 1247–1268.

Agarwal, A., and S. Narain. 1997. "Dying Wisdom: Rise, Fall and Potential of Traditional Water Harvesting Systems." *State of India's Environment* 4, Centre for Science and Environment, New Delhi.

Agrawal, A., and E. Ostrom. 2001. "Collective Action, Property Rights, and Decentralization in Resource Use in India and Nepal." *Politics & Society* 29 (4): 485–514.

Agrawal, A., E. Wollenberg, L. Persha. 2014. "Governing Agriculture-Forest Landscapes to Achieve Climate Change Mitigation." *Global Environmental Change* 29: 270–280.

Ahlers, R., and M. Zwarteveen. 2009. "The Water Question in Feminism: Water Control and Gender Inequities in a Neo-Liberal Era." *Gender, Place & Culture* 16 (4): 409–426.

Akinola, A. O. 2018. "Land Reform Question in South Africa: Rethinking the Feminization of Land." *Gender and Behaviour* 16 (3): 12130–12142.

Angelsen, A., and D. Kaimowitz. 1999. "Rethinking the Causes of Deforestation: Lessons from Economic Models." *World Bank Research Observer* 14 (1): 73–98.

Arora-Jonsson, S. and B. S. Basnett. 2018. "Disciplining Gender in Environmental Organisations: The Text and Practices of Gender Mainstreaming." *Gender, Work & Organization* 25 (3): 309–325.

Arts, B. 2014. "Assessing Forest Governance from a 'Triple G' Perspective: Government, Governance, Governmentality." Special Issue: Assessing Forest Governance. *Forest Policy and Economics* 49: 17–22.

Badstue, L., P. Petesch, S. Feldman, G. Prain, M. Elias, P. Kantor. 2018. "Qualitative, Comparative, and Collaborative Research at Large Scale: An Introduction to GENNOVATE." *Journal of Gender Agriculture and Food Security* 3 (1): 1–27.

Bakker, M., R. Barker, R. Meinzen-Dick, F. Konradsen. 1999. "Multiple Uses of Water in Irrigated Areas: A Case Study from Sri Lanka." System-Wide Initiative on Water Management Paper 8, World Bank, Washington, DC.

Balasubramanya, S. 2019. "Effects of Training Duration and the Role of Gender on Farm Participation in Water User Associations in Southern Tajikistan: Implications for Irrigation Management." *Agricultural Water Management* 216: 1–11.

Banana, A. Y., M. Bukenya, E. Arinaitwe, B. Birabwa, S. Ssekindi. 2012. *Gender, Tenure and Community Forests in Uganda*. Bogor: CIFOR.

Batliwala, S., and D. Deepa. 2004. "Gender Myths that Instrumentalise Women: A View from the Indian Frontline." *IDS Bulletin* 35 (4): 11–18.

Baviskar, A. 2005. "Adivasi Encounters with Hindu Nationalism in MP." *Economic and Political Weekly* 40: 5105–5113.

Bayrak, M., and L. Marafa. 2016. "Ten Years of REDD+: A Critical Review of the Impact of REDD+ on Forest-Dependent Communities." *Sustainability* 8 (7): 620.

Blundo-Canto, G., V. Bax, M. Quintero, G. Cruz-Garcia, R. Groeneveld, L. Perez-Marulanda. 2018. "The Different Dimensions of Livelihood Impacts of Payments for Environmental Services (PES) Schemes: A Systematic Review." *Ecological Economics* 149: 160–183.

Bose, P. 2017. "Land Tenure and Forest Rights of Rural and Indigenous Women in Latin America: Empirical Evidence." *Women's Studies International Forum* 65: 1–8.

Bromley, D. W. 1992. "The Commons, Common Property, and Environmental Policy." *Environmental and Resource Economics* 2 (1): 1–17.

Bromley, D. W., and M. M. Cernea. 1989. "The Management of Common Property Natural Resources: Some Conceptual and Operation Fallacies." Discussion Papers WDP 57, World Bank, Washington, DC.

Bruns, B., and R. Meinzen-Dick. 2000. *Negotiating Water Rights: Implications for Research and Action*. London: ITP.

CG Gender Platform. 2020. "Gender Equality in Food Systems." https://gender.cgiar.org/wp-content/uploads/2020/03/Gender-Platform-A4.pdf

Chappell, M. J. 2018. *Beginning to End Hunger: Food and the Environment in Belo Horizonte, Brazil, and Beyond*. Berkeley, CA: University of California Press.

CIFOR (Centre for International Forestry Research). 2016. *Gender and Forests: Climate Change, Tenure, Value Chains and Emerging Issues*. London: Routledge.

Clement, F., and E. Karki. 2018. "When Water Security Programmes Seek to Empower Women–A Case Study from Western Nepal." In *Water Security Across the Gender Divide*, edited by E. Fröhlich, G. Gioli, R. Cremades, R. Myrttinen, 151–169. Amsterdam: Springer.

Clement, F., W. Harcourt, D. Joshi, C. Sato. 2019a. "Feminist Political Ecologies of the Commons and Commoning (Editorial to the Special Feature)." *International Journal of the Commons* 13 (1): 1–15.

Clement, F., M. C. Buisson, S. Leder, S. Balasubramanya, P. Saikia, R. Bastakoti, E. Karki, B. van Koppen. 2019b. "From Women's Empowerment to Food Security: Revisiting Global Discourses through a Cross-Country Analysis." *Global Food Security* 23: 160–172.

Coleman, E. A., and E. Mwangi. 2013. "Women's Participation in Forest Management: A Cross-Country Analysis." *Global Environmental Change* 23 (1): 193–205.

Colfer, C. 2011. "Marginalized Forest Peoples' Perceptions of the Legitimacy of Governance: An Exploration." *World Development* 39 (12): 2147–2164.

Colfer, C., and D. Capistrano. 2005. *The Politics of Decentralization: Forests, People and Power.* London: Earthscan.

Colfer, C., R. Dennis, J. Mayer, G. Applegate, U. Chokkalingam, I. Kurniawan, H. Lachoski et al. 2005. "Fire, People and Pixels: Linking Social Science and Remote Sensing to Understand Underlying Causes and Impacts of Fires in Indonesia." *Human Ecology* 33 (4): 465–504.

Colfer, C. J. P., M. Elias, R. Jamnadass. 2015. "Women and Men in Tropical Dry Forests: A Preliminary Review." *International Forestry Review* 17 (2): 70–90.

Colfer, C. J. P, B. Basnett, M. Ihalainen. 2018. "Making Sense of 'Intersectionality': A Manual for Lovers of People and Forests." Occasional Paper 184, CIFOR, Bogor.

Coulibaly-Lingani, P., M. Tigabu, P. Savadogo, P.-C. Oden, J.-M. Ouadba. 2009. "Determinants of Access to Forest Products in Southern Burkina Faso." *Forest Policy and Economics* 11 (7): 516–524.

Datar, N. 2017. "The Legacy of Imperialism on Gender Law in India." *Historical Perspectives: Santa Clara University Undergraduate Journal of History* II: 22, Article 9.

Davis, C., L. Williams, S. Lupberger, F. Daviet. 2013. *Assessing Forest Governance.* Washington, DC: World Resources Institute.

De Soto, H. 1986. *The Other Path.* London: IB Tauris and Co. Ltd.

De Soto, H. 2001. *The Mystery of Capital.* London: Bantam Press.

Deere, D. C., G. Alvarado, J. Twyman. 2012. "Gender Inequality in Asset Ownership in Latin America: Female Owners Vs Household Heads." *Development and Change* 43 (2): 505–530.

Doss, C. R., and R. Meinzen-Dick. 2015. "Collective Action within the Household: Insights from Natural Resource Management." *World Development* 74: 171–183.

Doss, C., and R. Meinzen-Dick. 2018. "Women's Land Tenure Security: A Conceptual Framework." Resource Equity.

Doss, C., C. Kovarik, A. Peterman, A. Quisumbing, M. van den Bold. 2013. "Gender Inequalities in Ownership and Control of Land in Africa: Myths Versus Reality." Discussion Paper 1308, IFPRI, Washington, DC.

Doss, C., C. Kovarik, A. Peterman, A. Quisumbing, M. van den Bold. 2015. "Gender Inequalities in Ownership and Control of Land in Africa: Myths versus Reality." *Agricultural Economics* 46: 1–32.

Duchelle, A., C. de Sassi, P. Jagger, M. Cromberg, A. Larson, W. Sunderlin, C. Pratama et al. 2017. "Balancing Carrots and Sticks in REDD+: Implications for Social Safeguards." *Ecology and Society* 22 (3): 2.

Elias, M., R. Meinzen-Dick, and D. Joshi. Forthcoming. "Restoration for Whom, by Whom: Exploring the Socio-political Dimensions of Restoration."

Elmhirst, R. 2015. "Feminist Political Ecology." In *The Routledge Handbook of Gender and Development*, edited by A. Coles, L. Gray, J. Momsen, 519–531. London: Routledge.

Elmhirst, R., M. Siscawati, B. Basnett, D. Ekowati. 2017. "Gender and Generation in Engagements with Oil Palm in East Kalimantan, Indonesia: Insights from Feminist Political Ecology." *The Journal of Peasant Studies* 44 (6): 1135–1157.

Ely, R. 1918. "Private Colonization of the Land." *The American Economic Review* 8 (3): 522–548.

FAO (Food and Agricultural Organization). 2012. *The State of Food and Agriculture.* Rome.

Fortmann, L. 1990. "Locality and Custom: Non-Aboriginal Claims to Customary Usufructuary Rights as a Source of Rural Protest." *Journal of Rural Studies* 6 (2): 195–208.

Fortmann, L., and Bruce, J. W. 1988. *Whose Trees? Proprietary Dimensions of Forestry.* Boulder, CO: Westview Press.

Frankema, E. 2010. "The Colonial Roots of Land Inequality: Geography, Factor Endowments or Institutions?" *The Economy History Review* 63 (2): 418–451.

García-Fernández, C., M. Ruiz-Perez, S. Wunder. 2008. "Is Multiple-Use Forest Management Widely Implementable in the Tropics?" *Forest Ecology and Management* 256 (7): 1468–1476.

Ghebru, H. 2019. "Women's Land Rights in Africa." In *Securing Women's Land Rights in Africa*, edited by P. Robin, 1–8. Washington, DC: IFPRI.

Global Water Partnership. 2000. Towards Water Security: A Framework for Action. GWP, Stockholm and London.

Guha, R. 2000. *The Unquiet Woods: Ecological Change and Peasant Resistance in the Himalaya.* Berkeley, CA: University of California Press.

Gururani, S. 2002. "Forests of Pleasure and Pain: Gendered Practices of Labor and Livelihood in the Forests of the Kumaon Himalayas, India." *Gender, Place & Culture* 9 (3): 229–243.

Haraway, D. 1988. "Situated Knowledges: The Science Question in Feminism and the Privilege of Partial Perspective." *Feminist Studies* 14 (3): 575–599.

Haraway, D. 1991. *Simians, Cyborgs and Women: The Reinvention of Nature.* London: Free Association Books.

Harding, S. 1995. "Just Add Women and Stir?" In *Missing Links: Gender Equity in Science and Technology for Development*, edited by Gender Working Group, United Nations Commission on Science and Technology for Development, 295–307. Ottawa: ITDG.

Harvey, D. 1993. "The Nature of Environment: Dialectics of Social and Environmental Change." *Socialist Register 1993: Real Problems False Solutions* 29.

Hegde, N., M. Elias, H. A. H. Lamers, and M. Hegde. 2017. "Engaging Local Communities in Social Learning for Inclusive Management of Native Fruit Trees in the Central Western Ghats, India." *Forests Trees and Livelihoods* 26 (1): 65–83.

Hussain, I., and M. Giordano. 2003. *Water and Poverty Linkages. Case Studies from Nepal, Pakistan and Sri Lanka*. Colombo: IWMI.

IFPRI (International Food Policy Research Institute). 2012. *Women's Empowerment in Agriculture Index*. Washington, DC.

IWMI (International Water Management Institute). 2017. *Improving Gender Equity in Irrigation: Application of a Tool to Promote Learning and Performance in Malawi and Uzbekistan*. Colombo: CGIAR WLE.

Jackson, C. 1993. "Environmentalisms and Gender Interests in the Third World." *Development and Change* 24 (4): 649–677.

Jackson, C. 1995. "Radical Environmental Myths: A Gender Perspective." *New Left Review* 201: 124–143.

Jackson, C. 2003. "Gender Analysis of Land: Beyond Land Rights for Women?" *Journal of Agrarian Change* 3 (4): 452–480.

Joshi, D. 2011. "Caste, Gender and the Rhetoric of Reform in India's Drinking Water Sector." *Economic and Political Weekly* 46 (18): 56–63.

Joshi, D. 2015. "Like Water for Justice." *Geoforum* 61: 111–121.

Joshi, D., N. Donn-Arnold, M. Kamphuis. 2018. "Land and Water Reforms in S Africa: 'Men in White Coats.'" In *Water Security across the Gender Divide*, edited by C. G. Fröhlich, F. Gioli, F. Greco, R. Cremades, 83–100. Amsterdam: Springer.

Kabeer, N. 1999. "Resources, Agency, Achievements: Reflections on the Measurement of Women's Empowerment." *Development and Change* 30 (3): 435–464.

Kabeer, N. 2005. "Gender Equality and Women's Empowerment: A Critical Analysis of the Third Millennium Development Goal 1." *Gender & Development* 13 (1): 13–24.

Kabeer, N. 2017. "Economic Pathways to Women's Empowerment and Active Citizenship: What Does the Evidence from Bangladesh Tell Us?" *Journal of Development Studies* 53 (5): 649–663.

Kieran, C., K. Sproule, C. Doss, A. Quisumbing, S. M. Kim. 2015. "Examining Gender Inequalities in Land Rights Indicators in Asia." *Agricultural Economics* 46 (S1): 119–138.

Kieran, C., K. Sproule, A. Quisumbing, C. Doss. 2017. "Gender Gaps in Landownership Across and within Households in Four Asian Countries." *Land Economics* 93 (2): 342–370.

Knox, A., and R. S. Meinzen-Dick. 2001. "Collective Action, Property Rights, and Devolution of Natural Resource Management: Exchange of Knowledge and Implications for Policy." Workshop Summary Paper.

Lapniewska, Z. 2016. "Reading Elinor Ostrom through a Gender Perspective." *Feminist Economics* 22 (4): 129–151.

Larson, A. M. 2010. "Making the 'Rules of the Game': Constituting Territory and Authority in Nicaragua's Indigenous Communities." *Land Use Policy* 27 (4): 1143–1152.

Larson, A. M., D. Barry, G. Dahal. 2010. *Forests for People: Community Rights and Forest Tenure Reform*. London: Earthscan.

Larson, A. M., P. Cronkleton, J. Pulhin. 2015. "Formalizing Indigenous Commons: The Role of 'Authority' in the Formation of Territories in Nicaragua, Bolivia, and the Philippines." *World Development* 70: 228–238.

Larson, A. M., I. Monterroso, P. Canturias. 2019a. "Gender and Formalization of Native Communities in the Peruvian Amazon." InfoBrief 238, CIFOR, Bogor.

Larson, A. M., I. Monterroso, N. Liswanti, T. Herawati, A. Banana, P. Canturias, K. Rivera et al. 2019b. "Models for Formalizing Customary and Community Forest Lands: The Need to Integrate Livelihoods into Rights and Forest Conservation Goals." InfoBrief 253, CIFOR, Bogor.

Lastarria-Cornhiel, S. 1997. "Impact of Privatization on Gender and Property Rights in Africa." *World Development* 25 (8): 1317–1333.

Leach, M. 2007. "Earth Mother Myths and Other Ecofeminist Fables: How a Strategic Notion Rose and Fell." *Development and Change* 38 (1): 67–85.

Leach, M., R. Mearns, I. Scoones. 1999. "Environmental Entitlements: Dynamics and Institutions in Community-Based Natural Resource Management." *World Development* 27 (2): 225–247.

Lemos, M. C., and A. Agrawal. 2006. "Environmental Governance." *Annual Review of Environment and Resources*, 31: 297–325.

Li, T. M. 1998. "Working Separately but Eating Together: Personhood, Property, and Power in Conjugal Relations." *American Ethnologist* 25 (4): 675–694.

Locke, C., P. Muljono, C. McDougall, M. Morgan. 2017. "Innovation and Gendered Negotiations: Insights from Six Small-Scale Fishing Communities." *Fish and Fisheries* 18 (5): 943–957.

Lyons, K., and P. Westoby. 2014. "Carbon Colonialism and the New Land Grab: Plantation Forestry in Uganda and Its Livelihood Impacts." *Journal of Rural Studies* 36: 13–21.

Mai, Y. H., E. Mwangi, M. Wan. 2011. "Gender Analysis in Forestry Research: Looking Back and Thinking Ahead. *International Forestry Review* 13 (2): 245–258.

Malapit, H., A. Quisumbing, R. Meinzen-Dick, G. Seymour, E. Martinez, J. Heckert, G. Phase. 2019. "Development of the Project-Level Women's Empowerment in Agriculture Index (pro-WEAI)." *World Development* 122: 675–692.

McCulloch, A. K., R. Meinzen-Dick, P. Hazell. 1998. "Property Rights, Collective Action and Technologies for Natural Resource Management: A Conceptual Framework." SP-PRCA Working Paper No. 1, CAPRi, IFPRI, Washington, DC.

Meinzen-Dick, R. 2017. "Foreword." In *Redefining Diversity & Dynamics of Natural Resources Management in Asia*, edited by G. Shivakoti, U. Pradhan, H. Helmi, xix–xx. Amsterdam: Elsevier.

Meinzen-Dick, R., and E. Mwangi. 2009. "Cutting the Web of Interests: Pitfalls of Formalizing Property Rights." *Land Use Policy* 26 (1): 36–43.

Meinzen-Dick, R., and R. Pradhan. 2002. "Legal Pluralism and Dynamic Property Rights." CAPRi Working Paper 22, IFPRI, Washington, DC.

Meinzen-Dick, R., and M. Zwarteveen. 1998. "Gendered Participation in Water Management: Issues and Illustrations from Water Users' Associations in South Asia." *Agriculture and Human Values* 15 (4): 337–345.

Meinzen-Dick, R., L. R. Brown, H. S. Feldstein, A. R. Quisumbing. 1997. "Gender and Property Rights: Overview." *World Development* 25 (8): 1299–1302.

Meinzen-Dick, R., N. Johnson, A. Quisumbing, J. Njuki, J. Behrman, D. Rubin, E. Waithanji. 2011. "Gender, Assets, And Agricultural Development Programs: A Conceptual Framework." CAPRi Working Paper 99, IFPRI, Washington, DC.

Meinzen-Dick, R., C. Kovarik, A. Quisumbing. 2014. "Gender and Sustainability." *Annual Review of Environment and Resources* 39: 29–55.

Meinzen-Dick, R., A. Quisumbing, C. Doss, S. Theis. 2019. "Women's Land Rights as a Pathway to Poverty Reduction: Framework and Review of Available Evidence." *Agricultural Systems* 172: 72–82.

Mies, M. 1986. *Patriarchy and Accumulation on a World Scale: Women in the International Division of Labour.* London: Zed Books.

Mies, M., and V. Shiva. 1993. *Ecofeminism.* London: Zed Books.

Mitra, A., and N. Rao. 2019. "Gender, Water, and Nutrition in India: An Intersectional Perspective." *Water Alternatives* 12 (1): 169–191.

Mitra, A., and S. Balasubramanya. 2020. "Intra-household Decision Making and Participation in Water User Associations– Results from West Bengal." Working paper.

Mosse, D. 1997. "The Symbolic Making of a Common Property Resource: History, Ecology and Locality in a Tank-Irrigated Landscape in South India." *Development and Change* 28 (3): 467–504.

Mosse, D. 2002 "The Making and Marketing of Participatory Development." Participatory Action Research and Participation in Development, Seminar, Uppsala, April 11.

Mosse, D. 2008. "Epilogue: The Cultural Politics of Water–A Comparative Perspective." *Journal of Southern African Studies* 34 (4): 939–948.

Mwangi, E., and S. Dohrn. 2008. "Securing Access to Drylands Resources for Multiple Users in Africa: A Review of Recent Research." *Land Use Policy* 25 (2): 240–248.

Mwangi, E., R. Meinzen-Dick, Y. Sun. 2009. "Does Gender Influence Forest Management? Exploring Cases from East Africa and Latin America." Working Paper 40, CID, Harvard University, Cambridge, MA.

Mwangi, E., R. Meinzen-Dick, Y. Sun. 2011. "Gender and Sustainable Forest Management in East Africa and Latin America." *Ecology and Society* 16 (1): 17.

Nightingale, A. 2006. "The Nature of Gender: Work, Gender, and Environment." *Environment and Planning D: Society and Space* 24 (2): 165–185.

Nightingale, A. J. 2011. "Bounding Difference: Intersectionality and the Material Production of Gender, Caste, Class and Environment in Nepal." *Geoforum* 42 (2): 153–162.

Noble, I. R., and R. Dirzo. 1997. "Forests as Human-Dominated Ecosystems." *Science* 277 (5325): 522–525.

Notess, L., P. Veit, I. Monterroso, E. Sulle, A. Larson, A. Gindroz, A. Williams. 2018. *The Scramble for Land Rights: Reducing Inequity between Communities and Companies*. Washington, DC: World Resources Institute.

Ostrom, E. 1990. *Governing the Commons. The Evolution of Institutions for Collective Action*. New York: Cambridge University Press.

Ostrom, E. 1992. *Crafting Institutions for Self-Governing Irrigation Systems*. San Francisco, CA: ICS Press.

Ostrom, E. 1996. "Crossing the Great Divide: Coproduction, Synergy, and Development." *World Development* 24 (6): 1073–1087.

Ostrom, E. 1998. "A Behavioral Approach to the Rational Choice Theory of Collective Action: Presidential Address, American Political Science Association, 1997." *The American Political Science Review* 92 (1): 1–22.

Ostrom, E. 1999. "Coping with Tragedies of the Commons." *Annual Review of Political Science* 2 (1): 493–535.

Ostrom, E. 2000. "Collective Action and the Evolution of Social Norms. Journal of Economic Perspectives." *American Economic Association* 14 (3): 137–158.

Ostrom, E. 2007. "Institutional Rational Choice: An Assessment of the Institutional Analysis and Development Framework." In *Theories of the Policy Process*, edited by P. A. Sabatier, 21–64, Cambridge, MA: Westview Press.

Ostrom, E., R. Gardner, J. Walker. 1994. *Rules, Games, and Common-pool Resources*. Ann Arbor, MI: University of Michigan Press.

Ostrom, E., J. Burger, C. B. Field, R. B. Norgaard, D. Policansky. 1999. "Revisiting the Commons: Local Lessons, Global Challenges." *Science* 284 (5412): 278–282.

Ostrom, E., L. Schroeder, S. Wynne, H. Cummings. 1996. "Institutional Incentives and Sustainable Development: Infrastructure Policies in Perspective." *Canadian Journal of Agricultural Economics* 44 (1): 102–111.

Otsuka, K., and F. Place. 2001. *Land Tenure and Natural Resource Management: A Comparative Study of Agrarian Communities in Asia and Africa*. Baltimore, MD: Johns Hopkins University Press.

Perreault, T. 2014. "What Kind of Governance for What Kind of Equity? Towards a Theorization of Justice in Water Governance." *Water International* 39 (2): 233–245.

Porter-Bolland, L., E. Ellis, M. Guariguata, I. Ruiz-Mallen, S. Negrete-Yankelevich, V. Reyes-Garcia. 2011. "Community Managed Forests and Forest Protected Areas: An Assessment of Their Conservation Effectiveness across the Tropics." *Forest Ecology and Management* 268: 6–17.

Quisumbing, A. R., and K. Otsuka. 2001. "Land Inheritance and Schooling in Matrilineal Societies: Evidence from Sumatra." *World Development* 29 (12): 2093–2110.

Quisumbing, A. R., E. Payongayong, J. B. Aidoo, K. Otsuka. 2001a. "Women's Land Rights in the Transition to Individualized Ownership: Implications for the Management of Tree Resources in Western Ghana." *Economic Development and Cultural Change* 50 (1): 157–182.

Quisumbing, A. R., and K. Otsuka with S. Suyanto, J.B. Aidoo, and E. Payongayong. 2001b. "Land, Trees, and Women: Evolution of Customary Land Tenure Institutions in Western Ghana and Sumatra." IFPRI Research Report 121. Washington, DC: International Food Policy Research Institute.

Rao, N. 2010. "Migration, Education and Socio-Economic Mobility." *Compare: A Journal of Comparative and International Education* 40 (2): 137–145.

Ribot, J. C., and A. M. Larson. 2005. *Decentralization of Natural Resources: Experiences in Africa, Asia and Latin America*. London: Frank Cass.

Ribot, J. C., A. Agrawal, A. M. Larson. 2006. "Recentralizing While Decentralizing: How National Governments Reappropriate Forest Resources." *World Development* 34 (11): 1864–1886.

Rocheleau, D., Thomas-Slayter, B., Wangari, E. 1996. "Feminist Political Ecology: A Feminist Political Ecology Perspective." In *Feminist Political Ecology: Global Issues and Local Experience*, edited by D. Rocheleau, E. Wangari, B. Thomas-Slayter, 3–23. New York: Routledge.

Rockström, J., J. Williams, G. Daily, A. Noble, N. Matthews, L. Gordon, H. Wetterstrand et al. 2017. "Sustainable Intensification of Agriculture for Human Prosperity and Global Sustainability." *Ambio* 46: 4–17.

Rodríguez de Francisco, J. C, J. Budds, R. Boelens. 2013. "Payment for Environmental Services and Unequal Resource Control in Pimampiro, Ecuador." *Society & Natural Resources* 26 (10): 1217–1233.

Roth, D., R. Boelens, M. Zwarteveen. 2015. "Property, Legal Pluralism, and Water Rights: The Critical Analysis of Water Governance and the Politics of Recognizing 'Local' Rights." *The Journal of Legal Pluralism and Unofficial Law* 47 (3): 456–475.

Sayer, J. A., and J. R. Palmer. 1994. "Overview on Forest Research in Africa." CIFOR Working Paper 1, Invited Paper for International Symposium Supporting Capacity Building in Forestry Research in Africa AAS/IFS, Nairobi, June 28 –July 1.

Schlager, E, and E. Ostrom. 1992. "Property-Rights Regimes and Natural Resources: A Conceptual Analysis." *Land Economics* 68: 249–262.

Scott, J. C. 1998. *Seeing Like a State: How Certain Schemes to Improve the Human Condition Have Failed*. New Haven, CT: Yale University Press.

Sen, A. K. 1980. "Equality of What?" The Tanner Lecture on Human Values Delivered at Stanford University, May 22, 1979.

Seymour, F., T. La Vina, K. Hite. 2014. "Evidence Linking Community-Level tenure and Forest Condition: An Annotated Bibliography." Climate and Land Use Alliance.

Shah, R. B. 2016. "The Gender Friendly & Inclusive Community Participation to Promote Equity for the Successful Micro Hydro MUS Implementation & Postimplementation." International MUS Workshop, Kathmandu, February 23–26.

Shrestha, G., D. Joshi, F. Clement. 2019. "Masculinities and Hydropower in India: A Feminist Political Ecology Perspective." *International Journal of the Commons* 13 (1): 130–152.

Shiva, V. 1988. *Staying Alive: Women, Ecology and Survival in India*. New Delhi: Kali for Women.

Sun, Y., E. Mwangi, R. Meinzen-Dick. 2011. "Is Gender an Important Factor Influencing User Groups' Property Rights and Forestry Governance? Empirical Analysis from East Africa and Latin America." *International Forestry Review* 13 (2): 205–219.

Sunderland, T., R. Achdiawan, A. Angelsen, R. Babigumira, A. Ickowitz, F. Paumgarten, G. Shively. 2014. "Challenging Perceptions about Men, Women, and Forest Product Use: A Global Comparative Study." *World Development* 64: S56–S66.

Sunderlin, W., A. Angelsen, B. Belcher, P. Burgers, R. Nasi, L. Santoso, S. Wunder. 2005. "Livelihoods, Forests, and Conservation in Developing Countries: An Overview." *World Development* 33 (9): 1383–1402.

Sunderlin, W., A. Larson, A. Duchelle, I. Resosudarmo, T. Huynh, A. Awono, T. Dokken. 2014. "How Are REDD+ Proponents Addressing Tenure Problems? Evidence from Brazil, Cameroon, Tanzania, Indonesia, and Vietnam." *World Development* 55: 37–52.

Sunderlin, W., C. de Sassi, E. Sills, A. Duchelle, A. Larson, I. Resosudarmo, T. Huynh et al. 2018. "Creating an Appropriate Tenure Foundation for REDD+: The Record to Date and Prospects for the Future." *World Development* 106: 376–392.

Sweetman, C., and M. Ezpeleta. 2017. "Introduction: Natural Resource Justice." *Gender & Development* 25 (3): 353–366.

UNDP (United Nations Development Programme). 1997. *Human Development Report 1997: Human Development to Eradicate Poverty*. New York: UNDP.

UNDP (United Nations Development Programme) and SIWI (Stockholm International Water Institute). 2005. *Accountability in WASH*. Stockholm: SIWI and UNICEF.

Van Koppen, B. 1998. "Water Rights, Gender, and Poverty Alleviation. Inclusion and Exclusion of Women and Men Smallholders in Public Irrigation Infrastructure Development." *Agriculture and Human Values* 15 (4): 361–374.

Van Koppen, B. 1990. "Women and the Design of Farmer-Managed Irrigation Schemes-Experiences Provided by Two Projects in Burkina Faso." In *Design for Sustainable Farmer-Managed Irrigation Schemes in Sub-Saharan Africa*, vol 1, 2H. Proceedings of a workshop, Feb. 5–8. Wageningen: Wageningen Agricultural University.

Van Koppen, B., and I. Hussain. 2007. "Gender and Irrigation: Overview of Issues and Options. Irrigation and Drainage." *The Journal of the International Commission on Irrigation and Drainage* 56 (2–3): 289–298.

Van Koppen, B., and S. Mahmud. 1995. "Woman and Water-Pumps in Bangladesh: The Impacts of Participation in Irrigation Groups on Women's Status." Wageningen: Department of Irrigation and Soil and Water Conservation, Wageningen Agricultural University.

Van Koppen, B., and B. Schreiner. 2018. "A Hybrid Approach to Decolonize Formal Water Law in Africa." Research Report 173, IWMI, Colombo.

Van Koppen, B., and B. Schreiner. 2019. "A Hybrid Approach to Statutory Water Law to Support Smallholder Farmer-led Irrigation Development (FLID) in sub-Saharan Africa." *Water Alternatives* 12 (1): 146–155.

Van Koppen, B., and S. Smits. 2010. "Multiple-Use Water Services: Climbing the Water Ladder." *Waterlines* 29 (1): 5–20.

Van Koppen, B., A. Hellum, L. Mehta, B. Derman, and B. Schreiner. 2017. "Rights-Based Freshwater Governance for the Twenty-First Century: Beyond an Exclusionary Focus on Domestic Water Uses." In *Freshwater Governance for the 21st Century*, edited by E. Karar, 129–143. Amsterdam: Springer.

Van Noordwijk, M., and B. Leimona. 2010. "Principles for Fairness and Efficiency In Enhancing Environmental Services in Asia: Payments, Compensation, or Co-investment?" *Ecology and Society* 15 (4): 17.

Verzijl, A., J. Hoogesteger, R. Boelens. 2017. *Grassroots Scalar Politics in the Peruvian Andes: Mobilising Allies to Defend Community Waters in the Upper Pampas Watershed*. Water Text Series. London: Earthscan.

Von Benda-Beckmann, F. 1979. *Property in Social Continuity: Continuity and Change in the Maintenance of Property Relationships through Time in Minangkabau, West Sumatra*. The Hague: M. Nijhoff.

Von Benda-Beckmann, F., and K. von Benda-Beckmann. 2000. "Gender and the Multiple Contingencies of Water Rights in Nepal." *In Water, Land and Law: Changing Rights to Land and Water in Nepal,* 17–39, proceedings of a workshop, Kathmandu, March 18–20.

Von Benda-Beckmann, F., and K. von Benda-Beckmann. 2009. "Beyond the Law-Religion Divide: Law and Religion in West Sumatra". In *Permutations of Order: Religion and Law as Contested Sovereignties*, edited by T. Kirsch and B. Turner, 227–246. London: Routledge.

Von Benda-Beckmann, F., K. von Benda-Beckmann, J. Spiertz. 1996. "Water Rights and Policy." In *The Role of Law in Natural Resource Management*, edited by J. Spiertz, and M.G. Wiber, 7799. Gravenhage: VUGA.

Von Benda-Beckmann, F., K. von Benda-Beckmann, H. J. L. Spiertz. 1997. "Local Law and Customary Practices in the Study of Water Rights." In *Water Rights, Conflict and Policy*, edited by K. von Benda-Beckmann, H. J. L. Spiertz, R. Pradhan, F. von Benda-Beckmann, S. S. Kadka, K.A. Haq, 221–242. Colombo: IWMI.

Von Benda-Beckmann, F., K. von Benda-Beckmann, H. J. L. Spiertz. 1998. "Equity and Legal Pluralism: Taking Customary Law into Account in Natural Resource Policies." In *Searching for Equity*, edited by R. Boelens, G. Davila, 5769. Assen: Van Gorcum.

Von Benda-Beckmann, F., K. von Benda-Beckmann, M. Wiber, M. 2007. *Changing Properties of Property*. New York: Berghahn Books.

Von Benda-Beckman, F., K. von Benda-Beckman, A. Griffiths. 2009. "The Power of Law." In *The Power of Law in a Transnational World: Anthropological Enquiries*, 132. New York: Berghahn Books.

WEF (World Economic Forum). 2020. "Stakeholders for a Cohesive and Sustainable World." Annual Meeting, Davos, January 21–24.

Were, E., J. Roy, B Swallow. 2008. "Local Organisation and Gender in Water Management: A Case Study from the Kenya Highlands." *Journal of International Development* 20 (1): 69–81.

Whitehead, A., and N. Kabeer. 2001. "Living with Uncertainity: Gender, Livelihoods and Pro-Poor Growth in Rural Sub-Saharan Africa." Working Paper 134, Brighton, IDS.

Whitehead, A., and D. Tsikata. 2003. "Policy Discourses on Women's Land Rights in Sub-Saharan Africa: The Implications of the Re-turn to the Customary." *Journal of Agrarian Change* 3 (1–2): 67–112.

Wily, L. A. 2011. "'The Law Is to Blame': The Vulnerable Status of Common Property Rights in Sub⊠Saharan Africa." *Development and Change* 42 (3): 733–757.

World Bank. 2006. "India Water Supply and Sanitation: Bridging the Gap between Infrastructure and Service." Background Paper, Urban Water Supply and Sanitation, World Bank, Washington, DC.

World Fish. 2018. "Gender Strategy." CGIAR Research Program on Fish Agri-Food Systems (FISH), Penang.

Zwarteveen, M. 1994. *Gender Issues, Water Issues: A Gender Perspective to Irrigation Management.* Colombo: IWMI.

Zwarteveen, M. Z. 1997. "Water: From Basic Need to Commodity: A Discussion on Gender and Water Rights in the Context of Irrigation." *World Development* 25 (8): 1335–1349.

Zwarteveen, M. Z. 1998. "Identifying Gender Aspects of New Irrigation Management Policies." *Agriculture and Human Values* 15 (4): 301–312.

Zwarteveen, M. 2008. "Men, Masculinities and Water Powers in Irrigation." *Water Alternatives* 1 (1): 111.

Zwarteveen, M., and N. Neupane. 1995. "Gender Aspects of Irrigation Management: The Chhattis Mauja Irrigation System in Nepal." *Asia-Pacific Journal of Rural Development* 5 (1): 1–26.

Zwarteveen, M., and R. Meinzen-Dick. 2001. "Gender and Property Rights in the Commons: Examples of Water Rights in South Asia." *Agriculture and Human Values* 18 (1): 11–25.

Chapter 7

FROM VULNERABILITY TO AGENCY IN CLIMATE ADAPTATION AND MITIGATION

Sophia Huyer, Tatiana Gumucio, Katie Tavenner, Mariola Acosta, Nitya Chanana, Arun Khatri-Chhetri, Catherine Mungai, Mathieu Ouedraogo, Gloria Otieno, Maren Radeny, John Recha, and Elisabeth Simelton

Rising temperatures and more extreme weather associated with climate change are expected to exacerbate existing social and gender inequalities across the globe (Adger et al. 2014 , Dankelman 2010). The Intergovernmental Panel on Climate Change (IPCC) projects that the production of major crops such as wheat, rice, and maize will be affected across all regions and that climate change will progressively increase variability in crop yields. All aspects of food security are affected, including food access, utilization, and price stability. In rural areas, major impacts are expected on water availability and supply, food security, and agricultural incomes, with shifts in production of both food and non-food crops as growing zones change as a result of weather variability (Adger et al. 2014, Girvetz et al. 2017).

Climate-influenced migration is seen to be an intensifying factor in the feminization of agriculture in some regions, particularly South Asia and Central America, where out-migration for employment is predominantly male. Women left behind are required to manage farms and households with fewer resources and less family labor, further increasing their vulnerability (Gumucio et al. 2019a, Khatri-Chhetri et al. 2020).[1]

In a 2°C (or more)[2] world, gender equality will need to encompass women and men's increased resilience, as well as reduced vulnerability to climate change. Vulnerability is commonly defined as the extent to which a natural or social system is prone to damage. In the context of climate variability

1 See Chapter 8 of this volume for a broader discussion on the dynamics around the feminization of agriculture.

2 According to the IPCC, global warming is likely to reach 1.5°C if not beyond (2.0°C) between 2030 and 2052. This is expected to increase the risk of heatwaves, heavy rainfall events, crop productivity decline, reduction in water availability, undernutrition, habitat losses and others (Masson-Delmotte et al. 2018), and the effects get significantly worse at 2°C. The world has already witnessed about 1°C of temperature rise and is on track to exhaust the carbon budget for 1.5°C by 2030. Some projections put the world on track for 4°C of warming.

and change, vulnerability is a function of **exposure** to climate risks (such as extreme weather events, losses in agricultural productivity, and alterations in hydrological patterns), **sensitivity** to such risks, and **capacity to adapt**. It is characterized by interrelations between ecological and, increasingly, social systems, such as weather impacts, viability of natural resources, access to markets, and societal safety nets. Resilience is the capacity of communities or households to resist, cope with, or recover from shocks and stresses (Ulrichs et al. 2015) and arrive at a state of reduced or diminished vulnerability (see Perez et al. 2015). Women's agency in relation to climate resilience is the ability to access and act on (make choices based on) information and to participate in decisions that affect their lives. Women's collective action is an effective platform for achieving agency (Kabeer 1999, CCAFS and FAO 2013; see also case study on women's groups in South Asia, later in this chapter).

Climate change has varied effects on women and men, since they are exposed to different climate shocks and experience different impacts related to their gender-differentiated roles, rights, and opportunities (Dankelman 2010, Fisher and Carr 2015). Women and men's vulnerabilities vary according to gender, but also by ethnicity, religion, class, and age conditions. Less is known about men's gender-specific knowledge in relation to adaptation and mitigation but we do know that women's knowledge, networks, and assets are a significant aspect of resilience (Lane and McNaught 2009, Chanamuto and Hall 2015, McKune et al. 2015). Despite the significant roles that both women and men play as agents of change, however, gender discourses on development and climate change have centered overwhelmingly on women's greater vulnerability to climate change. This not only contributes to perpetuating stereotypes of women as victims but also prevents tackling the root causes of gendered vulnerabilities (MacGregor 2010, Arora-Jonsson 2011, Resurreccion 2011) by denying women their agency (Dankelman 2010, MacGregor 2010, Okali and Naess 2013).

Two approaches to climate adaptation and mitigation for sustainable rural development are climate-smart agriculture (CSA) and land use-based management. The goal of CSA is to help farmers adjust to climate change and manage climate risk by implementing strategies to sustainably increase productivity; build resilience of farming systems; and reduce greenhouse gas emissions. It builds on approaches that take into account the impact of land management decisions on ecosystem goods and services on a larger scale (Girvetz et al. 2017). Land use-based management initiatives focus on sustainable and inclusive interventions to reverse land degradation and biodiversity loss while promoting carbon sequestration. Delivering these environmental benefits also results in benefits of food security, nutrition, market access, and employment opportunities (Smith and Scherr 2002).

This chapter assesses how these climate adaptation and mitigation approaches can reduce women's and men's vulnerabilities, promote their capacities for resilience, support the ability of women to exercise their agency, and, consequently, increase gender equality. We review existing literature and regional case studies in relation to four dimensions of gender in/equality that are connected to these goals in climate-resilient agriculture and land use-based management: (1) participation in decision-making at different levels, (2) work burden, (3) access to and use of productive resources such as agroclimatic information, technology, livelihood incomes, and credit, and (4) collective action (see also Kabeer 1999, Dankelman 2010, CCAFS and FAO 2013).

For example, agricultural technologies and practices for adaptation that reduce workloads can increase production, reduce negative impacts on health, and allow women more time for other activities such as education or enterprises. This is critical since climate impacts such as drought in combination with deforestation are expected to significantly increase women's workload in rural areas (Dankelman 2010). Information and capacity-building to cope with and manage climate risk and variability, while important for both women's and men's capacity to adapt, can also promote women's participation in household decision-making and increase their agricultural production (Rengalakshmi et al. 2018, Huyer 2019b). Women's group organizing and collective action can engender capacity-building tailored to women's needs and constraints, and serve as platforms for women to exercise agency in implementing climate adaptation strategies (see South Asia case study later in this chapter).

We first summarize the gendered effects of climate change, then assess what we know so far about the potential of climate-resilient agriculture and land use-based management practices for gender equality. We next put the spotlight on cases from three different regions—South Asia, Southeast Asia, and West Africa—as potential models for increasing women's resilience and agency through the four dimensions of gender in/equality described above. The first case looks at women's groups as a platform for access, agency and voice in Madhya Pradesh, India, as a collective action approach to enabling women's agency and access to resources and information. The second case illustrates how women's economic empowerment and participation in community decision-making through non-timber forest product processing in Senegal can promote resilience. The third case sheds light on how participatory development of agro-advisories with women and men in Cambodia, Laos, and Viet Nam can promote agency and reduce work burdens. Finally, we draw out key insights on how these climate adaptation and mitigation approaches can contribute to gender equality, and propose areas for forward-looking research and action.

Effects of climate change on women and men

By 2050 a medium–high level of climate change is expected to increase the number of undernourished children by 4.8 million (IFPRI 2017). This will affect food availability and prices, which in turn will lead to a decrease in the amount of food consumed as well as its nutritional quality (Beuchelt and Badstue 2013, Bryan et al. 2017). Climate variability, including extreme events, can have significant impact on women's and men's health, well-being, and empowerment. Few studies consider the impacts of climate variability and change on men—yet climate stresses pose a significant health risk for them, including in rural areas. In rural Australia, suicide rates of men quadrupled over the 12-year Millennium Drought, which decimated the country's agriculture sector (Alston and Kent 2008). In India, suicide rates of men farmers have gone up over the past decade, owing to poverty and indebtedness from crop failures and water depletion (Reddy et al. 2019).

Rural women are at high risk of being negatively affected by climate change, particularly in relation to household responsibilities, agricultural activities, and male out-migration for employment—with resulting consequences on family nutrition and children's care and education (Kakota et al. 2011, Rao et al. 2017, Ylipaa et al. 2019). Women's nutrition levels are also affected: for example, when climate shocks affect food access, women tend to eat less to reserve food for the family (Nguyen et al. 2013). Natural disasters and their after-effects kill more women than men on average, for physiological reasons (such as pregnancy) or socio-cultural reasons (in the case of flooding, the clothes women wear or their responsibilities in caring for small children may restrict their ability to run or climb to avoid danger). Women are also at higher risk of physical, sexual, and domestic violence in times of climate shock and natural disaster (Correia 2001, Neumayer and Plumper 2007).

In many cases, women appear to be less able to adapt to climate change, even if they are aware of its effects. Gender inequalities in access to and control over resources, technology, and information, alongside less stable land tenure, restrict women's ability to act on and implement climate adaptation practices in agriculture (Fisher and Carr 2015, Huyer 2016, Jost et al. 2016, Assan et al. 2018). Gender norms may limit women's ability to respond to or make quick decisions in the face of climate events. In households where men are working off-farm in cities, women may lack the power to make timely farming decisions or to convince their husbands to agree to new practices (Goering 2015, World Bank et al. 2015).

Climate-resilient agriculture and land use-based management: can they advance gender equality?

Women have developed a range of coping strategies in response to risks and environmental impacts. Their assets, knowledge, and social networks related to their positions in the household and community are important pillars of resilience. Livestock-keeping is an important food safety net. Indigenous/local breeds and animals, including poultry, sheep, and goats, are inherently resilient to climate stress and can survive on crop and household residues (Kristjanson et al. 2014, Chanamuto and Hall 2015). Women's social networks for agricultural production can have a positive impact on household food security and productivity in the context of climate change (Tadesse et al. 2017, Violon et al. 2016). Community seed banks are repositories of local genetic diversity that can withstand climate stress, and are a useful resilience strategy. Women's role in informal seed networks and the related conservation of genetic resources is connected to dietary diversity and local knowledge. In addition, women often retain ties in their home village while creating new connections in their marriage community, suggesting they can be important avenues of seed distribution (Otieno et al. 2018). When CSA and land use-based initiatives interact with the four dimensions of gender in/equality for climate-resilient agriculture, they can promote gender equality and expand women's resilience strategies.

Climate-smart agriculture: can it exacerbate gender inequalities?

Specific approaches for integrating gender in diagnostics, prioritization, and impact assessment of CSA practices and technologies as well as climate services have been developed and tested (Jost et al. 2014, Nelson and Huyer 2016, Duong et al. 2016, 2017, Gumucio et al. 2018, Gumucio and Schwager 2019). Analysis to date shows that gender equality dimensions of participation in household decision-making, access to finance and information, and position in the household are factors associated with adoption of CSA. It also shows women and men may adopt differing adaptation practices depending on the gender division of labor, their awareness of climate impacts, and the type of information they have access to (Twyman et al. 2014; Jost et al. 2016, Mutenje et al. 2018, Aryal et al. 2020).

More recently, analysis of whether CSA can contribute to gender equality has started to emerge, indicating that: (1) CSA can exacerbate existing inequalities if implemented in a "gender-neutral" manner; (2) gender equality is in some cases a factor in the adoption of CSA (Huyer and Partey 2020); and (3) CSA can support gender equality if it integrates one or more of the four gender in/equality dimensions of climate-resilient agriculture. Most analysis to date has focused on how CSA can support gender equality. More research is needed to identify the enabling gender equality conditions for adoption of climate-resilient agriculture as well as potential for increased inequality as a result of CSA implementation.

1. CSA can exacerbate gender and social inequalities. Agricultural production is situated within broader societal structures and gender relations that affect the allocation of labor, resources, and other assets (Jordan 2018). CSA interventions will inevitably interact with these gender dynamics. To date, however, research on gender and CSA is relatively new, and has been concerned mostly with a male–female dichotomy that ignores power and social and political status stemming from gender, race, class, ethnicity, religion, and age (Djoudi et al. 2016, Mungai et al. 2017, Colfer et al. 2018).

As a result, the introduction of CSA technologies or practices may intensify inequality: prevailing power and gender relations within a community can be entrenched or solidified if questions are not asked about who is controlling the technology and who benefits (Haapala 2018, IFAD 2018). For example, in the western Indo-Gangetic plains of India, women adopt laser land-leveling (LLL) only through their children or a male relative. This is because of gender norms that prohibit public interaction between women and men: a woman farmer is not permitted to approach a male LLL owner or service provider either in person or by mobile phone (Aryal et al. 2015). However, an additional reason for differences in the adoption rate of LLL between women and men is the gender division of labor: the technology is aimed at men's agricultural domain exclusively; women's technology needs, related to their work and priorities, are overlooked.

Elsewhere, it has been noted that women can choose to adopt CSA technologies that in fact increase their labor load, counteracting other gender equality benefits such as increased production and control over income. In other cases, it was observed that women were not able to consider the adoption of CSA until barriers of access to family labor and lack of appropriate energy technology, transportation, and cultivation tools were addressed (Murray et al. 2016, Mutenje et al. 2019).

In the livestock sector, climate-smart interventions tend to assume impacts to be gender-neutral, and, as a result, neglect gendered power relations affecting men's and women's differential roles in livestock-keeping (Djoudi and Brockhaus 2011). These interventions can then intensify inequalities, particularly where they increase labor investment (Arora et al. 2017), overlook the roles of women (Gallina and Rozel Farnworth 2016), or ignore their activities in sectors such as informal dairy (Tavenner and Crane 2018). However, once women's roles in production are recognized, and they can access training, credit services, and technologies, the potential for gender equality outcomes improves markedly (see Gallina and Rozel Farnworth 2016). Since women's and men's responsibilities vary according to livestock type, program design should take into account the gender differences in a targeted livestock value chain. For example, cattle ownership is often a male activity because cows are a high-value commodity in comparison with chickens or goats. Gender dynamics like these may constrain women's participation. Participatory planning of livestock interventions can benefit women when it informs project components such as livestock breed selection (KIT et al. 2012).

2. CSA adoption is often contingent on gender equality. Gender equality can be a precondition in the adoption of CSA when women are able to make choices about their farming practices and to access the resources to implement them. In three countries in Southern Africa—Malawi, Mozambique, and Zambia—households with more female adults, and where women had greater bargaining power, were more likely to adopt CSA (Farnworth et al. 2018, Mutenje et al. 2019). In Bangladesh and Nicaragua, women's empowerment was a factor for increased crop diversification, an important element of dietary diversity and food security. Fruit and crop diversification tended to be a strategy in households where women had a say in decision-making in agricultural production (Gumucio et al. 2019a, de Pinto et al. 2019).

Group membership may increase CSA adoption and investment decisions by women. Mutenje et al. (2019) found that being a group member of a social/community platform (such as a savings group, cooperative, church, or mosque) positively influenced women's investment decisions on drought-tolerant maize with legume intercropping. This likely occurred as a result of group exchange of information and/or revolving credit.

3. CSA can contribute to gender equality. While gender equality can be a precondition for CSA adoption, emerging evidence indicates that CSA can also contribute to gender equality through the four gender in/equality dimensions:

Women in climate-smart villages (CSVs) in two regions of India (Bihar and Haryana) experienced empowerment as a result of CSA adoption, in the form of **increased participation in decision-making** over income from increased agricultural production as well as on farming practices and children's education, among other issues. Men in both regions also experienced increased empowerment, but to a lesser extent (Hariharan et al. 2020). In another region in India, access to agricultural information allowed women to discuss agricultural production with their husbands, increasing their role in decision-making on the farm (Mittal et al. 2016).

Reducing women's work burden by reducing time spent in labor, as well as increasing production, is an important aspect of gender equality in CSA, and often results in increased control over income (Mittal 2016). In East Africa, management of small ruminants (sheep and goats) and poultry that are hardy and adapted to climate stress promotes gender equality, since the livestock require less labor and women have greater control over the returns (Ojango et al. 2016). Climate services can also reduce women's workload by improving planting and harvesting, as a result lessening the risk of replanting later in the season (Simelton et al. 2019).

Gender-responsive CSA technologies reduce the labor required for tasks women engage in, while also increasing agricultural productivity. In Nepal, women's agricultural activities were matched with CSA technologies and practices that had been demonstrated, in other contexts, to reduce labor requirements in relation to efficiency, environment, and sustainability (Khatri-Chhetri et al. 2020). Some CSA technologies benefit both sexes directly. In a case study in Maharashtra, India, both women and men considered the rice drum seeder useful: men appreciated the increased production levels and farming income and women benefited from a reduced workload when seeding rice (Joshi et al. 2019).

The contribution of the third dimension, **access to resources, including information and income**, to gender equality is potentially significant. For example, in Viet Nam, participants in women-targeted training sessions on pest management and livestock-rearing experienced significant increases in self-confidence and participated more actively in household decision-making. Their incomes increased as a result of diversification of their production (they began to raise piglets for market with rice bran feed). Both the women themselves and family members felt their status in the household had increased (Chi et al. 2015). In East Africa, implementing different CSA options and participating in knowledge-sharing events allowed women to contribute to joint household decision-making on CSA technologies and practices (Radeny et al. 2018).

Agro- and climate information services can promote gender equality if they contribute to the needs and priorities of both women and men in rural areas, increase their resilience, and provide a means for women to challenge gender norms through public discussion and airing of their views (Sekabira and Qaim 2017, Huyer 2019b). In Kenya, hundreds of thousands of women benefited from the television show Shamba Shape-up, a show that presents real-life farm "renovations." Women and men farmers in approximately 428,566 households made changes in their farming practices as a result of watching the program. Production increased for both women and men, with the increase in women's dairy and maize production being proportionately greater than men's (AECF and University of Reading 2014). Radio Mang'elete in Kamba, Kenya, was an experiment with an interactive technology that enabled women to record their voices remotely for radio broadcast, without the need for intermediaries to facilitate content production. Use of the technology encouraged women to make suggestions for programming and express their views. They experienced increased agency and voice, adding to their self-confidence while increasing their recognition in the community (Sterling and Huyer 2010).

The gender equality potential of climate information services needs to be further investigated in line with these empowering results. Meeting rural women's service needs involves attention to communication channels that may differ from those used to reach men, depending on control and cost of technologies. Women may also need different information than men, depending on the gender division of labor in agriculture, including crop production. For example, in Kaffrine, Senegal, men control draft animals and plows, so that women must wait to prepare their fields until men have finished using them. As a result, information provided on rainfall onset benefited men but not women (Tall et al. 2014). Biased institutions and gender differences in group participation and networks can also constrain women's access to weather and climate information for managing risks and planning production (Gumucio et al. 2019b).

Despite these obstacles, women farmers who access climate information can use and benefit from it (Roncoli et al. 2009, Tall et al. 2014, Carr et al. 2016, Carr and Onzere 2017, Huyer et al. 2017, Gumucio et al. 2019b). In Tamil Nadu, India, for example, understanding farmers' perspectives on local weather and climate, and communicating gender-sensitive climate information and advisories, increased the resilience of women smallholder farmers.

Creating trust in the information accessed and understanding gendered needs within existing communication networks was critical for women to make informed agricultural decisions. Strengthening the social contract between climate experts and farmers in communicating climate information is also an important ingredient (Rengalakshmi et al. 2018). It should be emphasized, however, that climate information services risk reinforcing gender bias if they fail to account for women's concerns and priorities.

In relation to the fourth dimension, participation in **organizing and collective action**, women's organizations and community groups can be platforms for capacity development and agency, while acting as a vehicle for access to and control of resources. They can provide opportunities to share experiences, exchange information, and engage in group activities such as revolving credit, production and processing, and production cooperatives. They can act as mechanisms to magnify women's voice and shift the way they think of themselves and their entitlements, increasing their negotiating power in communities and households (Farnworth et al. 2017, Lecoutere 2017, Mello and Schmink 2017, Kumar et al. 2018). For example, the establishment of women's self-help groups (SHGs) in a highly drought-prone district of India allowed women farmers to interact more effectively with local government officials while improving their access to water (Desai and Olofsgård 2019).

Group membership can also support use of climate information. In Viet Nam, the national Women's Union trained women farmers across the country in low emissions development (LED) practices. It provided an avenue to navigate and overcome the discriminatory social and gender norms that were making it difficult for women to participate in LED, so that extension partners were able to reach women farmers more effectively (Farnworth et al. 2017). The inclusion of women's groups and networks in communication delivery channels can help address challenges women face to access weather and climate change information (Rengalakshmi et al. 2018).

Land use-based management approaches to resilience: can they promote gender equality?

Evidence as to whether land use-based management approaches contribute directly to gender equality is contested. For example, a recent assessment of the literature on gender and land degradation found that gendered biases related to land rights, access to resources and incentives, opportunities to participate in decision-making, and the distribution of costs and benefits hinder international efforts to promote gender equality in land improvement programming (Okpara et al. 2019).

To date, the largest land use management initiative is the program on Reducing Emissions from Deforestation and Forest Degradation (REDD+), that attributes financial value to the carbon stored in forests. As such, it provides incentives for developing countries to protect and manage forests to offset carbon emissions (Brown 2011). The funds gained through these incentives are meant to fund ecosystem services and pro-poor development. One step to empower rural women within the REDD+ framework is to ensure access rights to forest lands and resources (with commensurate carbon rights), thereby increasing their capacity to engage in decision-making and improving their economic and social status (Setyowati 2011).

However, **women have been largely marginalized** as stakeholders in the design and implementation of REDD+ (Larson et al. 2015, Wornell et al. 2015). A synthesis of lessons learned from REDD+ in Viet Nam found a variety of barriers to meaningful integration of gender, including: limited guidance on how to address inequality in forestry policy; limited representation of women in stakeholder groups, such as community organizations, non-governmental organizations, and government; inadequate understanding of women's priority needs; and lack of access by women to information (Thuy and Brockhaus 2015). A review of Indonesia's REDD+ policies found that less than 4 percent of documents substantively integrated gender (Wornell et al. 2015). A review by the Center for International Forestry Research (CIFOR) of 77 villages in 6 countries found that women were less involved in REDD+ design decisions and processes than men, even when they were the primary forest users. Program design focused on participation of women in meetings and training, ignoring deeper structural inequalities. As a result, implementation failed to take women's strategic interests into account in terms of land use and rights to forests (Larson et al. 2015, Evans et al. 2019).

REDD+ programming is also criticized for failing to take into account localized gendered power relations in pursuit of economic goals. Programs are designed to conserve forest resources and limit logging—areas that are largely under men's purview—with the result that access by women to non-timber forest products requires greater effort without commensurate benefits, in effect increasing their workloads. REDD+ initiatives in Burkina Faso perpetuated inequitable gender divisions of labor by shifting formal environmental decision-making "upward" to the state and development institutions and the burden of environmental labor downward onto the poor, in particular women (Westholm and Arora-Jonsson 2015). For policies such as REDD+, the challenge remains to integrate gender equality into implementation as well as design. Accountability for implementation requires tools such as gender budgeting, and the lack of this kind of incentive poses major challenges (Kristjanson et al. 2017, Ampaire et al. 2020).

Despite these critiques, land use-based management approaches have significant environmental, social, and gender equality benefits, including for resilience (Villamor et al. 2014). In agroforestry, trees or shrubs are grown between crops or around pastures for soil improvement; while fruit, nut, and vegetable production can improve nutrition and increase incomes. The cultivation of indigenous fruits and vegetables represents an important source of income for women in Benin, Burkina Faso, Cameroon, and Tanzania (Kiptot and Franzel 2012). Wood from on-farm woodlots can be sold or used in the household (Kiptot et al. 2014), decreasing women's and girls' workloads significantly. For example, in Malawi and Tanzania, improved access to woodlots decreased the time spent by women in collecting fuel by up to three hours per day (Oduol et al. 2006).

Strengthening women's collective action is an effective strategy for gender equality in agroforestry. In Tanzania, marketing by women's groups has increased revenue from indigenous fruit processing, while the use of marketing and extension services by women's groups in Kenya increased the value generated from fodder shrubs (Kiptot and Franzel 2011). Other strategies include improving women's access to information and extension services related to agroforestry (Kiptot and Franzel 2012) and targeting interventions to improve productivity and marketing of products considered to be women's provenance (for example, wild forest foods, fodder, and branches not considered economically important). Evidence is emerging to suggest that pairing diversification strategies with commercialization has a more positive effect for women smallholders than commercialization alone (Tavenner and Crane 2018, Tavenner et al. 2019).

Promising gender equality approaches in the context of climate change: regional experiences

Promising models for promoting gender equality in the context of agriculture and climate change are emerging. Three regional case studies presented here involve participatory approaches to climate adaptation and mitigation interventions as part of the CSV approach,[3] in the context of one or more of

---

3 A CSV is a platform for action research to test, develop, and support the scaling of CSA technologies and approaches and develop practical adaptation options for food security, resilience, and decreased carbon emissions (Aggarwal et al. 2018).

the four dimensions for gender in/equality. A common element is to promote women's agency and enhanced resilience through active partnerships with farmers organizations, both mixed-gender and women-only. Two of the case studies (India and Southeast Asia) are CSA-focused; the other (Senegal) combines both CSA and land use-based management approaches.

Women's groups as a platform for access, agency, and voice in South Asia: Madhya Pradesh, India

Collective action is a common strategy for women across the South Asian region, in the form of SHGs that serve as a platform for the adoption and spread of successful resilience strategies while promoting members' increased agency.

In 2017, a project was implemented by the CGIAR research program on Climate Change, Agriculture and Food Security (CCAFS) and the Madhya Pradesh, Uttar Pradesh, and Bihar states of India to improve women farmers' adaptive capacity. Components of the project included women-led groups, capacity-building, CSA technologies and practices, and climate information services, all placed within a larger initiative on climate-smart interventions for climate resilience. Preferences and needs of men and women were assessed and integrated into the design of the project. The representation of SHGs and other women-led groups in management and implementation was a strategic priority, as was strengthening women's agency and voice.

The women's collective action component was implemented[4] in a region identified as a "hotspot" of gender and climate risk in Madhya Pradesh. The area is characterized by significant involvement of women in agriculture, with high levels of drought risk (Chanana-Nag and Aggarwal 2020). The socio-cultural norms defining participation of women in public activities in Madhya Pradesh are less stringent in comparison with other regions (for instance women cover their faces when talking to strangers). During the project baseline assessment, it was observed that women's savings SHGs were well established in the region, and members of the community were familiar with and accepting of (to an extent) the idea of women's group-based activities. A participatory consultation process resulted in a focus on local institution-building, which involved leveraging the social capital of SHGs for project decision-making and implementation. Two mechanisms were used: participation of SHGs in Village Climate Management Committees to manage

4 Implemented by CCAFS, BAIF Development Research Foundation, and farmers' organizations in Betul district of Madhya Pradesh, funded by USAID and CCAFS.

and implement CSA; and setting up women-led Custom-Hiring Centers (CHCs) to rent out climate-smart technologies to farmers at affordable rates. While the two types of groups were part of a design implemented across all three project sites, in Madhya Pradesh the Village Climate Management Committee was women-only, in line with women's significant participation in agricultural production. The CHCs in all three project districts were women-led, to promote economic independence and group-based agency. This approach was used to counter local norms constraining male–female interaction, to ensure women farmers could dependably access the technologies made available.

Through information and communication technology (ICT)-based agro-advisory and weather/climate information services, women farmers became aware of new practices, even implementing some of them. Women farmers also participated equally—or in some cases more than men—in capacity-building events, technology prioritization, and community consultation, ensuring attention to their production preferences. Focus group discussions revealed that several technologies reduced women's work in the labor-intensive activities of weeding (cono-weeder), water management (solar pumps), and harvesting (harvesting machine).

In this case, collective action was a platform for gender equality, agency, and resilience. Women gained better access to information; their production is increasing; their workloads have decreased; and they participate in community decision-making. They no longer rely solely on their social networks for information, and regularly access the project's ICT-based agro-advisories for information on weather, markets, crop cultivation practices, and technology use. Survey results indicate that the percentage of household income generated from rice and wheat production has increased from 44 percent to 50 percent. Women are also actively sharing their new knowledge with other farmers.

Promoting resilience through processing of non-timber forest products in West Africa: Daga-Birame CSV, Senegal

CGIAR and partners implemented a collective action and technology training approach in the CSV of Daga-Birame in Senegal. The purpose was to promote women's resilience through increased agricultural production and participation in community management of natural resources. Activities included (1) planting fruit trees for improved vegetation cover and income generation, (2) gardening to increase nutrition and market sales, and (3) baobab (*Adansonia digitata*) processing. In West Africa, it is traditionally women who lead non-timber forest product processing. This offered a significant opportunity to increase incomes for women, to make up for decreasing agricultural production resulting from climate change.

A component of the project involved the establishment of a community-based Innovation Platform (IP). As one of its actions, the IP set up a committee responsible for processing baobab powder; the committee subsequently founded a women-run micro-enterprise for processing and selling the powder. The micro-enterprise was specifically developed with and for women members of the community.

Members of the micro-enterprise were trained in fruit processing and financial management, while environmental restoration activities such as increased vegetation cover and use of indigenous trees were introduced. To demystify beliefs about the length of time required for fruit production, and to provide training, five tree species were tested on community land as well as women farmers' individual plots. This approach was used for selecting fruit tree options and products: farmers were convinced to plant baobab once they learned that the grafted varieties produced fruit in a shorter time period (Sanogo et al. 2016).

Their involvement in the demonstration trials meant that the women farmers were trained in tree planting and grafting. They now grow baobab trees on their own plots, using the fruit, leaves, and bark for household purposes. A mixed-gender committee, also set up by the IP, manages community protected areas, including trees in the field and forest, and makes decisions on when to harvest fruit. Marketing and income management are carried out exclusively through the women-run enterprise (Sanogo et al. 2016). A village savings pool invests in community resilience activities.

Overall, the project has increased women's access to and control over forest resources, while the women's group controls the funds used for community improvement. Incomes have increased through fruit powder sales, and, as members on the different committees, women participate in community decision-making on tree management. Mixed- and single-gender collective action approaches have been used in combination to promote gender equality and resilience by increasing incomes, and agency through participation in community decision-making.

Despite these gender equality gains, the question of the long-term sustainability of the fruit processing enterprise remains. The product is sold locally, where there is limited demand. New commercialization initiatives need to be explored that can link the enterprise to subnational, national, and international markets, requiring investment in quality control and certification.

Promoting agency and reducing work burdens through participatory development of agro-advisories: Cambodia, Laos, and Viet Nam

The Agro-Climate Information Services project (ACIS) in Cambodia, Laos, and Viet Nam explored both mixed- and single-gender participatory approaches for integrating the preferences and activities of women and men in agro- and climate advisories (Le et al. 2018, Simelton et al. 2018b). The approach to gender equality went beyond provision of information in itself to enabling both women and men to co-develop the design, timing, and format of the agro-advisories. Unintended gender equality results of the approach included increased agency of women in the form of public speaking and joint household decision-making.

Across Southeast Asia, climate services value chains[5] are generally top-down. The forecast is disseminated via TV whereas advice is transferred via extension. In this context, gender equality is influenced by differential access to information (extension workers are men and male trainers are often the norm); time poverty (women work long hours every day); gender division of labor (women tend to be responsible for small livestock and home gardens, and men for forestry); unequal decision-making (while many couples say that they take shared decisions, women more often than men will consult with their spouse before spending their own or joint money); and migration for employment (increasingly changing both household composition and farming itself).

The core of the project was to provide a seasonal forecast to groups made up of women and men farmers, who then developed their own agro-advisories, in their preferred format. The basic package of weather forecast plus participatory agro-advisory development was either integrated into community development research projects (such as in My Loi CSV, Viet Nam) and rural development projects (CARE Viet Nam) or included as an add-on to already-existing projects (CARE in Cambodia and Laos; Ekxang CSV, Laos). The agro-advisory design component revealed gender differences in uses and formats: women wanted both text and visuals so that their children could read the advisory to them while they worked in the kitchen, whereas men relied more on visuals. The group work helped participants understand the underlying reasons for gender differences in agro-advisory needs and preferences.

---

5 A climate services value chain approach looks at improving the way decision-makers and users understand and use climate information (WMO 2018).

Local gender dynamics and differing project contexts led to project modifications in the different sites. In three sites, the project was combined with rural development projects targeting women-only village savings and loans associations (VSLAs) managed by CARE. VSLAs constitute a regular group activity for women in this region, and were considered an effective platform for ACIS, in addition to being empowering for women. The three sites all included ethnic minority populations, where local socio-cultural norms made women-only groups a better option: Cambodia, for example, has a high rate of male out-migration, with women left behind to manage the farm. In the CSV site in Viet Nam, the project utilized mixed-gender groups because both women and men participated significantly in household farming decisions.

Project activities were intended to facilitate social learning, collaboration, and trust between women and men through different methods. The format of mixed-gender group activities (including gender awareness training) encouraged discussion and exchange of information between household members and among group members. Both women and men said they had learned from each other. A combination of female and male trainers was used whenever possible, to facilitate women's active involvement. An unintended result was that the female trainers became role models for women in the community, who began to feel more confident to speak in public. Husbands and wives participated interchangeably in the mixed groups and both joined the savings group—ensuring they kept each other informed and encouraging greater communication at home.

The localized format of the agro-advisories improved timeliness and relevance for the agricultural planning needs of both women and men. More efficient resource use resulted, with fewer wasted inputs and less replanting, reducing women's work hours. The participatory design of the agro-advisories increased accessibility of content for both women and men farmers, which they shared with others, often through social media. Collective action approaches were successful in promoting women's participation and facilitated joint decision-making. In sites where gender roles were strictly defined and women's participation in mixed-gender groups was constrained, the project leveraged already existing women-led groups—the VSLAs—for gender equality results; in the site where more flexible gender norms allowed for collaboration between women and men in farming, work with mixed-gender groups was possible and encouraged joint decision-making in the household.

Despite these positive results, it should be noted that significant changes in gender relations do not occur overnight. Building mutual trust behind husband and wife—including adoption of new roles—takes time, as does the development of relationships between development practitioners and project participants.

From learning to action: lessons from the regions

All three projects, despite responding to different economic, environment, and socio-cultural contexts, exhibit common trends and approaches. The focus on women's participation in collective action supported resilience and promoted women's enhanced agency in all three communities. Resilience improved in relation to one or more dimensions of gender in/equality in climate-resilient agriculture. Increasing women's access to resources, technologies, and practices for climate-resilient production increased household incomes and resilience in two cases. The use of CSA technologies and practices to reduce women's work burden was a component of two projects, in South and Southeast Asia. Integral to any labor reduction outcomes was the meaningful involvement of women farmers in consultations to identify their technology needs and preferences.

All three cases are examples of how approaches that promote women's collective action (either mixed-gender or women-only), in combination with one or more of the other three dimensions for gender in/equality, can redress gender inequalities. The diversity of successful approaches to collective action demonstrates the importance of designing approaches that fit the local context.

Toward gender equality in climate-resilient agriculture and land management systems

Climate change has different effects on women and men farmers, in relation to their access to resources and income, stemming from gender norms around decision-making in different cultural contexts, and as a result of male out-migration for employment. Climate adaptation and mitigation strategies can perpetuate or, at worst, exacerbate gender inequalities when implemented in a gender-neutral approach—by, for example, diminishing or diverting the resources to which women have access. Gender-responsive approaches can promote equality while increasing resilience; at the same time, when women experience some degree of gender equality, they can be more apt to adopt CSA. This is predicated upon one or more dimensions related to agency and resilience: (1) participation in decision-making at local, national, and global levels, (2) work burden, (3) access to and use of productive resources such as agro-climatic information, livelihood incomes, credit, and climate finance, and (4) collective action.

To date, research on gender equality and climate-resilient agriculture has focused largely on "diagnostic" analyses of women's vulnerability, and on differences in how women and men perceive, are affected by, and respond to climate change (Kristjanson et al. 2017). The critical priority now is to focus on *how* to promote gender equality and increase resilience. We need more action-oriented, participatory research that explores strategies to build on previous diagnostic research. We know that gender equality and women's agency constitute an integral aspect of this research agenda, and there is a need for further exploration of the long-term interconnections between resilience and gender equality. For example, what are the longer-term effects of decreased work burdens on resilience and agency? Research is also needed on the conditions promoting gender equality along with women's agency and voice in different contexts through collective action. Promising examples need to be identified and methodologies tested to better understand what constitute successful approaches.

The full range of structural factors underlying gender inequalities in climate adaptation and mitigation also needs to be explored. This includes recognizing power relations between women and men, within households, communities, and countries, as well as among different groups in a community (Colfer et al. 2018, Haapala 2018). Power relationships are expressed through a range of structures—judicial, economic, social, and political—so that overcoming power imbalances involves promoting greater equality in control over resources (physical, human, intellectual, intangible) and ideology (beliefs, values, attitudes), as well as changes in institutions and structures (Rao and Kelleher 2005, Twyman et al. 2015). At the global level, the term "climate justice" has been used in relation to mitigation and locus of responsibility for global warming (Richards et al. 2015), and can be expanded to include climate adaptation (Boyd 2002, Rossi and Lambrou 2008).

The intersection of men and masculinities with women's resilience and gender equality is another underlying structural factor that is under-researched in relation to climate-resilient agriculture—although studies on CSA adoption do carry out sex-disaggregated analyses (see Villaneuva et al. 2016). More research is needed on the gendered impacts of climate change on men, as well as their experience of and responses to climate change (see Correia 2001). The role of men in promoting women's resilience and agency is an important area for research in relation to their response to women's increased incomes or their attitudes toward women-targeted initiatives. Questions to consider here include: How do we engage with men and power-holders to overcome resistance to gender equality (Hearn 2001)?[6] As the case studies showed, participatory, collective action approaches in the context of CSA and land use management are one avenue to address entrenched norms and attitudes.

Three specific research areas critical to the promotion of gender equality and women's agency and resilience in climate adaptation and mitigation are (1) use of CSA technologies and practices to reduce women's work burden; (2) access to agro-climatic information; and (3) gender and climate policy.

More analysis is needed on how CSA technologies can promote gender equality by reducing women's work burden and increasing their incomes. The examples from the literature review and case studies suggest CSA technologies can contribute to labor-saving outcomes for women when women's activities and priorities are taken into account. Apart from more insight into which technologies are appropriate, better understanding is needed of women's access to them, and the reasons women use or don't use them. They may relate to cost, lack of information, or a design that is culturally or physiologically inappropriate. There is evidence that, when women are not consulted, or when the gender division of labor and localized power relations are not taken into account, CSA technologies may benefit some community members more than others, or inadvertently increase women's workloads (Beuchelt and Badstue 2013, Haapala 2018). How can CSA technology approaches ensure women retain control over them and the proceeds of increased production? It is also necessary to assess how new technologies and practices may displace women's paid work in different socio-economic contexts (Paris and Chi 2005, Khatri-Chhetri et al. 2020).

Evidence to date indicates that access to agro-climatic information in different formats can enable women to play a greater role in household decision-making, increase production, and improve household health and nutrition. Training sessions and workshops, joint agro-advisory development, and Farmers' Field Schools are successful models. In cases where market and advisory information have been relevant to women's economic activities, access to information has also been associated with increased income.

Considering the importance of agro-climatic information and capacity-building for managing climate and related risk (Hansen et al. 2019), targeted research is necessary to identify the conditions and approaches that

6 This occurs in the context of *resistance* of men to involvement in gender equality debates, policies, and activities; the need to recognize the *responsibility* of men (and power-holders) to take part in the promotion of gender equality; and the process of *reaching out* to those who are less interested and less involved (Hearn 2001).

promote gender equality in climate services (Gumucio et al. 2019b). What kind of information do women want and need? What are women's priorities for information content and formats? What are the empowering effects of the use of climate information (Mittal 2016, Huyer 2019b)? What partnerships and platforms, including women's and community-based organizations, mobile service providers, and others, will support generation, access to, and use of climate information?

Increasing reliance on ICT- and digital-based information services—as well as research and survey approaches—also poses new questions. What are the equality implications of the digital gender gap in rural areas and the predominance of men in the IT sector (see García et al. 2019, Huyer 2019a)? How accurate are digital surveys?

Research with policymakers at all levels, complemented by advocacy, is also needed in order to understand the types of support, partnerships, policy structure, coordination mechanisms, and climate finance instruments that are most effective for achieving gender equality goals. Policy has largely failed to recognize women as agents of change in climate adaptation and mitigation (Dankelman 2010, UNDP 2016, Evans et al. 2019), to acknowledge the root causes of gender inequalities, or to promote gender equality in implementation (Acosta et al. 2019, Ampaire et al. 2020). A research agenda on strategies to increase the participation of women and men in the formulation of climate policy is critical and largely neglected. Policy can lead to climate action, if women are recognized as key stakeholders and change agents for adaptation and recovery, in a low-carbon society (Michael et al. 2019).

Research is needed on several fronts: substantive intersectional gender analysis should inform the *content* of climate policy, while we also need better understanding of *how* to integrate gender into climate policy. This includes how to build capacity of policymakers to mainstream gender (UNDP 2016, Ampaire et al. 2020), as well as identifying key aspects of national instruments such as National Determined Contributions, National Adaptation Plans, climate finance, etc. Political will as well as support from the private sector is needed, to ensure the commitment of financial and other resources to empowering farmer and women's organizations. Coordinated multistakeholder efforts with government, civil society, women's organizations, farmers, and youth as well as the private sector are needed to ensure that rhetoric about climate action and gender equality becomes reality.

A final critical question is how the scaling-up of climate-resilient approaches can be consistent with gender equality. Limiting global warming to the 2°C target to which countries have committed in the 2015 Paris

Agreement, and reaching the Sustainable Development Goals by 2030, will require large-scale expansion of climate-resilient approaches in agriculture, food systems, and land management. How can such approaches integrate the conditions and mechanisms necessary for gender equality? Can small-scale, inclusive, and localized approaches that incorporate cultural and socio-economic differences be coordinated with the broad-brush approaches required to scale up? Can women's collective action around climate-resilient agriculture and land use management be a mechanism for scaling up successful equality approaches? Questions need to be asked as to what constitutes inclusive scaling: Who and what is it for? How do scaling strategies take into account the capacities, priorities, and goals of all groups in society? These are the challenges in a warming climate.

Geography of study sites for publications cited in Chapter 7

Number of cited studies by research methodology (N=125)

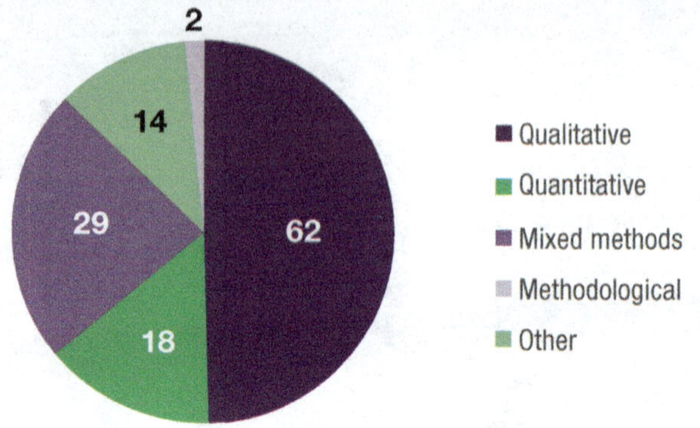

Timeline for references cited

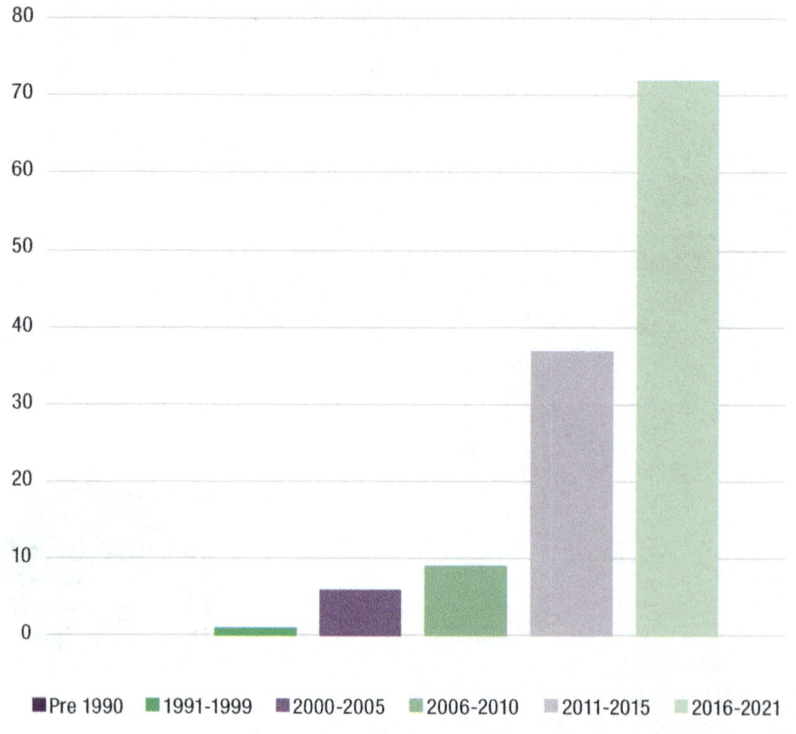

References

Acosta, M., S. van Bommel, M. van Wessel, E. Ampaire, L. Jassogne, P. Feindt. 2019. "Discursive Translations of Gender Mainstreaming Norms: The Case of Agricultural and Climate Change Policies in Uganda." *Women's Studies International Forum* 74: 9–19.

Adger W. N., J. M. Pulhin, J. Barnett, G. D. Dabelko, G. K. Hovelsrud, M. Levy, Ú. Oswald Spring et al. 2014. "Human Security." In *Climate Change 2014: Impacts, Adaptation, and Vulnerability. Part A: Global and Sectoral Aspects. Contribution of Working Group II to the Fifth Assessment Report of the Intergovernmental Panel on Climate Change*, edited by C. B. Field, V. R. Barros, D. J. Dokken, K. J. Mach, M. D. Mastrandrea, T. E. Bilir, M. Chatterjee et al., 755–791. Cambridge and New York: Cambridge University Press.

AECF (Africa Enterprise Challenge Fund) and University of Reading. 2014. *Assessing the Impacts of Shamba Shape Up*. Nairobi: AECF.

Aggarwal, P. K., A. Jarvis, B. Campbell, R. Zougmoré, A. Khatri-Chhetri, S. Vermeulen, S. Huyer. 2018. "The Climate-Smart Village Approach: Framework of an Integrative Strategy." *Ecology and Society* 23 (1): 14.

Alston, M., and J. Kent. 2008. "The Big Dry: The Link between Rural Masculinities and Poor Health Outcomes for Farming Men." *The Australian Sociological Association* 44: 133–147.

Ampaire, E. L., M. Acosta, S. Huyer, R. Kigonya, P. Muchunguzi, R. Muna, L. Jassogne. 2020. "Gender in Climate Change, Agriculture, and Natural Resource Policies: Insights from East Africa." *Climatic Change* 158 (1).

Arora, D., J. Arango, S. Burkart, N. Chirinda, J. Twyman. 2017. "Gender [Im]balance in Productive and Reproductive Labor among Livestock Producers in Colombia: Implications for Climate Change Responses." Info Note, CCAFS, Copenhagen.

Arora-Jonsson, S. 2011. "Virtue and Vulnerability: Discourses on Women, Gender and Climate Change." *Global Environmental Change* 21 (2): 744–751.

Aryal, J. P., M. B. Mehrotra, M. L. Jat. 2015. "Impacts of Laser Land Leveling in Rice–Wheat Systems of the North-Western Indo-Gangetic Plains of India." *Food Security* 7 (3): 725–738.

Aryal, J. P., C. Farnworth, R. Khurana, S. Ray, T. B. Sapkota, D. B. Rahut. 2020. "Does Women's Participation in Agricultural Technology Adoption Decisions Affect the Adoption of Climate-Smart Agriculture? Insights from Indo-Gangetic Plains of India." *Review of Development Economics* 24 (3): 973-990.

Assan, E., M. Suvedi, L. Schmitt Olabisi, A. Allen. 2018. "Coping with and Adapting to Climate Change: A Gender Perspective from Smallholder Farming in Ghana." *Environments* 5 (8): 86.

Beuchelt, T. D., and L. Badstue. 2013. "Gender, Nutrition- and Climate-Smart Food Production: Opportunities and Trade-Offs." *Food Security* 5 (5): 709–721.

Boyd, E. 2002. "The Noel Kempff Project in Bolivia: Gender, Power and Decision Making in Climate Mitigation." *Gender & Development* 10 (2): 70–77.

Brown, H. C. P. 2011. "Gender, Climate Change and REDD+ in the Congo Basin Forests of Central Africa." *International Forestry Review* 13 (2): 163–176.

Bryan, E., S. Theis, J. Choufani, A. de Pinto, R. Meinzen-Dick, C. Ringler. 2017. "Gender-Sensitive, Climate-Smart Agriculture for Improved Nutrition in Africa South of the Sahara." In *A Thriving Agricultural Sector in a Changing Climate: Meeting Malabo Declaration Goals through Climate-Smart Agriculture*, edited by A. de Pinto, and J. M. Ulimwengu, 114–135. Washington, DC: IFPRI.

Carr, E. R., and S. N. Onzere. 2017. "Really Effective (for 15% of the Men): Lessons in Understanding and Addressing User Needs in Climate Services from Mali." *Climate Risk Management* 22: 1–14.

Carr, E. R., G. Fleming, T. Kalala. 2016. "Understanding Women's Needs for Weather and Climate Information in Agrarian Settings: The Case of Ngetou Maleck, Senegal." *Weather, Climate and Society* 8 (3): 247–264.

CCAFS and FAO (Food and Agricultural Organization). 2013. *Gender and Climate Change Research in Agriculture and Food Security for Rural Development*. Second Edition. Rome: FAO.

Chanamuto, N. J. C., and S. J. G. Hall. 2015. "Gender Equality, Resilience to Climate Change, and the Design of Livestock Projects for Rural Livelihoods." *Gender & Development* 23 (3): 515–530.

Chanana-Nag, N, and P. K. Aggarwal. 2020. "Woman in Agriculture, and Climate Risks: Hotspots for Development." *Climatic Change* 158 (1): 13–27.

Chi, T. T. N., T. Paris, T. T. T. Anh, L. Duy, D. Loan. 2015. *Enhancing the Roles of Women in Rice Farming as an Adaptation Strategy to Climate Change Risks: A Case Study in Submergence Villages in Hau Giang Province, South Viet Nam*. Hanoi: CLRRI.

Colfer, C. J. P., B. Sijapati Basnett, M. Ihalainen. 2018. *Making Sense of "Intersectionality": A Manual for Lovers of People and Forests*. Bogor: CIFOR.

Correia, M. C. 2001. "Hurricane Mitch – the Gender Effects of Coping and Crises." PREM Notes 57, World Bank, Washington, DC.

Dankelman, I. 2010. "Introduction: Exploring Gender, Environment, and Climate Change." In *Gender and Climate Change: An Introduction*, edited by I. Dankelman, 1–20. London: Routledge.

De Pinto, A., G. Seymour, E. Bryan, P. Bhandari. 2019. *Women's Empowerment and Crop Diversification in Bangladesh: A Possible Pathway to Climate Change Adaptation and Better Nutrition*. Washington, DC: IFPRI.

Desai, R. M., and A. Olofsgård. 2019. "Can the Poor Organize? Public Goods and Self-Help Groups in Rural India." *World Development* 121: 33–52.

Djoudi, H., and M. Brockhaus. 2011. "Is Adaptation to Climate Change Gender Neutral? Lessons from Communities Dependent on Livestock and Forests in Northern Mali." *International Forestry Review* 13 (2): 123–135.

Djoudi, H., V. Locatelli, C. Vaast, K. Asher, M. Brockhaus, B. Basnett Sijapati. 2016. "Beyond Dichotomies: Gender and Intersecting Inequalities in Climate Change Studies." *Ambio* 45 (Suppl. 3): 248.

Duong, M. T., E. Simelton, V. H. Le. 2016. *Participatory Identification of Climate-Smart Agriculture Priorities*. Copenhagen: CCAFS.

Duong, M. T., A. Smith, T. T. Le, E. Simelton, M. Coulier. 2017. *Gender-Differences in Agro-Climate Information Services*. Copenhagen: CCAFS.

Evans, K., S. Flores, A. Larson. 2019. "Participatory Monitoring in Forest Communities to Improve Governance, Accountability and Women's Participation." *Small-Scale Forestry* 18: 165–187.

Farnworth, C. R., C. Stirling, T. Sapkota, M. Jat, M. Misiko, S. Attwood. 2017. "Gender and Inorganic Nitrogen: What Are the Implications of Moving towards a More Balanced Use of Nitrogen Fertilizer in the Tropics?" *International Journal of Agricultural Sustainability* 15 (2): 136–152.

Farnworth, C. R., C. M. Stirling, A. Chinyophiro, A. Namakhoma, R. Morahan. 2018. "Exploring the Potential of Household Methodologies to Strengthen Gender Equality and Improve Smallholder Livelihoods: Research in Malawi in Maize-Based Systems." *Journal of Arid Environments* 149: 53–61.

Fisher, M., and E. R. Carr. 2015. "The Influence of Gendered Roles and Responsibilities on the Adoption of Technologies that Mitigate Drought Risk: The Case of Drought-Tolerant Maize Seed in Eastern Uganda." *Global Environmental Change* 35: 82–92.

Gallina, A., and C. Rozel Farnworth. 2016. "Gender Dynamics in Rice-Farming Households in Viet Nam: A Literature Review." Working Paper 183, CCAFS, Copenhagen.

García, M. A., N. Orentlicher, J. Twyman, A. Eitzinger, O. Bonilla-Findji. 2019. *Reflections on the Use of Mobile Phones for Monitoring Gender Indicators Related to Climate Smart Agriculture Practices*. Wageningen: CCAFS.

Girvetz, E., C, Corner-Dolloff, C. Lamanna, T. Rosenstock. 2017. "'CSA-Plan': Strategies to Put Climate-Smart Agriculture (CSA) into Practice." *Agriculture for Development* 30: 12–16.

Goering, L. 2015. "From Second Jobs to New 'Stinginess', Women See Climate Change Differently." Thomas Reuters News, July 15.

Gumucio, T., and S. Schwager. 2019. *Checklist: Gender Considerations for Climate Services and Safety Nets.* Wageningen: CCAFS.

Gumucio, T., D. Arora, J. Twyman, A. Tickamyer, and M. Clavijo. 2019a. "Gender Equality and Trees on Farms: Considerations for Implementation of Climate-Smart Agriculture." In *Gender, Agriculture and Agrarian Transformations: Changing Relations in Africa, Latin America and Asia*, edited by C. E. Sachs, 203-220. New York: Routledge.

Gumucio, T., J. Hansen, S. Huyer, T. van Huysen. 2019b. "Gender-Responsive Rural Climate Services: A Review of the Literature." *Climate and Development* 12 (3): 241–254.

Gumucio, T., S. Huyer, J. Hansen. E. Simelton, S. Partey, S. Schwager. 2018. "Inclusion of Gender Equality in Monitoring and Evaluation of Climate Services." Working Paper. 249, CCAFS, Wageningen.

Haapala, A. 2018. "A Feminist Analysis of Oppressive and Emancipatory Potentialities within Technology-Prioritized Climate Change Adaptation Intervention: A Case Study from Phailom Community in Laos." In *Gender Dimension of Climate Change Research in Agriculture (Case Studies in Southeast Asia)*, edited by T. Paris, and M. Rola-Rubzen, 43–62. Wageningen: CCAFS.

Hansen, J., J. Hellin, T. Rosenstock, E. Fisher, J. Cairns, C. Stirling, C. Lamanna et al. 2019. "Climate Risk Management and Rural Poverty Reduction." *Agricultural Systems* 172 (June): 28–46.

Hariharan, V. K., S. Mittal, M. Rai, T. Agarwal, K. C. Kalvaniya, C. M. Stirling, M. L. Jat. 2020. "Does the Climate-Smart Village Approach Influence Gender Equality in Farming Households? A Case of Two Contrasting Ecologies in India." *Climatic Change* 158 (1): 77–90.

Hearn, J. 2001. "Men and Gender Equality: Resistance, Responsibilities and Reaching Out." Keynote Paper, Men and Gender Equality, Örebro, March 15–16.

Huyer, S. 2016. "Closing the Gender Gap in Agriculture." *Gender, Technology and Development* 20 (2): 105–116.

Huyer, S., J. Hansen, A. Rose, C. Vaughan, T. van Huysen. 2017. *What We Know about Gender and Rural Climate Services Preliminary Findings and Guidance.* Copenhagen: CCAFS.

Huyer, S. 2019a. "A Global Perspective on Women in Information Technology: Perspectives from the UNESCO Science Report 2015 towards 2030." In *Cracking the Digital Ceiling: Women in Computing Around the World*, edited by C. Frieze, and J. Quesenberry, 46–60. Cambridge: Cambridge University Press.

Huyer, S. 2019b. "ICT in a Changing Climate: A Path to Gender-Transformative Food Security." In *Taking Stock: Data and Evidence on Gender Digital Equality in Digital Access, Skills and Leadership*, edited by A. Sey, and N. Hafkin, 262–273. Macao: United Nations University Institute on Computing and Society/International Telecommunications Union.

Huyer, S., and S. Partey. 2020. "Weathering the Storm or Storming the Norms? Moving Gender Equality Forward in Climate-Resilient Agriculture." *Climatic Change* 158 (1): 1–12.

IFAD (International Fund for Agricultural Development). 2018. *How to Do: Design of Gender Transformative Smallholder Agriculture Adaptation Programmes*. Rome: IFAD.

IFPRI (International Food Policy Research Institute). 2017. *2017 Global Food Policy Report*. Extended Results of the IMPACT Model. Washington, DC: IFPRI.

Jordan, J. C. 2018. "Deconstructing Resilience: Why Gender and Power Matter in Responding to Climate Stress in Bangladesh." *Climate and Development* 11 (2): 167–179.

Joshi, P. K., M. T. Khan, A. Kishore. 2019. "Heterogeneity in Male and Female Farmers' Preference for a Profit-Enhancing and Labor-Saving Technology: The Case of Direct-Seeded Rice (DSR) in India." *Canadian Journal of Agricultural Economics* 67 (3): 303–320.

Jost, C., N. Ferdous, T. D. Spicer. 2014. *Gender and Inclusion Toolbox: Participatory Research in Climate Change and Agriculture*. Copenhagen: CCAFS, CARE International, and ICRAF.

Jost, C., F. Kyazze, J. Naab, S. Neelormi, J. Kinyangi, R. Zougmore, P. Kristjanson et al. 2016. "Understanding Gender Dimensions of Agriculture and Climate Change in Smallholder Farming Communities." *Climate and Development* 8 (2): 1–12.

Kabeer, N. 1999. "Resources, Agency, Achievements: Reflections on the Measurement of Women's Empowerment." *Development and Change* 30: 435–464.

Kakota, T., D. Nyariki, D. Mkwambisi, W. Kogi-Makau. 2011. "Gender Vulnerability to Climate Variability and Household Food Insecurity." *Climate and Development* 3 (4): 298–309.

Khatri-Chhetri, A., P. P. Regmi, N. Chanana, P. K. Aggarwal. 2020. "Potential of Climate-Smart Agriculture in Reducing Women Farmers' Drudgery in High Climatic Risk Areas." *Climatic Change* 158 (1): 29–42.

Kiptot, E., and S. Franzel. 2011. "Gender and Agroforestry in Africa: Are Women Participating?" Occasional Paper 13, ICRAF, Nairobi.

Kiptot, E., and S. Franzel. 2012. "Gender and Agroforestry in Africa: A Review of Women's Participation." *Agroforestry Systems* 84: 35–58.

Kiptot, E., S. Franzel, A. Degrande. 2014. "Gender, Agroforestry and Food Security in Africa." *Current Opinion in Environmental Sustainability* 6: 104–109.

KIT (Royal Tropical Institute), Agri-Profocus, IIRR (International Institute of Rural Reconstruction). 2012. *Challenging Chains to Change: Gender Equity in Agricultural Value Chain Development*. Amsterdam: KIT.

Kristjanson, P., A. Waters-Bayer, N. Johnson, A. Tipilda, J. Njuki, G. D. Baltenweck, D. MacMillan. 2014. "Livestock and Women's Livelihoods: A Review of the Recent Evidence." In *Gender in Agriculture and Food Security: Closing the Knowledge Gap*, edited by A. Quisumbing, R. Meinzen-Dick, T. Raney, A. Croppenstedt, J. A. Behrman, 209–233. Dordrecht: Springer Netherlands.

Kristjanson, P., E. Bryan, Q. Bernier, J. Twyman, R. Meinzen-Dick, C. Kieran et al. 2017. "Addressing Gender in Agricultural Research for Development in the Face of a Changing Climate: Where Are We and Where Should We Be Going?" *International Journal of Agricultural Sustainability* 15 (5): 482–500.

Kumar, N., S. Scott, P. Menon, S. Kannan, K. Cunningham, P. Tyagi, A. Quisumbing et al. 2018. "Pathways from Women's Group-Based Programs to Nutrition Change in South Asia: A Conceptual Framework and Literature Review." *Global Food Security* 17: 172–185.

Lane, R., and R. McNaught. 2009. "Building Gendered Approaches to Adaptation in the Pacific." *Gender & Development* 17 (1): 67–80.

Larson, A., T. Dokken, A. Duchelle, S. Atmadja, I. Resosudarmo, P. Cronkleton, M. Cromberg et al. 2015. "The Role of Women in Early REDD+ Implementation: Lessons for Future Engagement." *International Forestry Review* 17 (1): 43–65.

Le, T.T., R. Vidallo, E. Simelton, J. Gonsalves. 2018. "9 Steps to Scale Climate-Smart Agriculture: Lessons and Experiences from the Climate-Smart Villages in My Loi, Viet Nam and Guinayangan, Philippines." CCAFS, Hanoi.

Lecoutere, E. 2017. "The Impact of Agricultural Co-operatives on Women's Empowerment: Evidence from Uganda." *Journal of Co-operative Organization and Management* 5 (1): 14–27.

MacGregor, S. 2010. "Gender and Climate Change: From Impacts to Discourses." *Journal of the Indian Ocean Region* 6 (2): 223–238.

Masson-Delmotte, P., V. Zhai, H.-O. Po̊rtner, D. Roberts, J. Skea, P. R. Shukla, A. Pirani et al. 2018. "Global Warming of 1.5°C." An IPCC Special Report on the Impacts of Global Warming of 1.5°C above Pre-Industrial Levels and Related Global Greenhouse Gas Emission Pathways, in the Context of Strengthening the Global Response to the Threat of Climate Change. Cambridge and New York: IPCC.

McKune, S. L., E. C. Borresen, A. G. Young, T. D. Auria Ruley, S. L. Russo, A. Diao Camara, M. Coleman, E. P. Ryan. 2015. "Climate Change through a Gendered Lens: Examining Livestock Holder Food Security." *Global Food Security* 6: 1–8.

Mello, D., and M. Schmink. 2017. "Amazon Entrepreneurs: Women's Economic Empowerment and the Potential for More Sustainable Land Use Practices." *Women's Studies International Forum* 65: 28–36.

Michael, K., M. Shrivastava, A. Hakhu, K. Bajaj. 2019. "A Two-Step Approach to Integrating Gender Justice into Mitigation Policy: Examples from India." *Climate Policy* 20 (7): 800–814.

Mittal, S. 2016. "Role of Mobile Phone-Enabled Climate Information Services in Gender-Inclusive Agriculture." *Gender Technology and Development* 20 (2): 200–217.

Mungai, C., M. Opondo, G. Outa, V. Nelson, M. Nyasimi, P. Kimeli. 2017. "Uptake of Climate-Smart Agriculture Through a Gendered Intersectionality Lens: Experiences from Western Kenya." In *Climate Change Adaptation in Africa: Fostering Resilience and Capacity to Adapt*, edited by W. Leal Filho, S. Belay, J. Kalangu, P. Munishi, K. Musiyiwa, 587–601. Cham: Springer International Publishing.

Murray, U., Z. Gebremedhin, G. Brychkova, C. Spillane. 2016. "Smallholder Farmers and Climate Smart Agriculture: Technology and Labor-Productivity Constraints amongst Women Smallholders in Malawi." *Gender, Technology and Development* 20 (2): 117–148.

Mutenje, M. J., C. R. Farnworth, C. Stirling, C. Thierfelder, W. Mupangwa, I. Nyagumbo. 2019. "A Cost-Benefit Analysis of Climate-Smart Agriculture Options in Southern Africa: Balancing Gender and Technology." *Ecological Economics* 163: 126–137.

Nelson, S., and S. Huyer. 2016. *A Gender-Responsive Approach to Climate-Smart Agriculture: Evidence and Guidance for Practitioners*. Rome: FAO.

Neumayer, E., and T. Plumper. 2007. "The Gendered Nature of Natural Disasters: The Impact of Catastrophic Events on the Gender Gap in Life Expectancy, 1981–2002." *Annals of the Association of American Geographers* 97 (3): 561–566.

Nguyen, P. H., G. Strizich, A. Lowe, H. Nguyen, H. Pham, T. V. Truong, U. Ramakrishnan. 2013. "Food Consumption Patterns and Associated Factors among Viet Namese Women of Reproductive Age." *Nutrition Journal* 12 (1): 126.

Oduol, P., G. Nyadzi, R. Swai, M. Schueller, B. Gama, S. Matata, W. Mwageni et al. 2006. "Adoption and Impact of Agroforestry Technologies on Rural Livelihoods in Southern Africa." Second National Agroforestry and Environment Workshop, Mbeya, March 14.

Ojango, J., J. Audho, E. Oyieng, J. Recha, A. Okeyo, J. Kinyangi, A. Muigai. 2016. "System Characteristics and Management Practices for Small Ruminant Production in 'Climate Smart Villages' of Kenya." *Animal Genetic Resources* 58: 101–110.

Okali, C., and L. O. Naess. 2013. "Making Sense of Gender, Climate Change and Agriculture in Sub-Saharan Africa: Creating Gender-Responsive Climate Adaptation Policy." Working Paper 057, FAC, Brighton.

Okpara, U., L. Stringer, M. Akhtar-Schuster. 2019. "Gender and Land Degradation Neutrality: A Cross-Country Analysis to Support More Equitable Practices." *Land Degradation and Development* 30 (100): 1368–1378.

Otieno, G., H. Lacasse, J. Adokorach, J. W. Mulumba, J. W. Recha, T. W. Reynolds, C. Fadda. 2018. "Social Seed Networks for Climate Change Adaptation in Uganda: Strategies to Improve Access to Genetic Diversity and Information." CCAFS Info Note, CCAFS, Wageningen.

Paris, T. R., and T. Chi. 2005. "The Impact of Row Seeder Technology on Women Labor: A Case Study in the Mekong Delta, Viet Nam." *Gender, Technology and Development* 9 (2): 157–184.

Perez, C., E. Jones, P. Kristjanson, L. Cramer, P. Thornton, W. Foerch, C. Barahona. 2015. "How Resilient Are Farming Households and Communities to a Changing Climate in Africa? A Gender-Based Perspective." *Global Environmental Change* 34: 95–107.

Radeny, M., M. Ogada, J. Recha, P. Kimeli, E. Rao, D. Solomon. 2018. "Uptake and Impact of Climate-Smart Agriculture Technologies and Innovations in East Africa." Working Paper 251, CCAFS, Wageningen.

Rao, N., E. T. Lawson, W. Raditloaneng, D. Solomon, M. Angula. 2017. "Gendered Vulnerabilities to Climate Change: Insights from the Semi-Arid Regions of Africa and Asia." *Climate and Development* 11 (1): 14–26.

Rao, A., and D. Kelleher. 2005. "Is There Life after Gender Mainstreaming?" *Gender & Development* 13 (2): 57–69.

Reddy, K., S. Sravanth, N. Sundaram. 2019. "Agricultural Crisis and Farmers Suicides in India." *International Journal of Innovative Technology and Exploring Engineering* 8 (11): 1567–1580.

Rengalakshmi, R., M. Manjula, M. Devaraj. 2018. "Making Climate Information Gender Sensitive: Lessons from Tamil Nadu." *Economic and Political Weekly* 53 (17): 87–95.

Resurreccion, B. P. 2011. "The Gender and Climate Debate: More of the Same or New Pathways of Thinking and Doing?" Asia Security Initiative Policy Series Working Paper 10, NTS, Singapore.

Richards, M., T. Bruun, B. Campbell, L. Gregersen, S. Huyer, V. Kuntze, S. Madsen et al. 2015. "How Countries Plan to Address Agricultural Adaptation and Mitigation: An Analysis of Intended Nationally Determined Contributions." Info Note, CCAFS, Copenhagen.

Roncoli, C., C. Jost, P. Kirshen, M. Sanon, K. Ingram, M. Woodin, G. Hoogenboom. 2009. "From Accessing to Assessing Forecasts: An End-to-End Study Of Participatory Climate Forecast Dissemination in Burkina Faso." *Climatic Change* 92 (3–4): 433–460.

Rossi, A., and Y. Lambrou. 2008. *Gender and Equity Issues in Liquid Biofuels Production*. Rome: FAO.

Sanogo, D., D. Dayamba, M. Ouedraogo, R. Zougmore, J. Bayala, O. Ndiaye, M. Sall et al. 2016. *The Climate-Smart Village Approach: What Research and Insights from Current Implementation in Daga-irame CSV in Senegal?* Bamako: CCAFS and ICRISAT.

Sekabira, H., and M. Qaim. 2017. "Can Mobile Phones Improve Gender Equality and Nutrition? Panel Data Evidence from Farm Households in Uganda." *Food Policy* 73: 95–103.

Setyowati, A. 2011. "Ensuring that Women Benefit from REDD+." *Unasylva* (English ed.) 63 (No. 239): 57–63.

Simelton, E., M. Coulier, A. Carter, M. Duong, T. Le, T. Luu, E. Madsen. 2018a. "Actionability of Climate Services in Southeast Asia: Findings from ACIS Baseline Surveys in Viet Nam, Lao PDR and Cambodia." CCAFS, Wageningen.

Simelton, E, T. Le, M. Coulier, M. Duong, D. Le. 2018b. "Developing Participatory Agro-Climate Advisories for Integrated and Agroforestry Systems." In *Towards Low-Emissions Landscapes in Viet Nam*, edited by R. Mulia and E. Simelton, 129–144. Bogor: ICRAF.

Simelton, E., T. T. Le, T. M. Duong, H. D. Le, L. Gammelgaard. 2019. "From Sitting in the Back Rows to Sitting in the Front Row: How Climate Services Changed Gender Relations." CCAFS Gender Networking Meeting: Learning and Action for Gender-Transformative CSA, April 1–2, Wageningen.

Smith, J., and S. J. Scherr. 2002. "Forest Carbon and Local Livelihoods: Assessment of Opportunities and Policy Recommendations." Occasional Paper 37, CIFOR, Bogor.

Sterling, R., and S. Huyer. 2010. "89.1 FM: The Place for Development: Power Shifts and Participatory Spaces in ICTD." *The Journal of Community Informatics* 5 (3–4).

Tadesse, Y., C. J. M. Almekinders, R. P. O. Schulte, P. C. Struik. 2017. "Tracing the Seed: Seed Diffusion of Improved Potato Varieties through Farmers' Networks in Chencha, Ethiopia." *Experimental Agriculture* 53 (4): 481–496.

Tall, A., P. Kristjanson, M. Chaudhury, S. McKune, R. Zougmore. 2014. "Who Gets the Information? Gender, Power and Equity Considerations in the Design of Climate Services for Farmers." Working Paper 89, CCAFS, Wageningen.

Tavenner, K., and T. A. Crane. 2018. "Gender Power in Kenyan Dairy: Cows, Commodities, and Commercialization." *Agriculture and Human Values* 35: 701–715.

Tavenner, K., M. van Wijk, S. Fraval, J. Hammond, I. Baltenweck, N. Teufel et al. 2019. "Intensifying Inequality? Gendered Trends in Commercializing and Diversifying Smallholder Farming Systems in East Africa." *Frontiers in Sustainable Food Systems* 3–10.

Thuy, P. T., and M. Brockhaus. 2015. "Gender Mainstreaming in REDD+ and PES: Lessons Learned from Viet Nam." Brief for GSDR – 2016 Update, CIFOR, Bogor.

Twyman, J., M. Green, Q. Bernier, P. Kristjanson, S. Russo, A. Tall, E. Ampaire, et al. 2014. "Gender and Climate Change Perceptions, Adaptation Strategies, and Information Needs Preliminary Results from Four Sites in Africa." Working Paper 83, CCAFS, Copenhagen.

Twyman, J., J. Muriel, M. García. 2015. "Identifying Women Farmers: Informal Gender Norms as Institutional Barriers to Recognizing Women's Contributions to Agriculture." *Journal of Gender, Agriculture and Food Security* 1 (2): 1–17.

Ulrichs, M., T. Cannon, A. Newsham, and L. O. Naess. 2015. "Climate Change & Food Security Vulnerability Assessment." Working Paper 108, CCAFS, Copenhagen.

UNDP (United Nations Development Programme). 2016. *Gender Equality in National Climate Action: Planning for Gender-Responsive National Determined Contributions (NDCs).* New York: UNDP.

Villamor, G. B., M. van Noordwijk, U. Djanibekov, M. Chiong-Javier, D. Catacutan. 2014. "Gender Differences in Land-Use Decisions: Shaping Multifunctional Landscapes?" *Current Opinion in Environmental Sustainability* 6: 128–133.

Villanueva, A. B., Y. Jha, R. Ogwal-Omara, E. Welch, S. Wedajoo, M. Halewood. 2016. "Influence of Social Networks on the Adoption of Climate Smart Technologies in East Africa: Findings from Two Surveys and Participatory Exercises with Farmers and Local Experts." CCAFS Info Note, CCAFS, Copenhagen.

Violon, C., M. Thomas, E. Garine. 2016. "Good Year, Bad Year: Changing Strategies, Changing Networks? A Two-Year Study on Seed Acquisition in Northern Cameroon." *Ecology and Society* 21 (2): 34.

Westholm, L., and S. Arora-Jonsson. 2015. "Defining Solutions, Finding Problems: Deforestation, Gender, and REDD+ in Burkina Faso." *Conservation and Society* 13 (2): 189–199.

WMO (World Meteorological Organization). 2018. *Step-by-Step Guidelines for Establishing a National Framework for Climate Services.* Geneva: WMO.

World Bank, FAO (Food and Agricultural Organization), IFAD (International Fund for Agricultural Development). 2015. "Gender in Climate-Smart Agriculture." In *Gender in Agriculture Sourcebook*, edited by World Bank. Washington, DC: World Bank.

Wornell, E. J., A. R. Tickamyer, S. Kusujiarti. 2015. "Gender Mainstreaming Principles in Indonesia›s REDD+ Program: A Document Analysis." *Journal of Sustainable Development* (8): 159.

Ylipaa, J., S. Gabrielsson, A. Jerneck. 2019. "Climate Change Adaptation and Gender Inequality: Insights from Rural Viet Nam." *Sustainability* 11 (10): 2805.

7HP

Chapter 8

FROM THE "FEMINIZATION OF AGRICULTURE" TO GENDER EQUALITY

Cheryl Doss, Ayesha Qaisrani, Katrina Kosec, Vanya Slavchevska, Alessandra Galiè, and Nozomi Kawarazuka

The term "feminization of agriculture" is used to capture a wide range of gender dynamics and shifts in rural gender relations. Definitions range from the broadening and deepening of women's involvement in agriculture (Lastarria-Cornhiel 2006) to the increase in rural women's *measured* economic participation in agriculture and a *reported* visibility of women in agricultural activities (Deere 2005). The term may be used to imply an increase in women's labor in agriculture, in women's labor relative to that of men, or in women's roles in agricultural decision-making. Broadly speaking, however, literature on the feminization of agriculture has challenged researchers and development practitioners to consider how changing rural landscapes are affecting women, especially where men are moving out of agriculture. These changes have impacts on women's productive and reproductive workloads, both paid and unpaid, as well as their agency and decision-making, both within the farm and the household and in the community and a variety of institutions.

Literature on the feminization of agriculture has two distinct narratives. The first takes a negative view, suggesting that women's workload in agriculture is increasing at the same time as women continue to lack the resources needed to succeed (Song 1998, Padmaja et al. 2019, Tavenner et al. 2019, De Brauw et al. 2021). This narrative may refer to the women as being "left behind," implying a lack of agency and a worsening of their livelihoods. A second narrative argues that processes of rural transformation are providing opportunities for women's economic empowerment and gender equality (Khatri-Chhetri and Chanana 2017). As women move into paid employment or become more visible in smallholder commercial agriculture, they not only earn more income but also may gain greater visibility and voice within their household and community.

Uniting the narratives around the feminization of agriculture is evidence that the agriculture sector, and the rural sector more generally, are changing, and that these changes affect women and men differently. A number of different processes may be unfolding simultaneously—within as well as outside of agriculture—even within the same country. Some regions are experiencing growth in agriculture. Smallholder agriculture, including crops, livestock, forestry, and fisheries, may be commercializing, with changes in the products being produced and marketed and in the relative dependence on capital and labor. Newly created jobs provide new opportunities; these jobs may or may not be gender segregated, and they may differ from prior job opportunities in terms of both the location of the work and the skills required. New opportunities may also open up at various nodes along agricultural value chains for women and men (see Chapter 4 on value chains, this volume). Opportunities may also open up outside of the agriculture sector, changing patterns regarding who farms and how, as well as the returns to farming relative to other sectors. All of these changes are likely to influence gender equality.

As Chapter 1 introduced gender equality:

> "... refers to the equal rights, responsibilities and opportunities of women and men and girls and boys. Equality does not mean that women and men will become the same but that women's and men's rights, responsibilities and opportunities will not depend on whether they are born male or female. Gender equality implies that the interests, needs, and priorities of both women and men are taken into consideration, recognizing the diversity of different groups of women and men" (UN Women 2001).

Gender equality is multidimensional, and agrarian change will affect the various dimensions differently.

This chapter assesses research and literature on the feminization of agriculture—especially that of CGIAR centers—with respect to insights into the process of rural transformation and its effects on gender equality. We identify both the key insights provided as well as what is missing. To do so, the first section below considers how various drivers of agrarian change affect patterns of women's work and responsibilities. The second section then analyzes how these changes affect gender equality, primarily in terms of women's work and agency, and what mechanisms and conditions can ensure that agrarian change ultimately promotes gender equality. Finally, we identify research and data gaps, proposing a research agenda that can advance gender equality.

We draw primarily on research carried out through CGIAR. We carried out a comprehensive search for literature on gender and rural transformation, changes in employment, or the "feminization of agriculture" written by any of the CGIAR centers and CGIAR Research Programs—placing no restrictions on dates, and capturing both peer-reviewed work as well as gray literature. In addition, we drew on a broader set of literature review undertaken by a program supported by the CGIAR Collaborative Platform for Gender Research: "The 'Feminization' of Agriculture: Building Evidence to Debunk Myths on Current Challenges and Opportunities."

Drivers of change in the agriculture sector

The agriculture sector remains the backbone of the economy for many developing countries and employs much of the labor force, especially in rural Asia, Africa, and Latin America (Sachs 2019). However, the sector is undergoing dramatic transformations, shifting gender dynamics across geographical and cultural contexts. This section considers a number of drivers of change in gender relations in agriculture. These include the commercialization of both large-scale and smallholder agriculture, climate change, technologies and innovation, war and conflict, and the migration of women and/or men. For each driver, we discuss its effect on labor patterns, decision-making, and/or management roles. We note that it is not possible to describe all of the changes as the feminization of agriculture; in some areas women are migrating out of agriculture and rural areas.

Large-scale commercialization and wage work

Worldwide, neoliberal policies beginning in the 1980s focused on market-based approaches, reducing agricultural subsidies, and public investment in agriculture. These policies had a variety of impacts on agricultural producers depending, in part, on their scale of farming. Large-scale commercial farming, aimed at export markets, created a demand for wage labor, providing employment opportunities to many, including women (Dolan and Sorby 2003, Deere 2005, Bigler et al. 2017, Sachs 2019). It also led to visible change in the gendered division of labor globally (Deere 2005, Sugden et al. 2014, Abdelali-Martini and Dey de Pryck 2015, Najjar et al. 2018, Padmaja et al. 2019). Opportunities for employment in the sector offered an avenue for women to obtain remuneration for agricultural work and created a wave of women's participation in commercial farms (Lastarria-Cornhiel 2006). In Latin America, for example, women's participation as wage laborers in new emerging non-traditional export production and packing, such as flowers, fruits, and vegetables, increased over this period (Deere 2005).

Although commercial agriculture may open up income opportunities for women, tasks and working conditions often differ from those of men. For example, women often work in seasonal packing and distribution on commercial farms in Latin America whereas men typically hold permanent positions (Deere 2005). In Peru, women are employed cutting asparagus whereas men are generally responsible for field collection, packaging, pest control, and irrigation management (Bernardinie 2019). The details vary across countries, with some value chains offering year-long employment opportunities for both women and men (Dolan and Sorby 2003). Furthermore, wages typically differ across jobs, with men's jobs paying substantially more (Bernardinie 2019), even where women and men work together and perform the same tasks (Deere 2005, Sunderland et al. 2014, Najjar et al. 2018).

Changes in the structure of commercial agriculture may influence gender relations more broadly. For example, in West Kalimantan, Indonesia, married couples working on government-managed plantations were given permanent contracts that included medical care, childcare, and accommodation. However, as private firms took over the industry, the terms and conditions of employment became more informal. Women and men were offered different jobs and wages; childcare and accommodation benefits were withdrawn. Women became more likely to accept lower wages and poorer working conditions because of their lower bargaining power (Li 2015).

Overall, the job growth that the expansion of commercial agriculture has fueled is having gendered impacts, in terms of segregation in tasks as well as gendered differences in working conditions and wages—with important regional heterogeneity as well. Much of the evidence comes from relatively old and small-scale case studies, however.

Commercialization of smallholder agriculture

Increasing commercialization of smallholder agriculture has impacts on rural households, which are often very different for women and men. Smallholder households often combine multiple livelihood strategies, mixing production for home consumption with some degree of market-oriented farming in addition to participating in off-farm agricultural wage labor (Bigler et al. 2017). These mixed livelihood approaches, involving agricultural intensification and diversification, are heavily influenced by gender relations.

Men often handle the more commercial activities of smallholder farming or livestock-keeping whereas women are often largely responsible for subsistence agricultural activities (Ingabire et al. 2018, Tavenner et al. 2019) and activities related to household production. In some parts of sub-Saharan

Africa, women and men farm separate plots, with some crops seen as "men's crops" and others as "women's crops" (De Brauw et al. 2015, Ingabire et al. 2018;). Yet patterns change over time, particularly in response to market opportunities, and they are not absolute (Doss 2001). In South Asia and much of Latin America, it is mainly men who lead agricultural production activities, with women viewed as "helpers" rather than farmers themselves (Twyman et al. 2015). In Southeast Asia, especially in rice-based production systems, women and men often work together on the same tasks in the same fields (Akter et al. 2017). But in all of these situations, men are more likely to be seen as the farmer who is commercializing.

As smallholders move into more market-oriented agriculture, they may change the mix of crops grown, the provision of labor, and who controls the crops within the household. Crops traditionally grown by women may witness increased involvement of men when they become commercially viable. Depending on who dominates marketing and decision-making regarding production and access to and control over agricultural income may change for women.

In an example from Uganda, beans were traditionally considered a women's crop but when government policy prioritized their production as an export crop, men's participation in some production activities increased; they came to dominate site selection, fertilizer and insecticide application, and harvesting and women's activities came to include mainly winnowing, postharvest handling, sorting, planting, and weeding (Nakazi et al. 2017). The study does not offer insight into who controls the income from bean production but illustrates that the commercialization of crops may alter the participation rates of men and women in production activities. Similarly, in Zambia, men moved into groundnut production as shelling became mechanized and more profitable. In this situation, women welcomed men's involvement and were willing to give up some control over production as they were also released from drudgery (Orr et al. 2016).

Commercialization can also change the distribution of labor in other ways. For example, commercialization of livestock can result in new, gendered divisions of labor in which women assume physically demanding roles—specifically, tending livestock—while men take up less physically demanding work related to marketing. This can result in women feeling they have more work without receiving additional benefits (Njuki et al. 2016, Basu et al. 2019).

While examples abound of men moving into women's crops as market opportunities expand, or otherwise shifting less desirable work to women as more desirable work (for example marketing) arises, there is less research on examples of gender equality improving as a result of increased market opportunities.

Climate change

Climate change is a key driver of structural change within the agriculture sector. Not only does it directly affect cropping patterns and natural resource-based livelihood options (Sugden et al. 2014) but also it alters the existing division of labor via climate adaptation responses including migration (Djoudi and Brockhaus 2011, Mueller et al. 2014a, Kosec et al. 2018a, Khatri-Chhetri et al. 2019). A synthesis of relevant research by CCAFS, the CGIAR program on climate change, finds that women and men are exposed to different climate shocks and experience the impacts differently: few farmers adopt practices that will increase their resilience to climate change and women are even less likely to do so (Kristjanson et al. 2017).

Declining agricultural productivity as a result of climate change impacts may push smallholder farmers to seek alternative economic activities outside of their village. This may affect the gendered division of labor in agriculture (Sugden et al. 2014). In Morocco, rural–rural migration has increased for both women and men in response to droughts and climate change. This emerging pattern is breaking many stereotypes regarding the extent of women's involvement in agriculture in the region and the range of climate adaptation techniques they may employ (Najjar et al. 2017).

Climate change has visibly altered livelihoods, with gendered impacts in the north-western province of Mali (Djoudi and Brockhaus 2011). People have shifted from water- to forest-based livelihoods as a result of frequent droughts. Over the years, many men have abandoned livestock-herding and charcoal production and have migrated out of their villages. Women have moved out of crop agriculture and into forestry and livestock management. In smallholder farms in East Africa, climate adaptation through crop diversification increases women's control over income and foodstuffs, while adaptation measures that involve commercialization of crops have decreased women's control (Tavenner et al. 2019).

Thus, the literature confirms that the gender division of labor in agriculture is changing in response to climate change and subsequent household adaptation strategies. In some contexts, this leads to increases in women's work and responsibilities in agriculture. There is, however, insufficient evidence on whether and when these new roles and responsibilities contribute to reducing inequalities, particularly in economic opportunities. Policies and development interventions addressing climate change (for example climate-smart agricultural practices) have the potential to advance or disrupt gender equality. For more on gender equality approaches to climate change adaptation, see Chapter 7, this volume.

Technologies and innovation

Technologies introduced in the agriculture and rural sectors may also alter the labor patterns of women and men (Farnworth et al. 2019). The technologies may reduce labor inputs, such as agricultural machinery, or increase productivity, such as improved seed varieties. Mobile telephones and the internet can reduce price dispersion and misallocation across markets (Jensen 2007), increase farmers' access to weather data and thus influence planting decisions, serve as a platform on which to receive extension services (van Campenhout et al. 2018), permit access to services like picture-based weather insurance (Ceballos et al. 2019), and even increase communication between spouses—including when one of them has migrated. Information technology also affects individuals' influence over policymakers (Kosec and Wantchekon 2020).

Because agricultural labor, control over outputs, and access to technology itself are deeply gendered in most rural settings, the introduction of new technologies is likely to affect women and men differently. Agricultural technologies and gender relations are interrelated, and adoption of technology shapes gender relations. The introduction of new labor-saving technologies, such as mechanization, may have both positive and negative gendered impacts, depending on the extent to which men's or women's tasks are mechanized (Johnson et al. 2016, Khatri-Chhetri and Chanana 2017).

Abundant evidence suggests that women often have less access to and information about new technologies that would increase productivity, and fewer resources needed to purchase and use such technologies (Meinzen-Dick et al. 2010, Ragasa et al. 2012, Peterman et al. 2014, Kondylis et al. 2016). Women's ability to adopt technologies is constrained by the fact they are also much less likely than men to own land (Doss et al. 2015, Kieran et al. 2015) or to have secure tenure, which is often required for financing or long-term investment. As a result, women tend to focus on low-investment and low-profit innovation developed through their own social networks instead of via extension workers (Kawarazuka and Prain 2019). Poor targeting of extension services often neglects women farmers (Farnworth et al. 2018, Balasubramanya 2019). Thus, when men migrate, while remittances may allow for the purchase of some new technologies, the origin household may have other challenges in accessing them.

Technologies to increase productivity may have unanticipated impacts on the allocation of men's and women's labor. For example, the introduction of an improved breed of goats into a pastoral community in Tanzania changed the household distribution of livestock labor: traditionally, men and boys had taken the household herd grazing on the savannah, but, because the improved goats were not adapted to the long journeys, they remained with the women. The women welcomed the increased work burden because they also gained more milk for home consumption (Galiè and Kantor 2016). Technology may also substitute for men's labor. For instance, solar irrigation pumps installed in a village in Nepal with high rates of male outmigration have meant that de facto women household heads can irrigate their farms without relying on men (Khatri-Chhetri and Chanana 2017). Technologies that allow women to complete certain tasks on their own may reduce their dependence on men and increase their decision-making power in such tasks.

The impact technology has on gender dynamics also depends on what activities it aims to address and whose work it affects. New technologies adopted only by men may increase gender inequality. New technologies may also increase women's overall workload. Merely providing technology is not the answer; it is necessary to identify women's needs for labor-saving technology and provide appropriate training, and women must be able to benefit in terms of declining drudgery or increased income and sense of achievement. Further evidence is needed on how to use labor-saving technologies effectively to support women's work in the context of the feminization of agriculture.

Conflict

Conflict also often changes labor patterns. CGIAR research in conflict-affected areas such as Syria and the larger Middle East and North Africa region does not specifically talk about conflict as the main driver of structural shifts in agricultural labor. However, evidence shows that, with the increasing migration of men from such areas, women are providing more labor in smallholder agriculture (Abdelali-Martini and Dey de Pryck 2015, Galiè et al. 2017). In addition, research within Syrian refugee camps in Lebanon shows that women often undertake agricultural wage labor as a means of survival whereas men are more likely to take jobs in construction (Al Zoubi et al. 2019).

While research on the relationship between conflict and crises and the feminization of agriculture is limited, the patterns of agricultural labor certainly change during such critical moments. Conflict rarely results in welfare gains for those it affects. It disrupts economic systems, including the agriculture sector, and changes labor patterns. More research is needed to understand how conflict and crises—which disrupt supply and demand and may create security concerns that limit women's mobility—more broadly shift labor opportunities for men and for women, on and off the farm. Research is also required on how women's changing work and responsibilities in agriculture can facilitate gender equality in conflict and post-conflict settings.

Migration

One important way that many of these drivers change labor patterns in agriculture is by influencing migration. The term "feminization of agriculture" is even occasionally defined as a situation in which women take on more agricultural responsibilities because men have migrated (Khatri-Chhetri et al. 2019). While seasonal and circular migration have historically been common livelihood strategies for rural households, changes in the economy, economic crises, decreases in landholding size, and declining profitability of smallholder agriculture are pushing people, especially men and youth, to seek economic opportunities outside of rural villages (Deere 2005, Basnett 2013, Mueller et al. 2014a, Kosec et al. 2018a, Mueller et al. 2018). Large-scale commercialization, land reforms, and privatization of land have created a landless or land-poor labor force in some areas, which has contributed to migration (Mukhamedova and Wegerich 2014, Najjar et al. 2018), especially in settings where land rental markets are thin or otherwise do not function well (Kosec et al. 2018a). Climate change may make smallholder agriculture riskier and less profitable. Migration may also be a response to new economic opportunities in both rural and urban areas, as well as abroad, pulling people, especially youth, away from farms (De Brauw et al. 2014, Mueller et al. 2018).

The reasons for migration will influence the impacts in the rural communities. In Kyrgyzstan, men's outmigration is often a result of negative income shocks and economic stress facing agriculture-dependent households (Kosec et al. 2020). This predominately male outmigration is accompanied by reduced consumption and deteriorations in young child health (Kosec and Song 2019). A failure to control for economic conditions when considering the impacts of changes in rural labor patterns may skew the interpretation of their impacts. It may be more difficult for women to increase their agency and decision-making when prevailing economic conditions are bad.

If we consider only male outmigration, we may miss some key dimensions of the changes in rural areas. In some parts of Southeast Asia, women migrate and men are "left behind" in rural areas to farm (Elmhirst 2007, Hoang and Yeoh 2011). In some parts of Latin America too, women have a long history of migration, initially to cities within their own country and, since the 1990s, more frequently to developed countries for care work (Deere 2005, Deere et al. 2015). Women's migration has implications for the availability of labor in the sending community, but the empirical literature does not cover the impacts of female outmigration from agricultural areas in any depth.

While marriage is often cited as a reason for women's migration (Mueller et al. 2015), it is often categorized as a social rather than an economic reason (Rao 2017, De Brauw et al. 2018). Yet migration for marriage will influence whether or not women work, and in which sector (Mazumdar and Agnihotri 2014). Once married, women may also migrate with their husbands, again affecting their patterns of work.

One trend that has surged is the seasonal rural–rural migration of women to harvest crops. In Latin America, women may migrate seasonally alone or with their families (Deere 2005). In Morocco, women may migrate to nearby villages to work as hired laborers, since social norms discourage them from doing paid agricultural work in their own villages (Najjar et al. 2018). Similarly, women from rural Tajikistan migrate seasonally to other villages within Tajikistan or to nearby villages of Kyrgyzstan for casual agricultural work (Mukhamedova and Wegerich 2014). These cases show that migration trends are altering the pool of agricultural laborers as well as the set of economic opportunities for both women and men. Whether this is good or bad for gender equality depends on the work conditions and wages for both women and men as well as on who controls income.

Lessons from the feminization of agriculture research for gender equality

This section identifies key insights from the research on the feminization of agriculture regarding gender equality in the context of agrarian change. As discussed above, much of the literature focuses on the drivers of changes in women's work and agency without explicitly framing the discussion in terms of gender equality. We look at gender equality in terms of work and women's voice and agency. We consider this in four domains: commercial large-scale agriculture, smallholder farming, governance and natural resource management, and the role of development interventions in agrarian change.

The different aspects of gender equality are often not perfectly correlated: improvements on one indicator may correspond with a worsening on others. A nuanced view of gender equality that takes men into account and considers multidimensional impacts is thus critical. Importantly, we consider what other factors and gender dimensions we can draw from the existing studies that affect how and the extent to which changes in women's roles and responsibilities improve gender equality.

Equal employment opportunities in commercial agriculture

The review in the previous section confirms that the commercialization of agriculture can open up new employment opportunities for women yet it does not necessarily lead to gender equality related to wages, employment opportunities, and time use. Women often participate in these new activities and earn an income, but they tend to earn less than men and work under less secure contracts and in poorer working conditions. Thus, while there are new opportunities for women, they may not necessarily lead to gender equality. Broader structural changes that provide opportunities for women to earn the same wages as men and progress into both permanent and management jobs will be necessary for the sector to be a positive force toward gender equality.

In addition, patriarchal gender norms are one of the key underlying constraints to women's ability to take advantage of these new positions. Changes in men's and women's perceptions about the responsibilities around reproductive work appear to be a precondition for women to join commercial agricultural enterprises without a significant additional labor burden. Women are more likely to seize new opportunities when they are accompanied by a loosening of gender-biased norms that limit women's mobility and livelihood options, and when they do not merely increase women's overall work burden but rather create a more balanced division of labor between women and men. As women become visible in leadership positions, more households may become comfortable with the idea of their daughters, sisters, and wives earning money by working in the agriculture sector.

In Syria, some men assumed domestic responsibilities or purchased labor-saving home appliances when women were getting paid jobs in agriculture. Women's increased involvement in agricultural wage work brought about shifts in gender relations and women's empowerment. Labor contract work allowed women to come together as a community. Coming from a culture of seclusion, the work brought them into contact with women from other families. This offered them a space to share common problems, joys, and aspirations, opening their narrow worlds to new information, experiences, perspectives, networks, and friendships, and led to increased control over household expenditure and investment (Abdelali-Martini and Dey de Pryck 2015).

Other social norms also may limit women's ability to take advantage of new wage opportunities. For example, in South Asia, women's communication with outside men is often discouraged, limiting their opportunities to work in agriculture (Sugden et al. 2014). Norms that allow men to disrespect women may also limit women's choices. Women agricultural workers in Morocco often self-select to work in women-only groups to avoid harassment, even though this leads to lower wages (Najjar et al. 2018). Thus, deeply entrenched norms are powerful barriers to making new employment opportunities women-friendly and to bridging gender gaps in economic opportunities.

To summarize, evidence confirms that women's increased employment in agriculture does not necessarily yield gender equality in terms of wages and working conditions. Second, women's paid employment or income-generating activities tend to translate to a higher workload, particularly when the gender division of reproductive labor does not shift. Third, women's increased income-earning does not necessarily or directly translate into autonomy over that income and an increased say in household decision-making. Existing gender inequalities as well as gender-biased social norms affect the nature and impact of the employment and income-earning opportunities that women are seizing. Shifts in norms, in women's voice, in the gender division of labor, and in decision-making can and do occur, but not automatically. Improvements in women's opportunities in the commercial agriculture sector are welcome but are only one aspect in a broader complex of gender relations and dynamics. To move toward gender equality, it will be necessary for the sector to make good jobs available for women and for the social norms to change to allow women to take up these positions.

Changing opportunities within smallholder agriculture

In smallholder agriculture, several drivers of change with potentially diverse or contrasting effects are at play and have differing impacts on gender equality. Increased commercialization here opens up new opportunities for women but, without access to markets and control over the income earned, the changes may worsen gender equality. It is well documented in numerous contexts that women have less access to land and other resources needed to increase productivity (Deere and León 2003, Garikipati 2009, Ragasa et al. 2012, Peterman et al. 2014, Doss et al. 2015, Kieran et al. 2015). Increased market opportunities in the smallholder sector may also increase women's overall work burden. Thus, women's limited access to input and output markets, weak control over agricultural income, and double burden of reproductive and on-farm work (Ingabire et al. 2018) stand as barriers to gender equality. Often, these factors interact. As groundnut-shelling in Zambia was mechanized and men got more involved in production, women saw an increase in their decision-making around management and use of the income from the crop but men owned the land and retained decision-making power over what crops were grown (Orr et al. 2014).

Migration, particularly the outmigration of men, has had significant impacts with regard to patterns of work and responsibility within smallholder agriculture. The impacts on gender equality may be positive or negative. Men's off-farm migration may create space for women to engage in agricultural management (Abdelali-Martini et al. 2003, Mukhamedova and Wegerich 2014, Stanley 2015, Farnworth et al. 2018, Kar et al. 2018, Padmaja et al. 2019). This includes not only primary agricultural production but also marketing of agricultural products, negotiating labor contracts, supervising hired male labor, and interacting with extension and municipality agents. The shift in perceptions of women as "farm helpers" to farm managers has the potential to contribute to gender equality in terms of women's control of income. For some women, men's absence translates into more decision-making power, freedom of movement, and financial freedom (Farnworth et al. 2019).

The balance of own-farm work and employment may also shift. In Viet Nam, women are increasingly farming rice in smallholder households as men and some younger women migrate from villages. Although women now have more autonomy over managerial decisions on the farm, they often substitute this unpaid on-farm work for their salaried work (Bacud et al. 2019). Other studies point to a relocation of women's labor from non-farm to farm activities when a family member migrates (Binzel and Assaad 2011 for Egypt, Mu and van de Walle 2011 for China, Mendola and Carletto 2012 for Albania). Unless men's outmigration is coupled with gender-sensitive changes in terms of access to and control over productive resources and decision-making platforms, we may not see improvements in gender equality.

To better understand what is changing and how this affects gender equality, more research needs to focus not only on how male outmigration and commercialization of smallholder agriculture affect women but also on what kinds of institutional and societal changes are required so that these changes move us toward gender equality. Situations are diverse across regions as well as within communities. Studies need to move beyond documenting current labor situations toward understanding changes in decision-making and control over resources. From a gender equality perspective, it is equally relevant to reflect beyond changes that the commercialization of smallholder agriculture and the outmigration of men have brought on, to also consider other changes in the rural landscape that influence relationships between women and men, which are a result of changing opportunities within smallholder agriculture, such as women moving to cities; rural women's participation in the non-farm sector; and in-migration of other men.

Gender equality in governance and natural resource management

Women's involvement in the governance and management of agricultural enterprises and community resources is a critical dimension in achieving gender equality that addresses the interests, needs, and priorities of women as well as men (Li 2015, Elmhirst et al. 2017, Galiè et al. 2017). Evidence is emerging on how changes in the agriculture sector affect women's voice in their community and in local institutions. CGIAR has ongoing work to look at issues of governance of local institutions in the context of male outmigration, for instance water management institutions in Tajikistan and Nepal.[1] These have the potential to provide key insights on how to move toward gender equality as new governance structures develop. This type of research merits more attention, given that women's voices have the potential to influence the mix of policies and interventions affecting the agriculture and rural sector.

Women's respect and involvement in their community may be linked with their higher-earning opportunities and resulting greater self-confidence. For example, in one village in Nepal with high rates of male outmigration, solar irrigation pumps together with trainings and better support from the district agriculture offices contributed to women farmers attaining higher productivity from their farming and increasing their incomes. As a result, they gained more prestige in the community, which is one dimension of gender equality (Khatri-Chhetri and Chanana 2017). In South Asia, women gain more confidence in making decisions regarding their farms when they come together via informal or formal groups. These women also gain more access to rural institutions such as credit services and extension offices through collective effort (Padmaja et al. 2019).

The changing nature of rural institutions in the face of agrarian change will also influence gender equality. In Sughd province in Tajikistan, the privatization of many collective farms has pushed many women out of agriculture into conventional and domestic roles (Mukhamedova and Wegerich 2014). At the same time, the existing collective farms have become female-dominated as men have migrated out. Women have been emerging as leaders who organize other women workers and negotiate on payment and

---

1 See https://wle.cgiar.org/influencing-water-investments-support-women-tajikistan and https://gender.cgiar.org/selected-research-proposals-feminization-of-agriculture/

workload with the contractual (*dehkan*) farmers. Gender relations are shifting; women are actively taking over agriculture as primary farmers and gaining more confidence, leadership skills, and bargaining power as they organize themselves in cooperative labor units.

Yet commercialization and privatization can also negatively affect women's voice and their control over (community) resources. A critical concern here is women's voice in institutions that address issues of land tenure and land management. In Indonesia, changes in the rural sector have led to commercialized oil palm farming replacing women's farms where they previously grew vegetables, fruits, and other crops for subsistence. The land women farmed was given to oil palm companies, whose governance and management decisions rarely included women (and often young men) (Li 2015, Elmhirst et al. 2017). Such policies of exclusion of women from decision-making platforms reinforce historic practices of gender injustice, or introduce new inequalities.

Women are rarely compensated for their losses when they lose access to land as a result of a government policy or commercialization project; this is often because their involvement in agriculture is considered an "interest" rather than a "right." Compensation policies are often skewed either toward men or toward the household as a unit—in either case neglecting women's rights and usage of land, and often ignoring customary laws in the process (Doss and Meinzen-Dick 2019). In Burkina Faso, for example, the government's land expropriation policy has treated households as a single unit and given land to men, even though women are the main rice cultivators (von Koppen 2008). In Malawi, a resettlement program that otherwise improved households' access to land, tenure security, and food security actually jeopardized the land rights of women in male-headed households as matrilineal customs were abandoned in the new resettlement villages (Mueller et al. 2014b). Thus, while the government policy could have actively bridged gender gaps in decision-making and control, it actually worsened the situation.

Interventions to improve gender equality

Development interventions by government or non-governmental organizations (NGOs) may influence rural transformation and the resulting impacts on gender equality. These interventions can introduce new awareness of rights and responsibilities and challenge existing inequalities. They have the potential to increase the likelihood that rural transformation will benefit women. For instance, through trainings and support from extension officers in Nepal, women are increasingly participating in making decisions about wheat farming and are emerging as autonomous wheat growers. Women consult their husbands and discuss decisions with them but, with their increased success and economic incentives, their families are more encouraging of their participation as commercial farmers (Farnworth et al. 2019).

Yet development projects may also reinforce harmful gender and socioeconomic inequalities. For instance, land restoration efforts in Kenya may also reinforce gender-biased norms where women are involved in low-paying and more labor-intensive tasks with limited control over land and productive resources; and minimal tenure rights and cultural factors reinstate men's dominance in decision-making (Ihalainen 2018). In addition, development projects that ignore intersectionality may benefit some women while making others worse off. For example, international NGOs in India often find it easier to target upper-caste women from landholding households, which increases inequality among women of different castes and socioeconomic backgrounds (Farnworth et al. 2018).

For development programs to do good rather than harm to gender relations, insights into gender dynamics must inform gender-responsive interventions. To offset some of the negative effects of agrarian transformation on gender equality, a number of studies call for gender-responsive mechanisms. For example, creative solutions that bundle direct payments to women with non-financial incentives, such as increased access to veterinary services, can help offset the impacts of women's loss of control over milk incomes under commercialization (Tavenner and Crane 2018). In some cases, women experience more control over the incomes they receive individually or through women's groups, as opposed to through mixed gender groups (Ihalainen 2018).

Other responses use gender transformative approaches to engage explicitly with social norms to influence gender equal outcomes (see also Chapter 10, this volume). A number of CGIAR projects are attempting to change negative attitudes that exist toward women's engagement in agriculture. A project in Uganda encouraged couples to register at least one of their sugarcane blocks in the wife's name. The result was significant increases in women's involvement in the value chain and increased access of women to bank accounts, since these were a prerequisite for having a block registered in their name (Ambler et al. 2018). In Ethiopia, men as well as women were provided trainings on women's rights in agricultural value chains, benefiting women through increased access to resources, technologies, and knowledge (Gebremedhin et al. 2016).

One main takeaway from such projects is that it is important to target both women and men. This also counts for interventions that seek to raise aspirations of farmers. In Kyrgyzstan, for example, interventions aimed at raising women's aspirations can increase their involvement in household decision-making; interestingly, interventions that make their husbands more ambitious are associated with more egalitarian gender attitudes among both women and men, but no greater involvement of women in decision-making (Kosec et al. 2018b).

A research agenda to promote gender equality

Having explored the existing body of research on the feminization of agriculture, we now reflect on how it contributes to enhancing understanding of gender equality in agrarian change processes and develop a research agenda. A first observation relates to the size, scope, and generalizability of existing studies. The research on the feminization of agriculture, in its myriad dimensions, tends to be based on relatively small-scale qualitative and quantitative studies, often in a single location for which findings may not generalize to other locations. A few quantitative studies analyze nationally representative datasets but the variables available often limit the scope. Even if findings do not generalize, it would be useful to understand the context in which these individual studies are taking place and what they imply for the types of settings in which their findings should (and should not) hold. Thus, it will be critical to build a body of research in which the individual analyses more closely speak to each other, drawing on common framings and terms. The current CGIAR research project on the feminization of agriculture is in the process of developing such a framework.[2]

The literature discussed here both supports and challenges the two narratives that we presented in the introduction. The rural transformation that is occurring may have multiple effects, some of which may be empowering for women and move toward gender equality whereas others may be disempowering and disadvantage women relative to men. The findings confirm that, in many places, women's labor force participation in the agriculture sector is increasing for a variety of reasons—including commercialization of value chains, climate change, technology, conflict, and migration—but women are often concentrated in certain domains, such as seasonal, casual, or unpaid work.

Furthermore, women's increased paid work in agriculture does not necessarily result in gender equality in terms of wages, positions, and overall time

2 See https://gender.cgiar.org/feminization-of-agriculture/

use; prevailing patriarchal structures often create unequal distributions of both costs and benefits. Women's working conditions and pay often continue to be worse than those of men doing similar jobs, and women often remain responsible for productive and reproductive work in the home. This means entrance into the paid workforce increases their overall workload. Rural transformation, however, can move us toward gender equality, particularly when women are offered management or supervising roles, when labor-saving technologies and training are provided to women, and when women gain greater access to and control over resources and incomes.

While this review of literature has enhanced our understandings of gendered labor patterns in agriculture and provides insight into their implications for gender equality, it is necessary to address a number of key gaps. There is a need for greater evidence to identify the conditions under which rural transformation does increase gender equality. A future research agenda on rural transformation, the changing patterns of men's and women's work, and gender equality emerges when we look at the conceptual and methodological weaknesses and gaps.

Reconceptualizing the "feminization of agriculture"

Our review of existing studies points to the importance of looking at the existing institutional framework in order to understand the effect of rural transformation and the feminization of agriculture on gender equality. Social norms on gender emerge as a key factor in this. These include norms around women's paid employment and behavior outside the household, perceptions of what women can and should do, expectations of women's unpaid work within the household, and the extent of acceptance of violence against women and harassment.

Another need that emerges is for more information on the macro- and institutional-level policy levers that can affect labor patterns and increase their potential to bring about gender equality. Since most research focuses on the micro level, it tells us little about relationships with national-level policies. There is evidence on how policies regarding property rights or wage employment affect women but these analyses usually do not consider the rapidly changing situation in rural areas. In addition, many of the institutions that affect agricultural production operate at the community level, such as producer cooperatives or water user associations. When are these institutions able to adapt to the changes and incorporate women not only as members but also as agenda-setters and leaders? Considering subnational differences in policies within a single country, changes in policies within a single country over time, or differences in policies across countries can help researchers better assess which policies levers are most effective in bringing about gender equality in the context of rural transformation and other drivers of change in the rural sector.

Finally, there are dimensions of gender equality that are missing from current research and literature on the feminization of agriculture, and this is a gap that may need filling. In particular, how these rural transformations affect violence against women is critically important. CGIAR has carried out some innovative work on intimate partner violence in the context of social policy programs (Hidrobo et al. 2016, Roy et al. 2018) but less has been done on the relationship of agrarian change and violence against women.

There is a need for additional research on how agrarian change is unfolding in conflict and post-conflict situations. Severe policy and governance disruptions have the potential to promote rapid changes in women's and men's roles. Appropriate agricultural and labor market policies may support movement toward gender equality.

The institutional and policy environments in which women and men work affect both manifestations of rural transformation and the changing patterns of women's work and responsibilities, but they are also themselves affected by rural transformation. For example, as more women take on work outside the household and contribute to household income, social norms around what paid work women can and should do and norms around women's mobility and behavior outside the household are likely to change. Norms and perceptions around who is seen as a farmer and, therefore, a potential beneficiary of agricultural policies, may also change as more women take on the primary responsibilities for the household farm and men are absent from rural areas. And women may even help policies promote gender equality if agrarian change permits their greater involvement in decision-making processes at the local and national levels.

In analyzing the interactions between these different drivers, manifestations, and factors, it will be important to place these in the context of broader shifts in order to be able to assess whether change is moving toward or away from gender equality. There is a strong need for research on the broader trends across time and space in terms of rural women's and men's work within and outside of agriculture. As people move out of agriculture, where are they moving? To what extent are women moving into dynamic agricultural sectors rather than stagnating ones? And how do patterns vary with different drivers of change and transformation? New approaches, such as geo-spatial economic analyses and maps, may permit greater understanding and visualization of these changes. They may also serve as a valuable policy tool for explaining broad patterns succinctly.

Finally, the research demonstrates the importance of considering the differentiated impacts of these changing labor patterns on different groups of women and men. Who is providing more agricultural labor and who is providing less? How are the shifts affecting men's and women's total labor burdens and responsibilities? How are the poor and the landless particularly affected? While we may be interested in the impacts on gender equality at a national or regional level, there may be differences within specific contexts and groups of people, such as groups based on age, caste, ethnicity, or other socioeconomic characteristics. Taking intersectionalities seriously can lead to a more nuanced understanding of the trends—and thus more appropriate policy prescriptions.

Future research should more consistently provide evidence on what is changing and how drivers of rural transformation affect manifestations of the feminization of agriculture. A critical step in this is to investigate how these changes and these drivers affect different dimensions of gender relations. Many of the studies reviewed describe women's increased control over an asset, or their involvement in decisions as achievements in and of themselves. It is pertinent to go beyond such a narrow focus and to more systematically and comprehensively measure the domains in which gender equality has (and has not) been advanced.

Addressing data gaps with improved measures

An improved conceptual framework needs to be accompanied by improved ways of measuring changes in the dimensions of gender relations and in the patterns in agricultural and rural sectors. New quantitative work needs to take measurement seriously, and consider how to build upon existing studies through the collection of high-quality detailed data. These include data that allow us to better understand "jointness" in decision-making; data that capture the nature of asymmetric information between spouses, especially following male outmigration; data on women's and men's time use and how each goes about multitasking; and data on the policies or norms in a given setting that may affect the potential for agrarian change to bring about gender equality. We consider some of these data gaps in more detail here.

To understand shifting labor burdens, we need better data on time use in agriculture and domestic and unpaid care work, as well as off-farm work for women and men, girls and boys. The development of new methodologies on time use are currently underway but challenges remain as to how to best collect such data in rural settings in order to be able to better analyze patterns over time and space. In addition, while qualitative work, such as the GENNOVATE project (Petesch et al. 2018), has explored how to collect qualitative data across sites on *gendered power relations,* less quantitative work has been done to collect such data. Substantial work has been done, however, both qualitative and quantitative, on *measuring women's empowerment,* much of it through CGIAR collaborative projects.[3] See also Chapter 9, this volume, on measuring women's empowerment.

Related to the issue of measurement, a current body of research, again much of it within CGIAR, is exploring methods to better understand household decision-making. New approaches have involved analyzing data from multiple respondents within the same household to understand different perspectives regarding who within the household is making the decisions—and how different respondents' answers correlate with key outcomes (Kosec and Song 2019, Ambler et al. forthcoming). Other work seeks to understand how husbands and wives understand joint decision-making (Acosta et al. 2020). It would be useful to extend this literature to consider how gender dynamics affect household decision-making when one person has migrated or is spending time away from the farm or household.

In addition, since we are interested in how these changes in labor patterns and decision-making influence the well-being of people living in rural communities, it would be useful to identify additional measures of well-being, as well as of stress. Beyond increasing workloads, women may experience significant stress, despair, and other mental health issues associated with the increased pressures from new roles as primary household providers (for example in early stages of a husband's migration). Men may experience significant stress because new patterns in the division of labor and decision-making may challenge traditional masculinities. Men who are migrants themselves may face stress and unhappiness (Chen et al. 2019), affecting both the level of remittances they send home and relationships with their families.

We need new approaches to collecting well-being information from both family members who remain in rural areas and those who migrate. Comparative studies of how well-being, decision-making, and other indicators vary by migrant and non-migrant members are rare. We also need more studies analyzing the asymmetric information resulting from migration and the implications for the welfare of non-migrant women—including how the explosion of information technology has influenced these asymmetries. Research by Ambler (2015) uses experimental methods to identify these

3 See https://gaap.ifpri.info/; https://www.ifpri.org/project/weai

asymmetries based on interviews with both the sending and the receiving transnational households. The lack of intra-household data and research that covers both migrants and non-migrant family members makes assessing the impacts of migration on gender equality especially challenging.

Conclusion

The term "feminization of agriculture" is used to refer to the broadening and deepening of women's involvement in agriculture—in terms of workload, decision-making, or visibility. Some see this increased involvement of women in agriculture as empowering because it creates new opportunities for women; others fears disempowerment for women "left behind in agriculture," with few agency and livelihood opportunities. This chapter points to the need for more insight into how rural transformation is changing gender relations vis-à-vis progress toward gender equality. The emerging picture of changes in women's work and responsibilities has significant implications for CGIAR system-wide agricultural research and interventions.

First, the traditional gendered divisions of labor in agriculture are changing and labor patterns are increasingly diverse and variable. Agricultural interventions need to provide appropriate technologies, trainings, and policy recommendations to ensure that women as well as men benefit from the innovations and trends that are affecting rural labor markets, especially in agriculture. A failure to address these issues not only reduces adoption rates of technologies but also runs the risk of perpetuating gender inequality. Research and interventions need to question common assumptions about men's and women's roles and responsibilities; they must consider how the labor patterns as well as gender relations are changing, often in response to either new opportunities or setbacks. Context is critically important in this—and this raises the importance of qualitative work that sheds a light on why particular policies or interventions work in some settings but not in others.

Geography of study sites for publications cited in Chapter 8

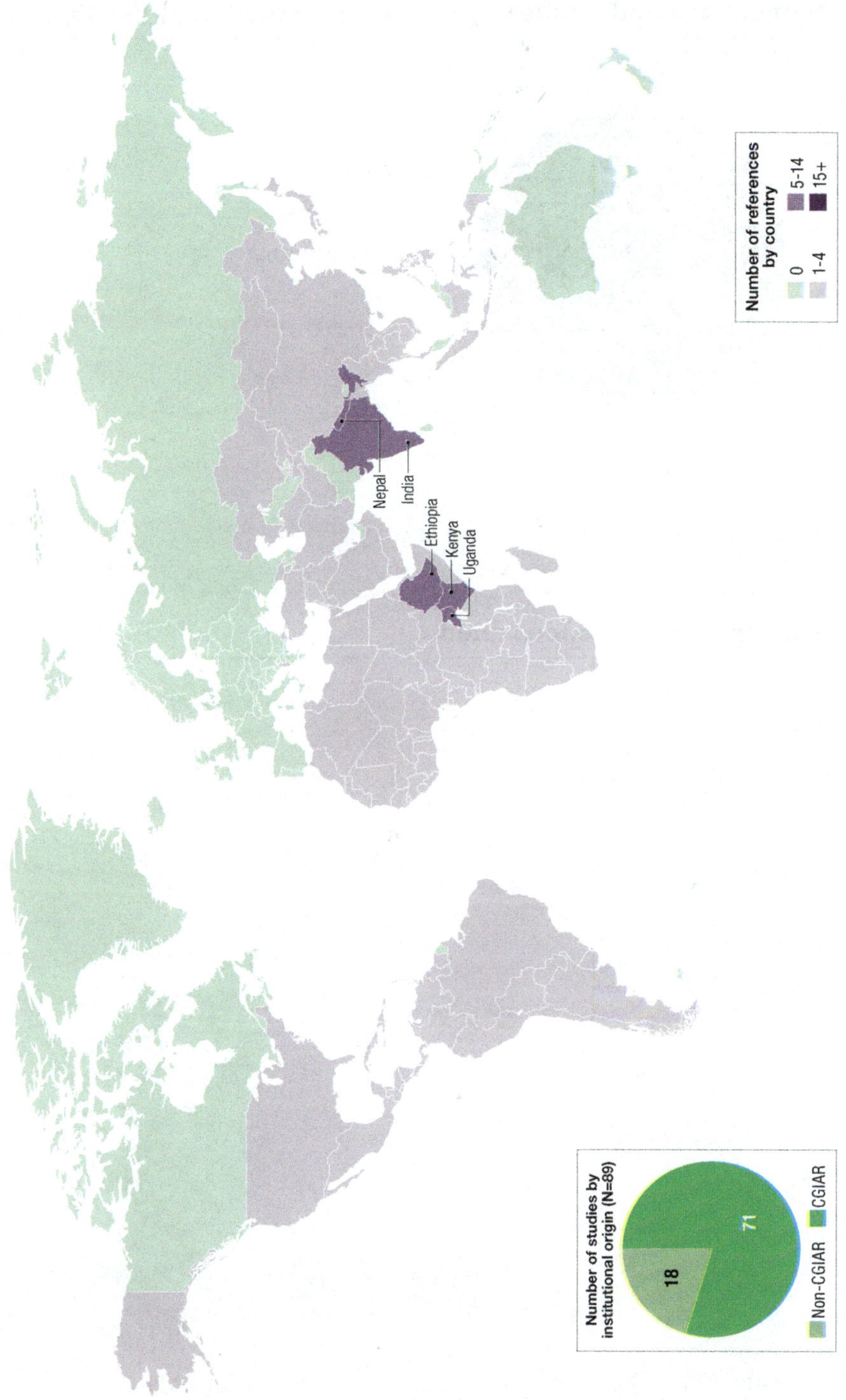

Number of cited studies by research methodology (N=89)

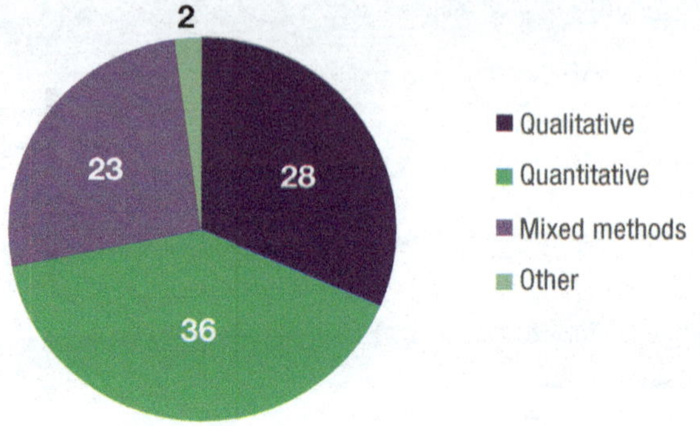

Timeline for references cited

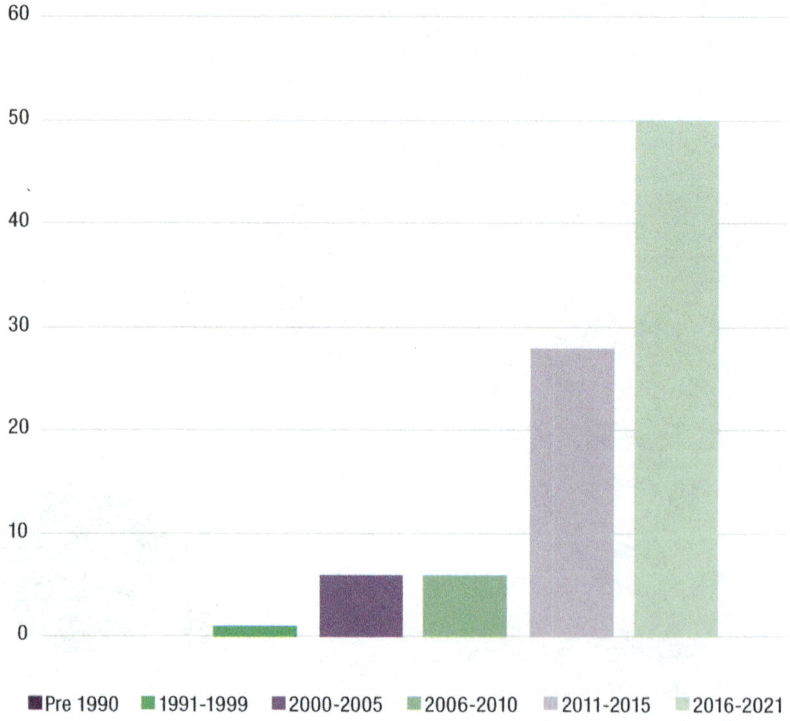

References

Abdelali-Martini, M., and J. Dey de Pryck. 2015. "Does the Feminisation of Agricultural Labour Empower Women? Insights from Female Labour Contractors and Workers in Northwest Syria." *Journal of International Development* 27 (7): 898–916.

Abdelali-Martini, M., P. Goldey, G. Jones, E. Bailey. 2003. "Towards a Feminization of Agricultural Labour in Northwest Syria." *Journal of Peasant Studies* 30 (2): 71–94.

Acosta, M., M. van Wessel, S. van Bommel, E. L. Ampaire, J. Twyman, L. Jassogne, P. H. Feindt. 2020. "What Does It Mean To Make A 'Joint' Decision? Unpacking Intrahousehold Decision Making In Agriculture: Implications For Policy And Practice." *The Journal of Development Studies* 56 (6): 1210–1229.

Akter, S., P. Rutsaert, J. Luis, N. M. Htwe, S. S. San, B. Raharjo, A. Pustika. 2017. "Women's Empowerment and Gender Equality in Agriculture: A Different Perspective from Southeast Asia." *Food Policy* 69: 270–279.

Al Zoubi, S. T., A. Aw-Hassan, B. Dhehibi. 2019. *Enhancing the Livelihood and Food Security of Syrian Refugees in Lebanon*. Amman: ICARDA.

Ambler, K. 2015. "Don't Tell on Me: Experimental Evidence of Asymmetric Information in Transnational Households." *Journal of Development Economics* 113: 52–69.

Ambler, K., C. Doss, C. Kieran, S. Passarelli. Forthcoming. "He Says, She Says: Spousal Disagreement in Survey Measures of Bargaining Power." *Economic Development and Cultural Change.* 69 (2): 765–788.

Ambler, K., K. Jones, M. O'Sullivan. 2018. "What Is the Role of Men in Connecting Women to Cash Crop Markets? Evidence from Uganda." Discussion Paper 01762, IFPRI, Washington, DC.

Bacud, E. S., R. Puskur, T. N. L. Duyen, B. O. Sander, J. Luis. 2019. "Rural Outmigration–Feminization–Agricultural Production Nexus: Case of Viet Nam." *Migration and Development.* https://doi.org/10.1080/21632324.2019.1679962

Balasubramanya, S. 2019. "Effects of Training Duration and the Role of Gender on Farm Participation in Water User Associations in Southern Tajikistan: Implications for Irrigation Management." *Agricultural Water Management* 216 (1): 1–11.

Basnett, B. S. 2013. "Taking Migration Seriously: What Are the Implications for Gender and Community Forestry?" *CIFOR Information Brief* 32 (65): 1–6.

Basu, P., A. Galiè, I. Baltenweck. 2019. "Presence and Property: Gendered Perspectives on Participation in a Dairy Development Program in Kenya and Uganda." *Women's Studies International Forum* 74: 68–76.

Bernardinie, M. D. 2019. "Permanently Seasonal Workers: Gendered Labor Relations and Working Conditions of Asparagus Agricultural Workers in Ica, Perú," In *Gender, Agriculture and Agrarian Transformations*, edited by C. E. Sachs, 183–202. New York: Routledge.

Bigler, C., M. Amacker, C. Ingabire, F. Birachi. 2017. "Rwanda's Gendered Agricultural Transformation: A Mixed-Method Study on the Rural Labour Market, Wage Gap and Care Penalty." *Women's Studies International Forum* 64: 17–27.

Binzel, C., and R. Assaad. 2011. *Egyptian Men Working Abroad: Labor Supply Responses by the Women Left Behind.* Bonn: Institute of Labour Economics.

Ceballos, F., B. Kramer, M. Robles. 2019. "The Feasibility of Picture-Based Insurance (PBI): Smartphone Pictures for Affordable Crop Insurance." *Development Engineering* 4: 100042.

Chen, J., K. Kosec, V. Mueller. 2019. "Moving to Despair? Migration and Well-Being in Pakistan." *World Development* 113: 186–203.

De Brauw, A., V. Mueller, H. Lee. 2014. "The Role of Rural-Urban Migration in the Structural Transformation of Sub-Saharan Africa." *World Development* 63: 33–42.

De Brauw, A., D. O. Gilligan, J. Hoddinott, S. Roy. 2015. "The Impact of Bolsa Familia on Schooling." *World Development* 70: 303–316.

De Brauw, A., V. Mueller, T. Woldehannah. 2018. "Does Internal Migraiton Improve Overall Well-Being in Ethiopia?" *Journal of African Studies* 27 (3): 347–365.

De Brauw, A., B. Kramer, and M. Murphy. 2021. "Migration, Labor and Women's Empowerment: Evidence from an Agricultural Value Chain in Bangladesh." *World Development* 142: 105445.

Deere, C. D. 2005. *The Feminization of Agriculture? Economic Restructuring in Rural Latin America.* Geneva: UNRISD.

Deere, C. D., and M. León. 2003. "The Gender Asset Gap: Land in Latin America." *World Development* 31 (6): 925–947.

Deere, C. D., G. Alvarado, A. D. Oduro, L. Boakye-Yiadom. 2015. *Gender, Remittances and Asset Accumulation in Ecuador and Ghana.* New York: UN Women.

Djoudi, H., and M. Brockhaus. 2011. "Is Adaptation to Climate Change Gender Neutral? Lessons from Communities Dependent on Livestock and Forests in Northern Mali." *International Forestry Review* 13 (2): 123–135.

Dolan, C., and K. Sorby. 2003. *Gender and Employment in High-Value Agriculture Industries.* Washington, DC: World Bank.

Doss, C. 2001. "Men's Crops? Women's Crops? Gender Patterns of Cropping in Ghana." *World Development* 30 (11): 1987–2000.

Doss, C., and R. Meinzen-Dick. 2019. *Land Tenure Security for Women: Conceptual Framework.* Seattle: IFPRI.

Doss, C., C. Kovarik, A. Peterman, A. Quisumbing, M. van den Bold. 2015. "Gender Inequalities in Ownership and Control of Land in Africa: Myth and Reality." *Agricultural Economics* 46 (3): 403–434.

Elmhirst, R. 2007. "Tigers and Gangsters: Masculinities and Feminised Migration In Indonesia." *Popultion, Space and Place* 13 (2): 225–238.

Elmhirst , R., M. Siscawati, B. S. Basnett, D. Ekowati. 2017. "Gender and Generation in Engagements with Oil Palm in East Kalimantan, Indonesia: Insights from Feminist Political Ecology." *Journal of Peasant Studies* 44 (6): 1137–1159.

Farnworth, C. R., T. Jafry, K. Lama, S. C. Nepali, L. B. Badstue. 2019. "From Working in the Wheat Field to Managing Wheat: Women Innovators in Nepal." *European Journal of Development Research* 31 (2): 293–313.

Farnworth, C. R., P. Bharati, T, Jafry, L. B. Badstue. 2018. "Strengthening Women in Wheat Farming in India: Old Challenges, New Realities, New Opportunities." GENNOVATE Resources for Scientists and Research Teams. Mexico City: CIMMYT

Galiè, A., and P. Kantor. 2016. "From Gender Analysis to Transforming Gender Norms: Using Empowerment Pathways to Enhance Gender Equity and Food Security in Tanzania". In *Transforming Gender and Food Security in Global South*, edited by J. Njuki, J. R. Parkins, A. Kaler, 189–216. London: Routledge.

Galiè, A., J. Jiggins, P. Struik. 2013. "Women's Identity as Farmers: A Case Study from Ten Households in Syria." *NJAS: Wageningen Journal of Life Sciences* 14 (1): 25–33.

Galiè, A., J. Jiggins, P. C. Struik, S. Grando, S. Ceccarelli. 2017. "Women's Empowerment through Seed Improvement and Seed Governance: Evidence from Participatory Barley Breeding in Pre-War Syria." *NJAS: Wageningen Journal of Life Sciences* 81: 1–8.

Garikipati, S. 2009. "Landless but Not Assetless: Female Agricultural Labour on the Road to Better Status, Evidence from India." *Journal of Peasant Studies* 36 (3): 517–545.

Gebremedhin, B., E. Tesema, A. Tegegne, D. Hoekstra, S. Nicola. 2016. *Value Chain Opportunities for Women and Young People in Livestock Production in Ethiopia : Lessons Learned.* Nairobi: ILRI.

Hidrobo, M. 2016. "The Effect of Cash, Vouchers, and Food Transfers on Intimate Partner Violence: Evidence from a Randomized Experiment in Northern Ecuador." *American Economic Association* 8 (3): 284–303.

Hoang, L. A., B. S. Yeoh. 2011. "Breadwinning Wives and 'Left-Behind' Husbands: Men and Masculinities in the Viet Namese Transnational Family." *Gender & Society* 25 (6): 717–739.

Ihaleinen, M. 2018. *Landscape Restoration in Kenya: Addressing Gender Equality.* Nairobi: CIFOR.

Ingabire, C., M. Mshenga, M. Amacker, J. Langat, C. Bigler, E. Birachi. 2018. "Agricultural Transformation in Rwanda: Can Gendered Market Participation Explain the Persistence of Subsistence Farming?" *Gender and Women's Studies* 2 (1): 1–18.

Jensen, R. 2007. "The Digital Provide: Information (Technology), Market Performance, and Welfare in the South Indian Fisheries Sector." *The Quarterly Journal of Economics* 122 (3): 879–924.

Johnson, N., C. Kovarik, R. Meinzen-Dick, J. Njuki, A. Quisumbing. 2016. "Gender, Assets and Agricultural Development: Lessosn from Eight Projects." *World Development* 83: 295–311.

Kar, A., V. Slavchevska, S. Kaaria, S. L. Taivalmaa, E. Mane, R. Ciacci, Y. T. Hoberg et al. 2018. *Male Outmigration and Women's Work and Empowerment in Agriculture: The Case of Nepal and Senegal.* Washington, DC: World Bank.

Kawarauka, N., and G. Prain. 2019. "Gendered Processes of Agricultural Innovation in the Northern Uplands of Viet Nam. " *International Journal of Gender and Entrepreneurship* 11 (3): 210–226.

Khatri-Chhetri, A., and N. Chanana. 2017. *Women's Groups Reaping the Benefits of Solar Energy for Irrigation in Nepal's Climate-Smart Villages.* Copenhagen: CCAFS.

Khatri-Chhetri, A., P. P. Regmi, N. Chanana, P. K. Aggarwal. 2019. "Potential of Climate-Smart Agriculture in Reducing Women Farmers' Drudgery in High Climatic Risk Areas." *Climate Change* 158: 29–42.

Kieran, C., K. Sproule, C. Doss, A. Quisumbing, S. Kim. 2015. "Examining Gender Inequalities in Land Rights Indicators in Asia." *Agricultural Economics* 46 (S1): 119–138.

Kondylis, F., V. Mueller, G. Sheriff, S. Zhu. 2016. "Do Female Instructors Reduce Gender Bias in Diffusion of Sustainable Land Management Techniques? Experimental Evidence from Mozambique." *World Development* 78: 436–449.

Kosec, K., and J. Song. 2019. *The Effects of Income Fluctuations on Rural Health and Nutrition.* Washington, DC: IFPRI.

Kosec, K., and L. Wantchekon. 2020. "Can Information Improve Rural Governance and Service Delivery?" *World Development* 125: 10346.

Kosec, K., H. Ghebru, B. Holtemeyer, V. Mueller, E. Schmidt. 2018a. "The Effect of Land Access on Youth Employment and Migration Decisions: Evidence from Rural Ethiopia." *American Journal of Agricultural Economics* 100 (3): 931–954.

Kosec, K., K. Akramov, B. Mirkasimov, J. Song. 2018b. "Aspirations and Women>s Empowerment: Evidence from Kyrgyzstan." Discussion Paper 01786, IFPRI, Washington, DC.

Kosec, K., J. Song, B. Holtemeyer. 2020. *The Gendered Impacts of Income Fluctuations on Migration and Employment Choices.* Washington, DC: IFPRI.

Kristjanson, P., E. Bryan, Q. Bernier, J. Twyman, R. Meinzen-Dick, C. Kieran, C. Ringler, C. Jost, C. Doss. 2017. "Addressing Gender in Agricultural Research for Development in the Face of a Changing Climate: Where Are We and Where Should We Be Going?" *International Journal of Agricultural Sustainability* 15 (5): 482–500.

Lastarria-Cornhiel, S. 2006. "Feminization of Agriculture: Trends and Driving Forces." Working Paper 41367, World Bank, Washington, DC.

Li, T. M. 2015. *Social Impacts of Oil Palm in Indonesia: A Gendered Perspective from West Kalimantan.* Bogor: CIFOR.

Mazumdar, I., and I. Agnihotri. 2014. "Traversing Myriad Trails: Tracking Gender and Labour Migration across India." In *Migration, Gender and Social Justice: Perspectives on Human Insecurity,Hexagon Series on Human and Environmenal Security and Pease,* edited by T. D. Truon, D. Gasper, J. Handmaker, 123–151. New York: Springer.

Meinzen-Dick, R., A. Quisumbing, J. Behrman, P. Biermayr-Jenzano, V. Wilde, M. Noordeloos, C. Ragasa et al. 2010. "Engendering Agricultural Research." Discussion Paper 973, IFPRI, Washington, DC.

Mendola, M., and C. Carletto. 2012. "Migration and Gender Differences in the Home Labour Market: Evidence from Albania." *Labour Economics* 19 (6): 870–880.

Mu, R., and D. van de Walle. 2011. "Left Behind to Farm? Women's Labor Re-Allocation in Rural China." *Labour Economics* 18 (1): 1–48.

Mueller, V., C. Gray, K. Kosec. 2014a. "Heat Stress Increases Long-Term Human Migration in Rural Pakistan." *Nature Climate Change* 4: 182–185.

Mueller, V., A. Quisumbing, H. L. Lee, K. Droppelmann. 2014b. *Resettlement for Food Security's Sake: Insights from a Malawi Land Reform Project.* Washington, DC: IFPRI.

Mueller, V., C. Kovarik, K. Sproule, A. R. Quisumbing. 2015. *Migration, Gender, and Farming Systems in Asia: Evidence, Data, and Knowledge Gaps.* Washington, DC: IFPRI.

Mueller, V., C. Doss, A. R. Quisumbing. 2018. "Youth Migration and Labour Constraints in African Agrarian Households." *Journal of Development Studies* 54 (5): 875–894.

Mukhamedova, N., and K. Wegerich. 2014. *Land Reforms and Feminization of Agricultural Labor in Sughd Province, Tajikistan.* Colombo: IWMI.

Najjar, D., B. Dhehibi, A. Aw-Hassan, A. Bentaibi. 2017. "Climate Change, Gender, Decision Making Power, and Migration into the Saiss Region of Morocco." Economic Research Forum, CGIAR, Giza.

Najjar, D., B. Baruah, A. Aw-Hassan, A. Bentaibi, G. Kassie. 2018. "Women, Work, and Wage Equity in Agricultural Labour in Saiss, Morocco." *Development in Practice* 28 (4): 525–540.

Nakazi, F., J. Njuki, M. A. Ugen, P. Aseete, E. Katungi, E. Birachi, R. Kabanyoro, I. J. Mugagga, G. Nanyonjo. 2017. "Is Bean Really a Women's Crop? Men and Women's Participation in Bean Production in Uganda." *Agriculture and Food Security* 6 (22).

Njuki, J. M., A. Wyatt, I. Baltenweck, K. Yount, C. Null, U. Ramakrishnan, A. Webb Girard et al. 2016. "An Exploratory Study of Dairying Intensification, Women's Decision Making, and Time Use and Implications for Child Nutrition in Kenya." *The European Journal of Development Research* 28 (4): 722–740.

Orr, A., T. Tsusaka, S. H. Kee-Tui, H. Msere. 2016. "What Do We Mean by 'Women's Crops'? Commercialisation, Gender and the Power to Name." *Journal of International Development* 28 (6): 919–937.

Padmaja, R., S. Pramanik, P. Pingali, C. Bantilan, K. Kavitha. 2019. "Understanding Nutritional Outcomes through Gendered Analysis of Time-Use Patterns in Semi-Arid India." *Global Food Security* 23: 49–63.

Peterman, A., J. A. Behrman, A. R. Quisumbing. 2014. "A Review of Empirical Evidence on Gender Differences in Nonland Agricultural Inputs, Technology and Services in Developing Countries. In *Gender in Agriculture: Closing the Knowledge Gap*, edited by A. R. Quisumbing, R. Meinzen-Dick, T. L. Raney, A. Croppenstedt, J. A. Behrman, A. Peterman, 145–186. New York: Springer.

Petesch, P., L. Badstue, G. Prain. 2018. *Gender Norms, Agency, and Innovation in Agriculture and Natural Resource Management: The GENNOVATE Methodology.* Mexico City: CIMMYT.

Rao, S. 2017. "Women and the Urban Economy in India: Insights from the Data on Migration." In *Gender and Time-Use in Global Context*, edited by R. Connelly, E. Kongar, 231–257. New York: Palgrave Macmillan.

Ragasa, C., G. Berhane, F. Tadesse, A. S. Taffesse. 2012. "Gender Differences in Access to Extension Services and Agricultural Productivity." ESSP Working Paper 49, IFPRI, Washington, DC.

Roy, S., M. Hidrobo, J. Hoddinott, A. Akhtet. 2018. "Transfers, Behavior Change Communication, and Intimate Partner Violence: Postprogram Evidence from Rural Bangladesh." *Review of Economics and Statistics.* 101 (5): 865–877.

Sachs, C. E. 2019. "Gender, Agriculture and Agrarian Transformations: Overview." In *Gender, Agriculture and Agrarian Transformations*, edited by C. E. Sachs, 3–8. New York: Routledge.

Song, Y. 1998. "'New' Seeds in 'Old' China: Impact Study of CIMMYT's Collaborative Programme on Maize Breeding in Southwest China." Wageningen: Wageningen Agricultural University.

Stanley, V. 2015. *Migration and Women's Agency in Agriculture – Women in Agriculture: The Impact of Male Out-Migration on Women's Agency, Household Welfare and Agricultural Productivity.* Washington, DC: World Bank.

Sugden, F., N. Maskey, F. Clement , V. Ramesh, A. Philip, A. Rai. 2014. "Agrarian Stress and Climate Change in the Eastern Gangetic Plains: Gendered Vulnerability in a Stratified Social Formation." *Global Environment Change* 29: 258–269.

Sunderland, T., R. Achdiawan, A. Angelsen , R. Babigumira, A. Ickowitz, F. Paumgarten, F., V. Reyes-García, G. Shively. 2014. "Challenging Perceptions about Men, Women, and Forest Product Use: A Global Comparative Study." *World Development* 64 (S1): 56–66.

Tavenner, K., and T. A. Crane. 2018. "Gender Power in Kenyan Dairy: Cows, Commodities, and commercialization." *Agriculture and Human Values* 35: 701–715.

Tavenner, K., M. van Wijk, S. Fraval, J. Hammond. 2019. "Intensifying Inequality? Gendered Trends in Commercializing and Diversifying Smallholder Farming Systems in East Africa." *Frontiers in Sustainable Food Systems* 3 (10).

Twyman, J., M. Juliana, M. A. Garcia. 2015. "Identifying Women Farmers: Informal Gender Norms as Institutional Barriers to Recognizing Women's Contributions to Agriculture." *Journal of Gender, Agriculture and Food Security* 1 (2): 1–17.

UN Women. 2001. "Concepts and Definitions." Accessed September 3, 2019. https://www.un.org/womenwatch/osagi/conceptsandefinitions.htm

Van Campenhout, B., D. Spielman, E. Lecoutere. 2018. "The Role of Gender in ICT-Mediated Agricultural Information Campaigns." Presented at conference, International Association of Agricultural Economists, Vancouver, July 28–August 2.

Von Koppen, B. 2008. "Gender, Resource Rights, and Wetland Rice Productivity in Burkina Faso." In *Institutional Economics Perspectives on African Agricultural Development*, edited by C. Poulton, N. Vink, J. Kirsten, A. Dorward, 389–407. Washington, DC: IFPRI.

Chapter 9

ASSESSING WOMEN'S EMPOWERMENT IN AGRICULTURAL RESEARCH

Marlène Elias, Steven Michael Cole, Agnes Quisumbing, Ana Maria Paez Valencia, Ruth Meinzen-Dick, and Jennifer Twyman

The concept of empowerment has steadily made its way onto the international development agenda. Batliwala (2007) traces its equivalents back several hundred years and across geographies in struggles for social justice. Feminists brought the concept of women's empowerment to the 1995 Fourth World Conference on Women in Beijing, where it gained traction, with the Beijing Declaration referring to "enhancing further the advancement and empowerment of women all over the world" (UN 1995, 7). Then, it was about collective struggles to challenge patriarchal structures, and intersecting structures of class, ethnicity, caste, and race, that shape women's (subordinate) position in society (Batliwala 2007). Twenty years later, "empowerment" animates the fifth Sustainable Development Goal (SDG5): "Achieve gender equality and empower all women and girls."

The field of agricultural research for development (AR4D) recognizes women's empowerment for its instrumental value and its links with several desirable outcomes related to health and nutrition (Sraboni et al. 2014, Galiè et al. 2018, Heckert et al. 2019),[1] productivity (Diiro et al. 2018), and resource management (for example Sodhi et al. 2010). Its intrinsic value is also increasingly acknowledged as a goal in itself (Cornwall and Edwards 2014). Yet a lack of conceptual clarity around the term as mobilized in the international development agenda, along with the subversion of the term in neoliberal political agendas, has diluted the concept that social activists brought to the table in Beijing (Batliwala 2007, Cornwall and Rivas 2015, Nazneen et al. 2019).

The complex, intangible, political, and context-specific nature of empowerment renders its assessment a formidable task. In 1999, Kabeer provided an in-depth discussion of the difficulties operationalizing the concept for measurement; today, the ethical, political, and epistemological debates that

1 A systematic review urges caution regarding links between women's empowerment and child nutrition, however, as many studies reviewed demonstrate a lack of rigor (Santoso et al. 2019).

characterize such measurement continue to merit proper consideration (for example Newton et al. 2019). Despite these challenges, applied researchers and practitioners pursue their attempts at assessment, based on "the realization that we must devise ways of checking whether the policies, resources, and strategies applied toward building more equitable, sustainable, rights-affirming, inclusive and peaceful societies are working effectively or not—whether they are producing the changes we wish to see" (Batliwala and Pittman 2010, 3). On the global agenda and in AR4D initiatives, which increasingly define women's empowerment as a goal, such assessments—however imperfect—are important for advancing women's empowerment and gender equality.

In this chapter, we ask: "How is women's empowerment assessed in AR4D, and how can such assessments advance women's empowerment and gender equality?" In so doing, we challenge those working in the field of agriculture to return to the foundational concepts, to move from instrumental to more political and transformative engagements as implied in the original concept of empowerment. We further bring recent developments in assessing empowerment in agriculture into the fold of the broader literature on the concept. This is relevant not only to strengthen assessments but also for the framing of empowerment in AR4D and in the agriculture and natural resource management (NRM) sectors, as "what is measured—and not measured—influences discourse and confers legitimacy to certain categories of intervention or institutional change" (O'Hara and Clement 2018, 112).

We begin by defining the concept of women's empowerment as used in AR4D and how it relates to gender equality. We then argue that assessing women's empowerment in the context of AR4D can advance gender equality—although we highlight that tensions and challenges accompany such an effort. Next, we examine different methodologies, with a focus on tools, for assessing women's empowerment in agriculture and NRM in and beyond CGIAR. Finally, we raise critical questions related to assessing women's empowerment for a future AR4D agenda.

Conceptualizing empowerment

It is perhaps unsurprising that multiple definitions of empowerment exist in the literature, given the term's use by scholars and practitioners from different disciplines, theoretical-epistemological backgrounds (for example Narayan 2005), and regional contexts (Ibrahim and Alkire 2007). This multiplicity of definitions reveals healthy debates and evolving thinking about the concept. In AR4D, many studies (including this book, see Chapter 1) refer to empowerment as "the processes by which those who have been denied the ability to make strategic life choices acquire such an ability" (Kabeer 1999, 435, 2017). This conceptualization draws from Sen's (1985, 1999) capability approach, and emphasizes people's freedom to define and lead the life they have reason to value. "Strategic" life choices are those that hold significance over one's life direction, such as those related to whether, who, and when to marry, family formation, or which type of livelihood strategy one will pursue. These defining choices set the parameters for practical, day-to-day decisions, with historical and structural conditions influencing the range of options people see before them and value (Kabeer 2005).

We can conceptualize the ability to exercise choice over strategic decisions along three interconnected dimensions: "resources (defined broadly to include not only access, but also future claims, to both material and human and social resources); agency (including processes of decision making, as well as less measurable manifestations of agency such as negotiation, deception and manipulation); and achievements (well-being outcomes)" (Kabeer 1999, 435). Resources are the preconditions that enhance people's abilities to exercise choice—although women's strengthened agency can also unlock access to resources (Farnworth et al. 2019). Formal and informal rules, including norms,[2] mediate access to these resources in different institutional domains of society (for example the household, community, or market). Achievements are realized when people have agency and access to resources that enable them to define and act upon their goals. Achievements cannot be predefined, as in any given context different people may value and seek different ways of being and doing (Sen 1985).

Agency—a person's ability to define and act upon one's goals—is at the heart of the concept of empowerment. It is often operationalized as decision-making but also takes the form of bargaining, negotiation, resistance, and critical reflection and analysis. Agency is exercised at individual and group levels, through collaborative relations and collective action[3] (collective agency or "power with"), and can be framed in both positive and negative terms in

---

2 Norms are socially constituted rules that "govern social relations and establish expectations as to how we are to act in our everyday affairs" (Knight and Ensminger 1998, 105).

3 Collective action entails women "gaining solidarity and taking action collectively on their interests, to enhance their position and expand the realm of what is possible. It mobilises and strengthens women and girls' collective power, enabling them to have more influence than when they act individually and in isolation" (van Eerdewijk et al. 2017, 32). Collective action develops "power with" and is linked to "power within," as coming together can change women's perceptions of power inequalities and sense of self. It also influences "power to," "by amplifying voice and exercising choice in decision-making processes" (ibid., 32).

relation to power (as per Rowlands 1997). In positive terms, agency is when people recognize their self-worth and the purpose they bring to their actions (intrinsic agency, or "power within") and are able to act to realize their goals (instrumental agency, or "power to"), even when opposed by others or by social norms. In negative terms, it refers to actors superseding the agency of others, and exercising control or "power over" their lives and resources (Kabeer 1999). Empowerment, then, is about changes in these multiple manifestations of power, which interconnect and are mutually reinforcing to create unequal outcomes (Hillenbrand et al. 2015). Such changes in oppressive power relations can occur at the individual or group level (Eyben et al. 2008).

There are fundamental contestations as to whether the expansion of individual women's agency represents empowerment, or whether empowerment is about something more—a critical consciousness[4] of women's rights, women's solidarity, and the collective challenge to patriarchal structures and power relations that curtail their freedoms. Feminist scholars and activists adhere to the latter perspective (Kabeer 1994, Cornwall and Rivas 2015, Ewerling et al. 2017), and critique mainstream development practice for treating empowerment as an individual pursuit focused on entrepreneurship and self-reliance (Nazneen et al. 2014). This framing reflects a co-optation of the concept in the neoliberal international development agenda that divests the state of its responsibilities by "empowering" local women to look after themselves (for example Batliwala 2007, Nazneen et al. 2019).

Empowerment is generally considered a process, although it is sometimes treated as both a process and an outcome or as an outcome (Carr 2003). As a process, it refers to the changes in institutional structures, access to resources, critical consciousness, and so on that facilitate people's abilities to make, act upon, and achieve their strategic life choices. As an outcome, it embodies the degree of freedom people have to control and have positive impacts on their lives and futures (van Eerdewijk et al. 2017). Empowerment is relative: people are empowered (or disempowered) in comparison with others or with themselves at another point in time (Mosedale 2005). Importantly, empowerment necessarily requires women to be the prime movers. As such, interventions may "be conceived not as empowering women but as clearing some of the obstacles from the path and providing sustenance for women as they do empowerment for themselves" (Cornwall and Rivas 2015, 405).

---

4 Critical consciousness refers to a person's awareness of her or his ability to make life-changing choices, including by challenging oppressive socio-political structures (Freire 1970).

The plural definitions of empowerment imply the use of various frameworks to explain its multiple and interrelated dimensions. For example, van Eerdewijk et al. (2017) frame empowerment in terms of resources, agency, and institutional structures. Hillenbrand et al. (2015) argue that considering these three dimensions together is important to maintain a focus on collective responsibility and political engagement, rather than placing the burden of change on individual women. Narayan (2005) identifies key factors facilitating or constraining empowerment and broader development outcomes—namely, institutional climate, social and political structures, individual assets and capabilities, and collective assets and capabilities.

Lombardini et al. (2017) (see also Lombardini and McCollum 2018) focus on measuring changes in empowerment at the personal, relational, and environmental levels. Personal empowerment relates to changes taking place within the person—in a woman's beliefs about her own worth, capacities, and actions. The focus here is on the immaterial, related to power within, self-perception, and critical consciousness, rather than on individual-level material elements. Relational empowerment refers to changes taking place in a person's relationships and in the power relations within which she or he

FIGURE 9.1 Framework for assessing women's empowerment

Source: Lombardini et al. (2017, 6).

is embedded—in a woman's position relative to others, such as her partner, family, community, local authorities, or social networks. Changes at the environmental level occur in broader societal institutions and structures. These can be formal (such as in political and legislative frameworks) or informal (such as in social norms, attitudes,[5] and beliefs). Changes at one level will stimulate changes at others, although these changes do not necessarily move at the same pace or in the same direction (Figure 9.1).

We draw upon this framing to structure our analysis of tools for assessing women's empowerment; at the personal level, we also consider whether tools support an exploration of changes in material resources that can affect women's empowerment. The relational and environmental levels of the framework are of particular relevance for highlighting the power-laden and political nature of empowerment, and the fact that transformative change toward gender equality must go far beyond only "changing women."

Assessment to "move the needle" on women's empowerment and gender equality

Assessing empowerment in AR4D can play an important role in advancing women's empowerment and gender equality in at least four ways. First, quantitative and qualitative assessments of empowerment can **support holistic design of projects, programs, and policies**. Multidimensional measures can support the development and prioritization of interventions that address women's empowerment, gender equality, and other project objectives. For instance, the project-level Women's Empowerment in Agriculture Index (pro-WEAI, described below) allows projects to identify in which domains women are most disempowered, so they can develop and prioritize interventions that address these (Malapit et al. 2019). If no measures of empowerment are available, program implementers might concentrate on changes that can be measured and demonstrated, such as women's income, rather than less tangible changes that hold equal or greater importance for women's empowerment (Mosedale 2005).

Evidence on *how* to advance women's empowerment and gender equality is also needed to shed light on the causal pathways that lead to empowerment, and on how women's empowerment correlates with other development goals. This can contribute to evidence-based interventions and policy-influencing

5 In contrast with norms, which are held at the group level, attitudes refer to individual beliefs and emotions toward something, someone, or some occurrence (Ajzen 1991).

(Lombardini et al. 2017). Bangladesh's Agriculture, Nutrition, and Gender Linkages (ANGeL)[6] pilot project was designed based on results from the Women's Empowerment in Agriculture Index (WEAI), described below. Data from the WEAI demonstrated the extent of women's and men's disempowerment, the factors that contributed most to this, and the interrelationship between women's empowerment and household food security and dietary diversity of children (Sraboni et al. 2014).

Second, assessments are needed **to monitor whether and how initiatives such as projects, programs, policies, or social movements and efforts led by women's organizations are contributing—positively or negatively—to women's empowerment**. Nuanced assessments are important for adaptive learning, to identify areas of strength as well as weakness in the strategies they deploy (Carter et al. 2014). Galiè (2013) discusses how a participatory plant-breeding project in Syria actively sought to address the needs of women farmers. Efforts to assess effects on women's empowerment revealed the stigmatization a young woman experienced for having traveled alone to a conference. Thereafter, the project took steps to reduce the risk of social ostracism by involving a larger group of women. Having sound and concrete bearings with respect to empowerment can thus encourage efforts to broaden or deepen strategies within institutions and their programming.

Third, **measuring and/or assessing empowerment serves to build upward and downward accountability and credibility** (Batliwala and Pittman 2010).[7] For example, governments must be held downwardly accountable to their constituents, and in some cases upwardly accountable to international and regional organizations, for their commitments, such as reaching SDG targets. Most of the key strategic elements women's rights organizations advocated have been included as targets under SDG5 (Razavi 2016). Yet the SDG framework's weak accountability mechanisms, with no mandatory reporting requirements, essentially rely on the goodwill of governments to implement the agenda and track changes (Deere 2018). Close monitoring using adequate measures is needed to track progress and enable women's rights advocates and their allies to lobby for the agenda's proper implementation

6 This pilot project was developed by IFPRI and implemented at scale by the Bangladeshi Ministry of Agriculture to identify actions and investments in agriculture that would help increase farm household income, improve nutrition, and empower women (see https://www.ifpri.org/project/agriculture-nutrition-and-gender-linkages-angel).

7 Upward accountability refers to accountability to higher-level structures or institutions, such as from senior managers to boards or projects to donors; downward accountability is accountability to lower levels, such as from governments to citizens or projects to the local communities with which they work.

(Razavi 2016, Deere 2018). This imperative has given rise to initiatives such as Data2X, which uses gender data to support global efforts to achieve gender equality.[8] Failing to track or using inadequate or narrow measures to monitor women's empowerment can augment the risk of selectivity and dilution of policies in the process of implementation.

At a programmatic and project level, governments and donors use indicators in monitoring, evaluation, and impact assessments as the basis for judging performance and allocating resources. Inevitably, the things we measure are those that receive attention and on which we focus for change. Although a growing number of projects claim to advance women's empowerment, many such projects do not, in fact, make conscious efforts to define what empowerment means in their context, or to diagnose or address constraints to women's agency (Mosedale 2005). Danielsen et al. (2018) found that, out of a portfolio of 18 projects funded by the Canadian International Food Security Research Fund, which advanced gender integration as a key feature of its program, only about one third achieved "women's empowerment sub-outcomes," including changes in gender norms, and increased women's recognition, control over decisions, and formal leadership. Likewise, reviewing 13 AR4D projects with the stated goal of empowering women, Johnson et al. (2018) found that many had neither strategies that would be expected to increase women's abilities to make strategic life choices nor ways of measuring whether such changes take place. Hence, the authors highlight that it is important to be clear about whether project objectives are to reach, benefit, or empower women; and about what women's empowerment may consist of in the context of AR4D.

Assessments also hold programs and projects downwardly accountable. For example, in Galiè's (2013) study, women participants pushed to hold researchers accountable in supporting their empowerment, or in not pushing them too much if there was no support to be given.[9] Assuming that empowerment, as captured in certain measures, is necessarily what women want can be misleading, and highlights the importance of gathering perspectives from the women whose life experiences are being explored. In a normatively restrictive environment, women who are considered "empowered" can be frowned upon and socially shunned, and risk direct backlash in the form of intimate partner

8 Data2X is a partnership to "improve the quality, availability, and use of gender data in order to make a practical difference in the lives of women and girls worldwide" (see https://data2x.org/).

9 One of the project's women participants asked, "Why do you make us dream, then, if you can't do anything about it" (Galiè 2013, 87).

violence (for example Basu 1995, Jewkes 2002)—a risk not all women are willing to take without any safeguards.

Fourth and finally, the **assessment process itself can challenge power relations** (Hillenbrand et al. 2015). For instance, engaging participants in deciding what, how, and when to measure, as well as who does the measuring, can be empowering (Morgan 2014, Newton et al. 2019). When we apply participatory approaches to measurement in a transformative way, and women drive the assessment process, they can facilitate critical reflection and action on norms and power relations that disempower women and cause gender inequalities (Kantor 2013, Cole et al. 2014, Cornwall and Sardenberg 2014, Newton et al. 2019). Privileging the voices of marginalized groups in the assessment process can validate their knowledge, shift power into their hands, and lead to locally demanded actionable change (Holland and Reudin 2012). Newton et al. (2019, 4) note that, "Because empowerment is both an outcome and a process of transformative change it requires the participation of those being empowered to explain changes, as these may not be observed by others." Exploring local visions of empowerment and priorities of women and men should also be a key step in informing programming and assessment (Hillenbrand et al. 2015).

Tensions and challenges with assessing women's empowerment

Yet assessing women's empowerment is not necessarily empowering or desirable. Critical scholars and feminists flag the need to reflect on which measurements are meaningful and useful, at which conjuncture, and to challenge assumptions that it is possible, or should be, to assess abstract and intangible processes of social change (Batliwala and Pittman 2010). Difficulties associated with capturing "power within," coupled with neoliberal biases, result in assessments privileging some dimensions of empowerment (such as economic) over others (such as psychological) (Narayan 2005). There are challenges with identifying appropriate methods to situate women's empowerment processes within their spatial, temporal, and historical contexts (Nazneen et al. 2014), and with defining global indicators of empowerment, given that forms of agency or achievements that indicate empowerment in some contexts may not be relevant in others (Mahmud et al. 2012). Different local understandings of empowerment pose difficulties with translating the concept itself into different languages (or cultural equivalents) (Tsikata and Darkwah 2014, Meinzen-Dick et al. 2019), and mean that externally determined indicators may not correspond with what is valued by those whose empowerment is assessed (Kabeer 1999).

Measurement is a political process that privileges certain types of knowledge and knowing, and the priorities of some actors over others (Batliwala and Pittman 2010, Holland and Reudin 2012, Hillenbrand et al. 2015). There are thus ethical and epistemological issues related to why, and by whom, empowerment should be measured (Morgan 2014, Nazneen et al. 2014, Newton et al. 2019). The use of feminist methodologies to understand women's empowerment can flatten power hierarchies between researchers and participants, situate knowledge production within contexts and relationships, and foster the co-production of knowledge as part of a social change process (for example Cornwall and Sardenberg 2014). Yet current development and policy paradigms tend to favor quantifiable, "objective" indicators over qualitative analyses of trajectories of change in women's lives, expressed in their own words (Nazneen et al. 2014). Nonetheless, all methods make assumptions about what we can and cannot measure and the scale at which we can assess empowerment. For quantitative measures, this includes judgments about proxy indicators of empowerment, their validity, and their relative importance (weighting) (Box 9.1).

Lastly, assessing empowerment as a process is challenging because it is often attempted at one point in time but must capture forward and backward movements and trajectories. Ideally, assessments capture "different dimensions and sites of empowerment in a more holistic way, one that aims to understand the relational dynamics of power and positive change at a variety of levels, in different spaces and over time" (Cornwall 2016, 345). Many measures are cross-sectional snapshots and must be applied longitudinally to provide a sense of change over time. Others ask for retrospective data, which can yield faster results but entails limitations associated with recall. Panel data on empowerment outcomes are better suited for examining longitudinal trajectories of women's empowerment and can complement qualitative assessments that focus on trajectories.

Assessment approaches

Measuring empowerment requires a strong foundational understanding of the concept and its core dimensions, to guide the assessment, develop related indicators, and choose level(s) on which to focus (Narayan 2005, Ibrahim and Alkire 2007, Huis et al. 2017, Richardson 2018). Below, we review a selection of tools to measure empowerment in AR4D identified following a call to

BOX 9.1 Methodological choices in development of the WEAI

Most quantitative measures, recognizing the multidimensional nature of empowerment, use some form of aggregation to construct an empowerment scale or index. The WEAI measures women's empowerment across five domains in agriculture: 1) decisions about agricultural production; 2) access to and decision-making power over productive resources; 3) control over use of income; 4) leadership in the community; and 5) time use (Alkire et al. 2013). These domains, measured in 10 indicators, were based on the areas the United States Agency for International Development (USAID) Feed the Future Initiative could directly affect through its programming.

Two sections of the survey questionnaire proved difficult to administer in the field: the autonomy in production decisions module and the time use module. The desire to reduce survey administration time (and field costs) led to the development of the Abbreviated-WEAI (A-WEAI), with 6 instead of 10 indicators. Indicators that were controversial were removed, such as the "speaking in public" indicator, which was difficult to implement in areas that had experienced civil unrest.

The choice of cut-offs or thresholds for the WEAI and A-WEAI involved value judgments on what made sense for an individual to be considered "adequate" under that indicator, and in many cases was informed by qualitative research in the area. The 80 percent threshold in WEAI (to be empowered, a woman has to be "adequate" in 80 percent of the indicators) was chosen because too high a threshold meant that it would be very difficult to achieve and may not be sensitive to short-term policy changes; and too low a threshold would be too easy to achieve and may not work as a programmatic target (Alkire et al. 2013).

The WEAI co-developers opted for the use of fixed weights—an index rather than a scale—to facilitate comparability across a portfolio, as USAID wanted to compare countries in the Feed the Future Initiative. In WEAI, the five domains were equally weighted, but the indicators were not, as the domains did not have an equal number of indicators. This changed in pro-WEAI, which has 12 equally weighted indicators, equally distributed across the domains. Most agency indicators are instrumental (referring to "power to"), reflecting the areas that agricultural projects can affect directly. Collective agency indicators are few and in the early stages of development. Psychometric methods are being used for scale validation, including estimation of theoretically sound models that have good fit to the data (Yount et al. 2019).

CGIAR gender researchers and key partners and researchers. Some of these respondents also shared reflections on the strengths and limitations of their tools, and findings emerging through their use.

Our framework to analyze these tools comprises five components:

1. Dimensions of empowerment (resources, agency, and/or achievements);
2. Primary levels of inquiry (personal, relational, and/or environmental);
3. Participant focus (who participates in the assessment);
4. Attention (or lack thereof) to gender parity; and
5. Assessment perspective (etic versus emic).[10]

Table 9.1 presents a brief summary of our analysis of the tools across these components, with attention to the quantitative or qualitative nature of the tools. Oftentimes, tools cannot be exclusively labeled as quantitative or qualitative based on the way they are operationalized and on how the data collected are analyzed. Hence, we do not distinguish between quantitative and qualitative methods in a strict or binary way but rather surface some of the strengths and limitations that methods steeped in different epistemological traditions can offer for understanding and assessing empowerment, and the value of bringing these together for richer and more complete assessments.

Dimensions and levels of empowerment

We combine the first two components of our framework—dimensions and levels—in a light mapping of the tools to represent their relative placement along two axes (Figure 9.2). The horizontal axis indicates the multidimensionality of the measure and the vertical axis its multilevel character. Moving from the bottom left toward the top right, tools explore more dimensions and levels of empowerment.

The tools cluster roughly into four groups. First, tools that use a unidimensional approach to assessing empowerment at one level are located in the bottom left corner. In contrast, tools that focus on one empowerment dimension but at multiple levels are located in the upper left corner. Third, a group of measures that use a multidimensional approach to assessing empowerment at one or more levels are located in the center of the figure. A fourth cluster consists of tools that explore the three dimensions of empowerment at

10 Emic perspectives refer to perceptions of "insiders": people within a given social group. Etic perspectives are those of observers or "outsiders" to the given group.

TABLE 9.1 Tools to measure empowerment used in agricultural research for development

Tool name	Description	Empowerment dimension(s) focus$	Primary level(s) of inquiry*	Participant focus	Assessment of gender parity?	Measurement perspective
PRIMARILY QUANTITATIVE						
Women's Empowerment in Agriculture Index (WEAI) (Alkire et al. 2013)	Designed for implementation in population-based surveys, with a strong focus on women's productive roles. Comprises two sub-indices: 1) 5DE—women's empowerment across five domains in agriculture and 2) the Gender Parity Index—gender parity in empowerment within the household.	Agency, with some focus on resources	Personal and relational	Individual women and men within the same household	Yes	Etic
Abbreviated-WEAI (A-WEAI) (Malapit et al. 2017)	Shorter version of the WEAI for use in population-based surveys to measure women's empowerment.	Agency, with some focus on resources	Personal and relational	Individual women and men within the same household	Yes	Etic
Project level-WEAI (pro-WEAI) (Malapit et al. 2019, Meinzen-Dick et al. 2019)	Measures women's empowerment in project-specific contexts; includes optional modules tailored to livestock and/or nutrition and health programs.	Agency, with some focus on resources	Personal and relational, some focus on environmental	Individual women and men within the same household	Yes	Etic
Women's Empowerment in Livestock Index (WELI) (Galiè et al. 2018)	Adaptation of the WEAI to assess the empowerment of women in the livestock sector; complemented by two rounds of qualitative research pre- and post-application of the survey.	Agency, with some focus on resources	Personal and relational	Individual women from livestock-producing household	No	Etic and emic
Women's Empowerment in Fisheries Index (WEFI) (Cole et al. 2020)	Adaptation of the A-WEAI; includes a scale to assess gender attitudes, from which a score is created.	Agency, with some focus on resources	Personal and relational, some focus on environmental	Individual women and men value chain actors	Yes	Etic
Gender Empowerment Index for Climate-Smart Villages (GEI–CSV) (Hariharan et al. 2018)	Based on the Global Gender Gap Index1 and the WEAI. Constructed across four domains (political, economic, social, agricultural), each with a different weight, based on insights from focus group discussions (FGDs) conducted before the survey to inform design of the tool.	Agency, with some focus on resources	Personal and relational	Individual women and men within the same household	Yes	Etic

continued

Tool name	Description	Empowerment dimension(s) focus$	Primary level(s) of inquiry*	Participant focus	Assessment of gender parity?	Measurement perspective
IRRI's Women's Empowerment Index (WEI) (Achandi et al. 2018)	Index calculated on a scale from 1 to 5 based on the level of participation of women in decision-making within the household; adapted to rice farming systems.	Agency	Relational	Individual women	No	Etic
Empowerment profiles (Najjar et al. 2018)	Comprise 27 variables drawing from the WEAI's 5DE, and covering socioeconomic characteristics and asset ownership. Use of multivariate cluster analysis to identify homogenous groups of women and men (empowerment typology).	Resources, with some focus on agency	Personal and relational	Individual women and men from a stratified sample to cover diversity in land access, ownership, and use	No	Etic
Women's Decision-Making Index and Gender Attitudes Index (WDI–GAI) (Kosec et al. 2018)	Women's empowerment indices focused on decision-making and gender attitudes, constructed using Principal Component Analysis (PCA) and Anderson indices. Survey instruments are adapted to national contexts.	Agency	Relational, and some focus on environmental	Applied with all adult household members or with household head and spouse	No	Etic
CARE's Women's Empowerment Index (WEI) (CARE 2015)	Adaptation of the WEAI combined with other measures. Questions on mobility and use of FGDs to gather data on time use. Composite score with country-specific thresholds.	Agency, with some focus on resources	Personal and relational	Individual women and men in the same household	Yes	Etic
Oxfam's Women's Empowerment Index (WEI) (Lombardini and McCollum 2018)	Indicators of an "empowered woman" adapted to context. Questions derived mainly from Demographic and Health Survey toolkit questionnaires (USAID),2 WEAI, and Living Standards Measurement Survey (World Bank).3 Explicitly considers social norms or policies and laws at the environmental level that contribute to women's (dis)empowerment.	Agency, with some focus on resources	Personal, relational, and environmental	Individual women	No	Etic
Comparison of the Five Dimensions of Men's and Women's Empowerment (5 Dimensions) (Mayanja et al. 2018)	Indicators of empowerment in five domains, based on WEAI. In FGDs, respondents are asked about their own ability to make decisions within each particular domain, and about the ability of women and men in their community to make decisions in these domains.	Agency	Relational	Individual women and men, not necessarily from the same household	No	Etic

Tool name	Description	Empowerment dimension(s) focus[$]	Primary level(s) of inquiry*	Participant focus	Assessment of gender parity?	Measurement perspective
PRIMARILY QUALITATIVE						
Ladder of Power and Freedom (Petesch and Bullock 2018)	Uses scores and narrative data to understand women's and men's sense of their capacity to make strategic life decisions, or those of other women or men in their community; and to shed light on processes underpinning changes in their sense of agency over time.	Agency, but can capture changes in other dimensions	Relational, and some focus on environmental	Women and men, not necessarily from the same household	Possible; depends on the analysis	Etic and emic
Life histories and well-being timelines (Petesch et al. 2018)	Explores a person's occupational, economic, and social, psychological, and cultural histories. Participants identify key moments along their life trajectories in these spheres going back 10 years. They score significant moments and explain the reasons for their scores. An overall "well-being" trend line is developed based on the consolidated data.	Resources, agency, and achievements	Personal, relational, and environmental	Women and men, not necessarily from the same household	No	Etic and emic
Gender Indicator Monitoring Tool (GIMT) (CARE 2015, Hillenbrand et al. 2015)	Participatory outcome mapping to identify incremental indicators of behavior change toward the vision of gender equality outlined by community members. Through FGDs, evaluates behavior changes on a six-monthly basis around 1) household decision-making processes; 2) men's engagement and personal changes; and 3) community leaders' views and practices; as well as ascertaining women's own definitions of empowerment.	Agency and achievements, with some focus on resources	Personal, relational, and environmental	Women's group members, men's group members or spouses, community leaders	No	Etic and emic

Notes: [$] Empowerment conceptualized along three interconnected dimensions: resources, agency, and achievements (well-being outcomes) (Kabeer 1999, 435).

* Based on the framing of different empowerment levels by Lombardini et al. (2017).

[1] See https://www.weforum.org/reports/the-global-gender-gap-report-2018

[2] See https://www.k4health.org/toolkits/dhs

[3] See http://surveys.worldbank.org/lsms

FIGURE 9.2 Positioning tools based on their attention to dimensions and levels of empowerment

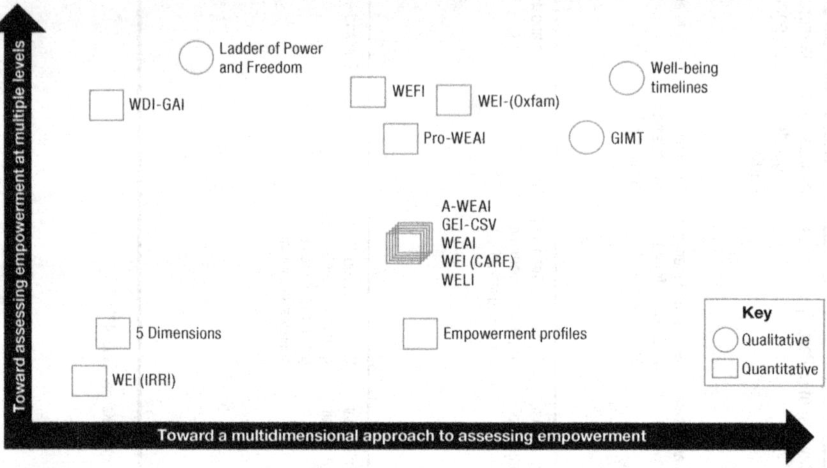

Note: Tool names are abbreviated here; refer to Table 9.1 for full names and descriptions of the tools.

the three levels of inquiry—personal, relational, and environmental—located in the upper right corner.

As mentioned above, the two tools situated in the bottom left corner measure only one dimension of empowerment at one point in time. Both measure one aspect of agency—decision-making within the household—to examine relational aspects of women's empowerment. The International Rice Research Institute (IRRI) **Women's Empowerment Index** (WEI) scores women's participation in significant decisions within their households along a five-point scale (Achandi et al. 2018). The two extremes indicate that either the husband (WEI = 1) or the wife (WEI = 5) makes all decisions in the household solely, whereas a score of 3 indicates that spouses have an equal say in intrahousehold decision-making. Similarly, the **Comparison of the Five Dimensions of Men's and Women's Empowerment** tool examines women's and men's perceived ability to make intrahousehold decisions in relation to five domains of empowerment drawn from the WEAI (Mayanja et al. 2018).

The tools located in the upper left corner remain focused on agency, and particularly on decision-making, but explore this dimension in relation to some of the structural (or "environmental") dimensions of (dis)empowerment: the norms that underpin gender inequalities and constrain women's abilities with regard to self-determination. The **Women's Decision-Making Index and Gender Attitudes Index** (WDI–GAI) maintain emphasis on women's decision-making, and examine gender attitudes to bring some of the beliefs that underlie decision-making patterns to light (Kosec et al. 2018). The qualitative GENNOVATE **Ladder of Power and Freedom** explores women's and men's sense of freedom to decide on important matters in their lives (Petesch and Bullock 2018).[11] Participants, either individually or in a focus group discussion (FGD), score this capacity along a metaphorical 5-step ladder, and reflect on changes and reasons for these over the past 10 years. These factors may include changes in resources, in formal or informal structures, in critical consciousness, or more. In this sense, although the tool explicitly asks about changes in agency, the number of dimensions of empowerment the tool addresses depends on participants' reflections on their experiences—as does the number of levels at which it captures changes in women's empowerment.

Most of the tools reviewed, many of which relate closely to each other, sit in the central area of the figure, providing a more multidimensional and multilevel examination of women's empowerment. The WEAI (Alkire et al. 2013) and related measures focus primarily on agency, but also touch upon aspects related to resources. They explore empowerment at the personal and relational levels but are less suited to capturing the environmental level. The **project-level WEAI** (pro-WEAI) takes a mixed-methods approach to examine women's empowerment within project-specific contexts (Malapit et al. 2019, Meinzen-Dick et al. 2019, Yount et al. 2019). Compared with its precursors, it further unpacks agency by looking at three domains: intrinsic agency ("power within"), instrumental agency ("power to"), and collective agency ("power with"). It expands upon the WEAI, including with indicators related to intrahousehold harmony, attitudes toward intimate partner violence toward women, and mobility. The accompanying qualitative tools assess elements related to the environmental level. In the fisheries sector, the **Women's Empowerment in Fisheries Index** (WEFI) combines the WEAI and elements from the framework proposed by van Eerdewijk et al. (2017) with a gender attitudes scale (Cole et al. 2020). It assesses change in agency (in terms of decision-making about income) and in exercising choice to partake in livelihood opportunities (an expression of agency), resources (in terms of control over value chain assets), and institutional structures (attitudes toward inequitable gender norms). The gender attitudes scale captures additional elements of empowerment at the environmental level.

11 GENNOVATE is a comparative qualitative research initiative designed to examine the relationship between gender norms, agency, and agricultural innovation (see https://gennovate.org/).

In the upper right of Figure 9.2 sit two qualitative tools that explore the three empowerment dimensions at personal, relational, and environmental levels: the **well-being timelines** of the GENNOVATE methodology and CARE's **Gender Indicator Monitoring Tool** (GIMT). The former explores occupational, economic, social, psychological, and cultural histories to identify and understand the most significant milestones in a person's life. The tool offers deep insights into diverse aspects of agency, resources, and achievements, and sheds light on how individual capacities, relations and interpersonal dynamics, and social institutions affect these. The GIMT, developed by CARE as part of its Pathways Program, uses participatory outcome mapping (Hillenbrand et al. 2015) to identify incremental indicators of behavior change that demonstrate progress toward a vision of gender equality outlined by community members. The tool evaluates behavior changes around household decision-making processes; men's engagement in projects and their personal changes; and community leaders' views and practices. It privileges women's own definitions of empowerment, as do the GENNOVATE well-being timelines.

Several key points emerge from this analysis. First, most of the reviewed tools recognize the multidimensional and multilevel nature of empowerment in assessments, which bodes well for bringing some of the complexity of the concept into AR4D thinking and practice. In this regard, the tools offer potential to consider the interactions among changes across dimensions and levels. Yet, studies based on these tools rarely perform such an analysis and, in general, the tools offer little guidance for analysis, such as interpreting how deep and broad, and of what scale, are the changes taking place.

Second, many tools fall short of carefully exploring changes at the environmental level, and thus of shedding light on structural causes of gender inequality. Assessments that focus on the individual and relational levels reflect, and can reinforce, a programmatic and project focus on change at these levels. Such an emphasis on individual capacities risks ignoring power relations and structures that (re)produce gender inequalities and constrain women's capacities to make purposive choices (Kabeer 2005, Batliwala 2007, Woodall et al. 2012, Hillenbrand et al. 2015).

Third, although many AR4D interventions focus on enhancing rural women's (and men's) resources—tangible and "countable" areas of change, such as income and assets, which are market-driven values (Narayan 2005, Cornwall 2014)—the tools we reviewed focus less on this dimension of empowerment. This may be because they are often integrated in larger monitoring and evaluation strategies, which include surveys that ask about changes in resources and (material) achievements. Within their focus on agency, most tools explore instrumental agency ("power to") rather than changes in "power within" and "power with." This may owe to the difficulty of assessing the multiple dimensions of agency. Interventions may also privilege efforts toward what they consider they can most directly affect, such as instrumental agency (as decision-making over resources), rather than intrinsic and collective agency. These latter are difficult to address with short-term projects and funding, and may be considered out of scope and of lower value in a neoliberal development agenda.

Placement of tools along the horizontal and vertical axes does not necessarily indicate the tools' ability to reveal the breadth of the changes taking place. Nor does it reflect the quality of the tools or the data they elicit per se. Data interpretation, on which most of the tools offer limited guidance, is also key to the quality of the assessments. Moreover, single tools can be integrated as part of a broader methodology that addresses other dimensions or levels of empowerment. For example, the Ladder of Power and Freedom tool constitutes one part of the GENNOVATE methodology, which combines different tools to study normative change and women's and men's empowerment in agriculture and natural resource management. The United Nations Food and Agriculture Organization and WorldFish combined the Ladder of Power and Freedom tool with five others in qualitative case studies in Bangladesh; the other tools assessed village characteristics and wealth distribution, gendered divisions of labor, factors affecting participation in aquaculture value chains, intrahousehold decision-making processes, and access to resources and services (Choudhury et al. 2017). This combination offered insight into the informal institutional structures that influenced empowerment. Nonetheless, tools that adopt a narrow assessment focus risk reinforcing a limited and counterproductive understanding of empowerment. This calls for clarity on the scope and limitations of each tool, and on the need to implement them reflexively, considering their appropriateness and possible need for adaptation.

Participant focus: gender parity and intersectionality

While all tools focus their assessments of the agency dimension of empowerment mostly at the personal and/or relational level, the majority situate the analysis within the household, a formerly often neglected domain, and particularly looking at relations among spouses. Some tools rely on interviews of women only, whereas others rely on interviews with both women and men, often, but not exclusively, within the same household (Table 9.1).

The WEAI and many of its related measures explicitly measure gender parity in empowerment between a woman and man within a household to understand if or how these relate to each other. Such a comparison can contribute to understanding whether and which facets of women's disempowerment result from oppressive gender regimes, and which owe to poverty and contextual constraints that disempower both women and men. Measures such as the WELI and WEI (IRRI) focus only on women, likely because of specific targets for projects set by institutions, donors, or researchers. In few cases, tools are administered to actors outside the household (such as to community leaders, in CARE's GIMT) or ask participants to reflect on empowerment at the community level (such as the 5DE and the GENNOVATE Ladder of Power and Freedom tools). Even when empowerment is assessed at the environmental level, it is done from the perspective of individuals who experience the (dis)empowering effects of societal structures.

Most of the tools reviewed do not provide explicit guidance on sampling beyond the household level, as such decisions depend on the purpose of the study. Discussions of intersectionality in relation to measuring women's empowerment in AR4D are surprisingly limited in the literature and tools reviewed. Some studies provide insights into how the tools can surface how gender interacts with other axes of social discrimination to create disempowerment and marginalization. For instance, socioeconomic and demographic data collected for the WEAI and related measures enable analyses of correlations between degrees of empowerment and sex, age, marital status, and other variables.

Bourdier's (2019) analysis of WEAI data in Ghana takes into account polygyny and highlights the importance of looking at which wives are sampled within a household. In Nepal, O'Hara and Clement (2018) iterate between the WEAI and subjective measures of critical consciousness, highlighting the importance of household structure (extended versus nuclear families) in affecting women's empowerment. In India, Hariharan et al. (2018) calculate the **GEI-CSV** in two states (Haryana and Bihar) and highlight the geographical unevenness of, and constraints to supporting, women's empowerment in different cultural contexts. In turn, Najjar et al. (2018) use 27 variables to create empowerment profiles of Egyptian women and men farmers with different land entitlements, to investigate the link between empowerment, sex of the farmer, and land access and ownership. Applied with women and men of different age and wealth groups, GENNOVATE tools enable comparative analyses that link empowerment processes with life cycle and socioeconomic status.

Measurement perspective, holistic assessments, and mixing methods

The choice of approach, tools, and methods for assessing women's empowerment depends on the motives for the assessment, as well as its scale. Monitoring empowerment at a global level, for instance, often calls for measures that enable comparative analyses. Yet, as the concept of empowerment holds meanings only within the specific contexts it inhabits, balancing between the ability to measure across countries and assessments that capture the contextual nature of empowerment is important (Richardson 2018).

Qualitative methods are particularly apt at providing contextual information and eliciting context-specific attributes of empowerment, and at grounding definitions of empowerment in the experiences of women of different backgrounds (see for example Newton et al. 2019 on participatory approaches). They are also valuable for shedding light on *processes* of change, including on when or how transformative change occurs (Morgan 2014, Elias and Morgan 2016). Qualitative narratives foreground the complex, emergent, and non-linear nature of empowerment, and how advances in empowerment in one area of life may, or not, be accompanied by advances (or setbacks) in another. Yet, compared with quantitative tools, qualitative tools offer less comparability, information on trends, and numerical information, which donors and decision-makers are often seeking. Quantitative tools can also be designed to be context-sensitive and comparable, if consistent guidelines and protocols for adaptation are developed.

The quantitative tools reviewed here use an etic perspective when defining or conceptualizing empowerment, with some exceptions. The **Women's Empowerment in Livestock Index** (WELI) (Galiè et al. 2019) adapts the WEAI and pro-WEAI to assess empowerment of women in the livestock sector. Two rounds of qualitative research—pre- and post-application of the survey—complement the mainly quantitative tool. A formative qualitative and participatory study captures universal dimensions of empowerment that allow for comparison across settings, and local meanings of empowerment that can be used for in-depth monitoring and assessment. Likewise, Oxfam GB's **Women's Empowerment Index**[12]—based on Lombardini et al.'s (2017) framework (Figure 9.1)—comprises a range of indicators that represent the characteristics of an "empowered woman." These indicators are adjusted to the socioeconomic context under analysis based on qualitative fieldwork on

12 This is not the same as the Women's Empowerment Index developed by Achandi et al. (2018) or that of CARE.

perceptions of what constitutes an empowered woman, thereby allowing context-specific signs of empowerment to surface.

The qualitative tools reviewed mostly, but not exclusively, use an emic perspective. The GENNOVATE Ladder of Power and Freedom tool takes an emic perspective in eliciting local understandings of what strategic decisions consist of and what influences ability to make them, but an etic perspective to analyzing the data. CARE's GIMT adopts a similar approach to defining empowerment using emic perspectives. The GENNOVATE methodology includes a module focused on life histories, which asks participants to identify, score, and explain the significance of key moments in different arenas of their life going back 10 years (Petesch et al. 2018). These well-being timelines reflect participants' emic understanding of the combinations, interactions, and sequencing of key events over their trajectories and their influence on subjective well-being. In general, open-ended, qualitative tools allow participants to express in their own words aspects related to resources, agency, and achievements; the different levels at which empowerment manifests itself; and their interrelationships. They also surface the relative importance of different factors in supporting or hindering empowerment.

Quantitative and qualitative methods for assessing empowerment both have their strengths and limitations. Combining and triangulating methods can be valuable in both measurement and analysis. Qualitative and quantitative methods can be combined to create measures, in the measures themselves, and in interpretation of results. Several tools have used qualitative methods to inform the development of quantitative measures. For instance, Oxfam uses FGD data to develop locally relevant thresholds and indicators for its Women's Empowerment Index (Lombardini and McCollum 2018). The pro-WEAI developed a suite of qualitative tools (including key informant interviews, FGDs, and life histories) to be used with the surveys. Together with past qualitative data from the project areas, these informed development of the domains, indicators, and thresholds of the index (Malapit et al. 2019, Meinzen-Dick et al. 2019). In particular, the negative views of women having "power over" others (notably men, but also over other women) informed the decision to exclude a domain on coercive power.

The qualitative data also revealed differences between societies depending on whether individual or joint asset ownership or decision-making was considered (more) empowering. Thus, these indicators in the index accepted both individual and joint as "empowered." Perhaps most importantly, the qualitative data reinforced the understanding that empowerment is relational and needs to be understood in the context of the entire family and community (Meinzen-Dick et al. 2019). This underscores the importance of collecting pro-WEAI survey data from men and women or multiple members of extended families—that is, co-wives in polygynous households or mothers- and daughters-in-law in extended families in South Asia.

Combining qualitative and quantitative methods in the measures themselves is less common. One example, noted above, is CARE Pathways' use of FGD data on women's time use along with survey data on other indicators of women's empowerment. In CARE's GIMT, the assessment is qualitative in nature but the data collected can be quantified to demonstrate the direction of change in certain broad categories of indicators. Similarly, the GENNOVATE Ladder of Power and Freedom offers a qualitative assessment as well as a quantitative figure to show a direction and relative magnitude of change. However, combining qualitative and quantitative often entails converting qualitative to quantitative data, during which much of the nuance in and advantage of collecting qualitative data is lost.

Joint or iterative use of qualitative and quantitative data for interpretation is one of the most valuable uses of mixed methods. CARE assesses women's empowerment by combining tools, such as the GIMT; the Women's Empowerment Index (Miruka et al. 2015) (see Table 9.1); the **Women's Empowerment—Multidimensional Evaluation of Agency, Social Capital and Relations** (WeMEASR) scale; and the **Social Norms Analysis Plot** (SNAP). The quantitative and qualitative data may contradict each other, but this is not a drawback of this method (ibid.). Rather, the creative tension between qualitative and quantitative findings should be anticipated and appreciated, and can be used to add nuance to the understanding and interpretation of results. For instance, by combining the WEAI, their constructed measure of critical consciousness, and qualitative data, O'Hara and Clement (2018) could better capture local understandings of empowerment within a broader cultural context that shapes values, meanings, and identities.

Galiè et al. (2019) illustrate the value of integrating methods in interpretation. They combine data from the WELI and quantitative indicators of food security with FGD data from pastoralist households in Tanzania. They find no significant association between women's empowerment and household food security in the quantitative analysis; yet, in FGDs, women identified mechanisms through which changes in their time use and control over livestock and land resources had influenced their ability to provide sufficient nutritious food for their families. Further analysis points to gender differences in who is in charge of securing food versus nutrition at household level—men and women, respectively, in this context. Analysis of the qualitative data associated with pro-WEAI also reveals interconnections among the quantitative indicators. For instance, burdens on women's time as well as relations with their husbands and in-laws limit women's mobilities and abilities to participate in groups (Meinzen-Dick et al. 2019). Projects that promote group membership should address such disempowering aspects, for instance by engaging with husbands and mothers-in-law.

Critical questions in a future research agenda

The growing number of tools and methods for assessing women's empowerment in agriculture and beyond reflects significant efforts to advance this field. Yet urgent political and ethical questions as well as substantive challenges remain. First, women's empowerment remains framed predominantly as a pathway for enhanced agricultural outcomes (for example FAO 2011, World Bank 2012). In this regard, women farmers are expected to adopt technologies to increase productivity and food and nutrition security, without questioning their roles and responsibilities. Accordingly, tools designed to assess such processes focus on individual women's access to material resources or visible forms of agency, such as decision-making. This reinforces flawed assumptions about how empowerment may be achieved through agriculture. Yet caution is needed: "Such forms of agency might not lead to social change and to collective action that would allow women to challenge oppressive economic, social and political structures, as long as women do not critically reflect on gender inequalities and its structural causes" (O'Hara and Clement 2018, 121). A renewed focus is thus necessary on critical consciousness and women's collective action, and their key role in empowerment and gender equality—and, more generally, meaningful social change.

Challenging apolitical and instrumentalist views of empowerment in AR4D will require refocusing methodologies to explore women's collective and intrinsic agency ("power with" and "power within") and identify the power relations and structures—the environmental-level elements and the "power over"—that underpin women's disempowerment and gender inequalities. The pro-WEAI takes an important step in this direction by incorporating domains related to intrinsic and collective agency. So, too, does the emerging body of research on gender transformative approaches (see Chapter 10, this volume) that builds on efforts to assess normative change in the field of AR4D (CARE 2017) and beyond (BMGF 2018).

Second, a focus on women's empowerment *in agriculture* should not lose sight of the possibility that agriculture itself is not always empowering for women (or men). The prospects agriculture can offer as a pathway toward empowerment depend on women's aspirations; and empowerment will ultimately require that women have the resources and agency to choose to pursue meaningful livelihoods within or beyond the sector. The agricultural focus of several tools presented offers important insights but may divert attention from other areas of rural women's lives that are at least as relevant for empowerment. More holistic measures that capture empowerment outside agriculture are needed, to avoid the risk of misclassifying women who have left agriculture as disempowered.

Third and related, there is a need to systematically document *how* shifts in empowerment and transformative change occur within agriculture and NRM and beyond, at what level(s), and for whom. Qualitative or mixed methods approaches can help us focus on the change mechanisms and trajectories that enable women to empower themselves. These methodologies will also be highly valuable for incorporating a meaningful intersectional perspective. Representative samples that are comparable across social groups can also add to capturing the diverse, lived realities of marginalized groups (Yount et al. 2018).

Fourth, measures of empowerment must be able to detect situations in which advances lead to backlash and setbacks in a change trajectory. Positive change in some dimensions can engender impediments in others. Women's economic empowerment does not necessarily correlate with familial, psychological, legal, political, and socio-cultural dimensions of empowerment (Bayissa et al. 2018). In fact, it can be a double-edged sword, leading to regressive change in certain dimensions. Serious challenges to social power structures can create resistance, which may be misinterpreted as a lack of effectiveness if assessments are not sensitive to this process. This potential backlash has implications from a programmatic perspective and has not yet been adequately resolved in measures of women's empowerment (for example Batliwala and Pittman 2010).

Fifth, measures must grapple with the nuances and complexities of decision-making and agency. The WEAI and related measures have made progress in recognizing different degrees of "jointness" in decision-making among spouses or household members, and preferences for joint decision-making in some conditions and cultures (Acharya et al. 2010, Belcher et al. 2011, Farnworth et al. 2019). It is equally relevant to recognize that, in some situations, women may not wish to be involved in certain types of decisions (Nazneen et al. 2014). An ideal measure of empowerment should be able to discern such scenarios of "choosing not to choose" (Kabeer 1999) as a sign of agency rather than lack thereof. Measures and interpretations should also be sensitive to how heightened self-awareness and critical consciousness can give rise to a decreased sense of agency (Freire 1970). Women's exposure to new spaces, information, and critical reflections on their lives can lead them to (downwardly) reassess their own knowledge and sense of empowerment (for example Galiè 2013).

The refocusing of methodologies, as outlined above, calls for significant changes within AR4D. Action-oriented research with multiple actors (such as researchers, diverse local groups, non-governmental organizations, local authorities, government, etc.) can be particularly well suited to understand and address some of the underlying causes of women's disempowerment and gender inequalities (see also Cornwall 2016). Such research, which engages with social hierarchies, is complex and messy, and can push AR4D researchers outside of their comfort zone. Yet it can also help unearth the structural barriers that create privilege and opportunity for some, and constraints, exclusions, and disempowerment for others.

There is also large potential in "using unconventional tools, creative use of conventional methods, and thinking outside the box for capturing the less researched aspects or developing a deeper understanding of women's empowerment" (Nazneen et al. 2014, 59). These include participatory photography (Cornwall and Sardenberg 2014), storytelling and creative writing (Ali 2014), intergenerational life history narratives (Tsikata and Darkwah 2014), and other methods informed by feminist ethics and epistemological considerations (Cornwall and Sardenberg 2014). Visual methods have shown their worth for exploring relational agency and aspirations, and surfacing emotions and feelings that are difficult to express verbally (Eger et al. 2018), but are used only marginally in AR4D.

Embracing less conventional and mixed-methods approaches to assessing empowerment will require new commitments from the AR4D community. It will mean learning to respect and dialogue across disciplines rooted in different epistemological traditions, and also adequate investment in strengthening capacities in (qualitative) research that demands a specific set of skills that is often in shorter supply in the sector. The AR4D ecosystem will need to move beyond a preference for quantitative data and experimental designs; reconsider assumptions that change follows a linear trajectory; open itself up to exploring unanticipated and negative outcomes; value changes in relationships that are often less visible, tangible, and thus measurable; and allow (and budget) for assessments (and changes) to take place well beyond short project cycles, considering that empowerment and social change can be lengthy processes (Morgan 2014). These efforts can bring us closer to the changes needed for women to empower themselves and advance gender equality through and beyond agriculture.

Geography of study sites for publications cited in Chapter 9

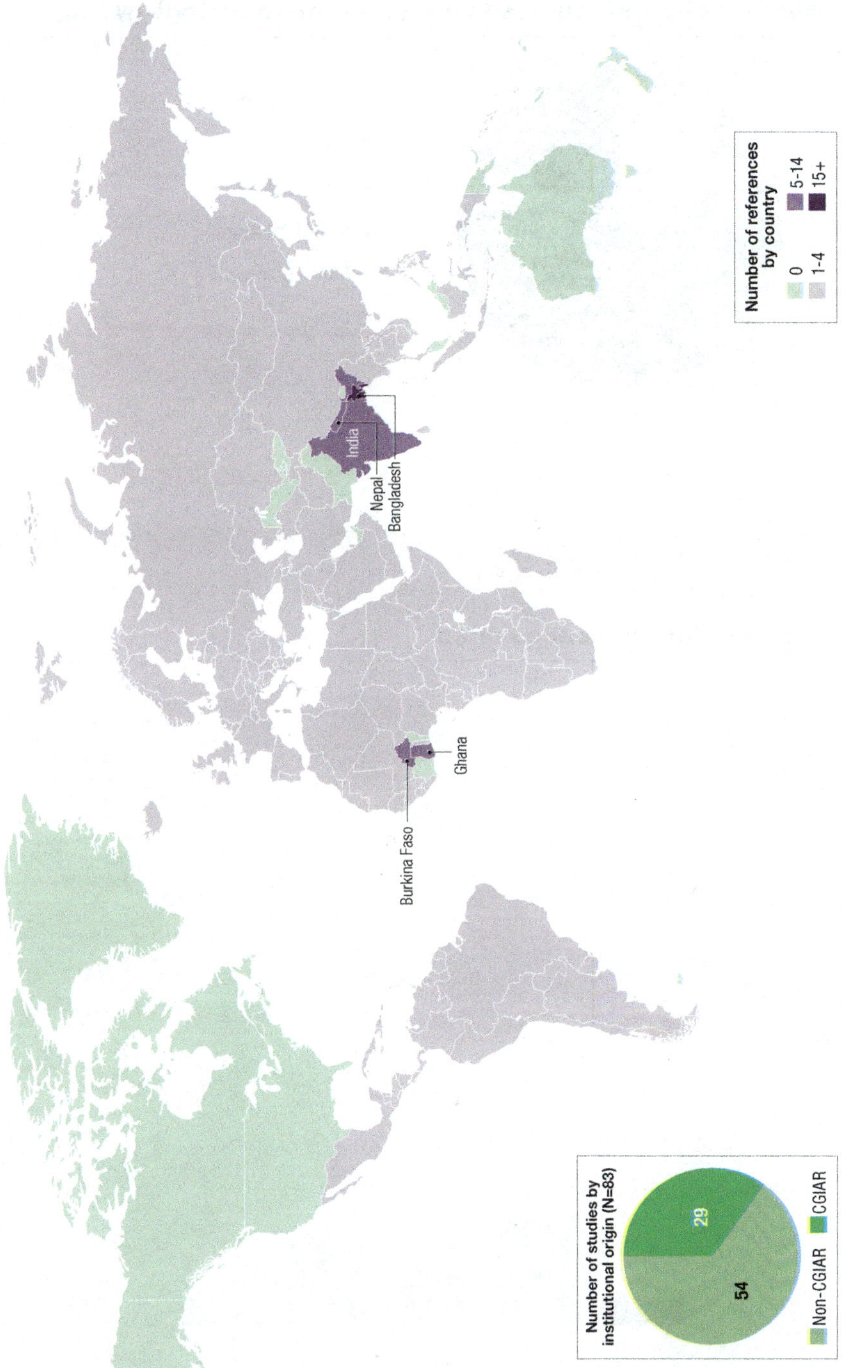

Number of cited studies by research methodology (N=83)

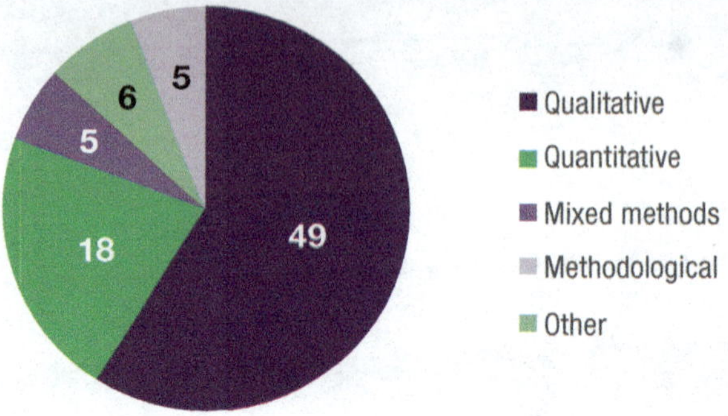

Timeline for references cited

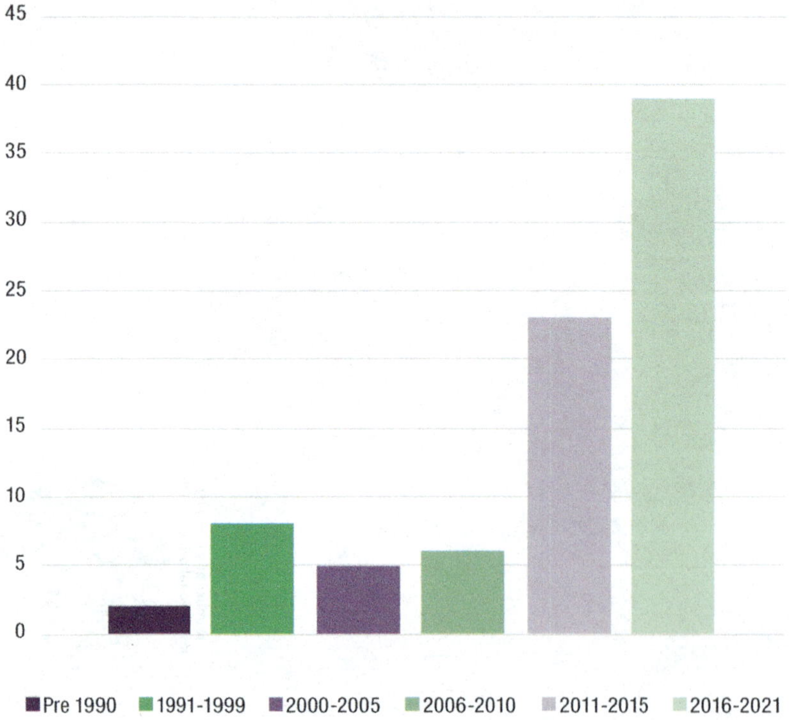

References

Achandi, E. L., G. Mujawamariya, A. Agboh-Noameshie, S. Gebremariam, N. Rahalivavololona, J. Rodenburg. 2018. "Women's Access to Agricultural Technologies in Rice Production and Processing Hubs: A Comparative Analysis of Ethiopia, Madagascar and Tanzania." *Journal of Rural Studies* 60: 188–198.

Acharya, D, J. Bell, P. Simkhada, E. van Teijlingen, P. Regmi. 2010. "Women's Autonomy in Household Decision-Making: A Demographic Study in Nepal." *Reproductive Health* 7 (1): 15.

Ajzen, I. 1991. "The Theory of Planned Behavior." *Organizational Behavior and Human Decision Processes* 50 (2): 179–211.

Ali, M. 2014. "Stories/Storytelling for Women's Empowerment/Empowering Stories." *Women's Studies International Forum* 45: 98–104.

Alkire, S., R. Meinzen-Dick, A. Peterman, A., Quisumbing, G. Seymour, A. Vaz. 2013. "The Women's Empowerment in Agriculture Index." *World Development* 52: 71–91.

Basu, A. 1995. "Introduction: The Challenges of Local Feminisms." In *Women's Movements in Global Perspective: The Challenge of Local Feminisms*, edited by A. Basu, 1–25. Boulder, CO: Westview Press.

Batliwala, S. 2007. "Taking the Power out of Empowerment: An Experiential Account." *Development in Practice* 17 (4–5): 557–565.

Batliwala, S., and A. Pittman. 2010. "Capturing Change in Women's Realities: A Critical Overview of Current Monitoring & Evaluation Frameworks and Approaches." Toronto: AWID.

Bayissa, F. W., J. Smits, R. Ruben. 2018. "The Multidimensional Nature of Women's Empowerment: Beyond the Economic Approach." *Journal of International Development* 30 (4): 661–690.

Belcher, J. R., E. V. Peckuonis, B. R. Deforge. 2011. "Family Capital: Implications for Interventions with Families." *Journal of Family Social Work* 14 (1): 68–85.

BMGF (Bill & Melinda Gates Foundation). 2018. "Advancing Social Norms Practice." Meeting Report, Social Norms Convening, Seattle, WA, February 12–13.

Bourdier, T. 2019. "Women's Empowerment and Child Nutrition in Polygynous Households of Northern Ghana." Discussion Paper 1809, IFPRI, Washington, DC.

CARE. 2015. *Designing a Common Gender Indicator Framework for CARE's Pathways Program.* Atlanta, GA: CARE USA.

CARE. 2017. *Applying Theory to Practice: CARE's Journey Piloting Social Norms Measures for Gender Programming.* Atlanta, GA: CARE USA.

Carr, E. S. 2003. "Rethinking Empowerment Theory Using a Feminist Lens: The Importance of Process." *Affilia: Journal of Women and Social Work* 18 (1): 8–20.

Carter, J., S. Byrne, K. Schrader, H. Kabir, Z. Uraguchi, B. Pandit, B. Manandhar et al. 2014. "Learning about Women's Empowerment in the Context of Development Projects: Do the Figures Tell Us Enough?" *Gender & Development* 22 (2): 327–349.

Choudhury, A., C. McDougall, S. Rajaratnam, C. M. Y. Park. 2017. *Women's Empowerment in Aquaculture: Two Case Studies from Bangladesh*. Rome: FAO.

Cole, S. M., P. Kantor, S. Sarapura, S. Rajaratnam. 2014. "Gender-Transformative Approaches to Address Inequalities in Food, Nutrition and Economic Outcomes in Aquatic Agricultural Systems." Working Paper AAS-2014-42, AAS, Penang.

Cole, S. M., A. M. Kaminski, C. McDougall, A. S. Kefi, P. Marinda, M. Maliko, J. Mtonga. 2020. "Gender Accommodative versus Transformative Approaches: A Comparative Assessment within a Post-Harvest Fish Loss Reduction Intervention." *Gender, Technology, and Development* 24 (1): 48–65.

Cornwall, A. 2014. "Women's Empowerment: What Works and Why?" WIDER Working Paper 2014/104, UN University, Helsinki.

Cornwall, A. 2016. "Women's Empowerment: What Works?" *Journal of International Development* 28 (3): 342–359.

Cornwall, A., and J. Edwards. 2014. "Negotiating Empowerment." In *Feminisms, Empowerment and Development: Changing Women's Lives*, edited by A. Cornwall, and J. Edwards, 1–22. London: Zed Books.

Cornwall, A., and A. M. Rivas. 2015. "From 'Gender Equality' and 'Women's Empowerment' to Global Justice: Reclaiming a Transformative Agenda for Gender and Development." *Third World Quarterly* 36 (2): 396–415.

Cornwall, A., and C. Sardenberg. 2014. "Participatory Pathways: Researching Women's Empowerment in Salvador, Brazil." *Women's Studies International Forum* 45: 72–80.

Danielsen, K., F. Wong, D. McLachlin, S. Sarapura. 2018. *Typologies of Change: Gender Integration in Agriculture and Food Security Research*. Amsterdam: KIT.

Deere, C. D. 2018. "Sustainable Development Goals, Gender Equality and the Distribution of Land in Latin America." *cadernos pagu* 52: e185206.

Diiro, G. M., G. Seymour, M. Kassie, G. Muricho, B. W. Muriithi. 2018. "Women's Empowerment in Agriculture and Agricultural Productivity: Evidence from Rural Maize Farmer Households in Western Kenya." *PloS ONE* 13 (5): e0197995.

Eger, C., G. Miller, C. Scarles. 2018. "Gender and Capacity Building: A Multi-Layered Study of Empowerment." *World Development* 106: 207–219.

Elias, M., and M. Morgan. 2016. "Mixing Methods for Holistic Project Evaluations: Revisiting Nepal's 'Home Garden Project' through a Qualitative Lens." Impact Assessment Brief 18, Bioversity International, Rome.

Ewerling, F., J. Lynch, C. Victora, A. van Eerdewijk, M. Tyszler, A. Barros. 2017. "The SWPER Index for Women's Empowerment in Africa: Development and Validation of an Index Based on Survey Data." *The Lancet Global Health* 5 (9): e916–e923.

Eyben, R., N. Kabeer, A. Cornwall. 2008. "Conceptualising Empowerment and the Implications for Pro Poor Growth." Paper for the DAC Poverty Network, IDS, Brighton.

FAO (Food and Agricultural Organization of the United Nations). 2011. *The State of Food and Agriculture. Women in Agriculture: Closing the Gender Gap for Development.* Rome: FAO.

Farnworth, C. R., T. Jafry, K. Lama, S. C. Nepali, L. B. Badstue. 2019. "From Working in the Wheat Field to Managing Wheat: Women Innovators in Nepal." *The European Journal of Development Research* 31 (2): 293–313.

Freire, P. 1970. *Pedagogy of the Oppressed.* New York: Herder and Hered.

Galiè, A. 2013. "Empowering Women Farmers: The Case of Participatory Plant Breeding in Ten Syrian Households." *Frontiers: A Journal of Women's Studies* 34 (1): 58–92.

Galiè, A., N. Teufel, L. Korir, I. Baltenweck, A. Webb Girard, P. Dominguez-Salas, Y. M. Yount. 2018. "The Women's Empowerment in Livestock Index." *Social Indicators Research* 142: 799–825.

Hariharan, V. K., S. Mittal, M. Rai, T. Agarwal, K. Kalvaniya, C. Stirling, M. Jat. 2018. "Does Climate-Smart Village Approach Influence Gender Equality in Farming Households? A Case of Two Contrasting Ecologies in India." *Climatic Change* 158: 77–90

Heckert, J., D. K. Olney, M. T. Ruel. 2019. "Is Women's Empowerment a Pathway to Improving Child Nutrition Outcomes in a Nutrition-Sensitive Agriculture Program? Evidence from a Randomized Controlled Trial in Burkina Faso." *Social Science & Medicine* 233: 93–102.

Hillenbrand, E., N. Karim, P. Mohanraj, D. Wu. 2015. "Measuring Gender Transformative Change: A Review of Literature and Promising Practices." Working Paper, CARE USA, Atlanta, GA.

Holland, J. and L. Ruedin. 2012. *Monitoring and Evaluating Empowerment Processes.* Stockholm: Sida.

Huis, M. A., N. Hansen, S. Otten, R. Lensink. 2017. "A Three-Dimensional Model of Women's Empowerment: Implications in the Field of Microfinance and Future Directions." *Frontiers in Psychology* 8: 1678.

Ibrahim, S., and S. Alkire. 2007. "Agency and Empowerment: A Proposal for Internationally Comparable Indicators." *Oxford Development Studies* 35 (4): 379–403.

Jewkes, R. 2002. "Intimate Partner Violence: Causes and Prevention." *The Lancet* 359 (9315): 1423–1429.

Johnson, N., M. Balagamwala, C. Pinkstaff, S. Theis, R. Meinzen-Dick, A. Quisumbing. 2018. "How Do Agricultural Development Projects Empower Women? Linking Strategies with Expected Outcomes." *Journal of Gender, Agriculture and Food Security* 3 (2): 1–19.

Kabeer, N. 1994. *Reversed Realities: Gender Hierarchies in Development.* London: Verso.

Kabeer, N. 1999. "Resources, Agency, Achievements: Reflections on the Measurement of Women's Empowerment." *Development and Change* 30 (3): 435–464.

Kabeer, N. 2005. "Gender Equality and Women's Empowerment: A Critical Analysis of the Third Millennium Development Goal 1." *Gender & Development* 13 (1): 13–24.

Kabeer, N. 2017. "Economic Pathways to Women's Empowerment and Active Citizenship: What Does the Evidence from Bangladesh Tell Us?" *Journal of Development Studies* 53 (5): 649–663.

Kantor P. 2013. "Transforming Gender Relations: Key to Positive Development Outcomes in Aquatic Agricultural Systems." Brief AAS-2013-12, AAS, Penang.

Knight, J., and J. Ensminger. 1998. "Conflict over Changing Social Norms: Bargaining, Ideology, and Enforcement." In *The New Institutionalism in Sociology*, edited by M. C. Brinton, and V. Nee, 105–126. Stanford, CA: Stanford University Press.

Kosec, K., K. Akramov, B. Mirkasimov, J. Song. 2018. "Aspirations and Women's Empowerment: Evidence from Kyrgyzstan." Discussion Paper 1786, IFPRI, Washington, DC.

Lombardini, S., and K. McCollum. 2018. 'Using Internal Evaluations to Measure Organisational Impact: A Meta-Analysis of Oxfam's Women's Empowerment Projects." *Journal of Development Effectiveness* 10 (1): 145–170.

Lombardini, S., K. Bowman, R. Garwood. 2017. "*A 'How To' Guide to Measuring Women's Empowerment: Sharing Experience from Oxfam's Impact Evaluations.*" Oxford: Oxfam.

Malapit, H. J., C. Pinkstaff, K. Sproule, C. Kovarik, A. Quisumbing, R. Meinzen-Dick. 2017. "The Abbreviated Women's Empowerment in Agriculture Index (A-WEAI)." Discussion Paper 1647, IFPRI, Washington, DC.

Malapit, H., A. Quisumbing, R. Meinzen-Dick, G. Seymour, E. Martinez, J. Heckert, D. Rubin et al. 2019. "Development of the Project-Level Women's Empowerment in Agriculture Index (pro-WEAI)." *World Development* 122: 675–692.

Mahmud, S., N. M. Shah, S. Becker. 2012. "Measurement of Women's Empowerment in Rural Bangladesh." *World Development* 40 (3): 610–619.

Mayanja, S., N. Mudege, D. Naziri. 2018. "Validation of the Gender in Value Chains Analytical Tools." Technical Report, IPC, Lima.

Meinzen-Dick, R., D. Rubin, M. Elias, A. Mulema, E. Myers. 2019. "Women's Empowerment in Agriculture: Lessons from Qualitative Research." Discussion Paper 1797, IFPRI, Washington, DC.

Miruka, M., J. Njuki, L. Starr, E. Kruger, E. Hillenbrand. 2015. "Measuring Women's Empowerment in Agriculture. Addressing the Multidimensional Nature of Gender Dynamics in Agriculture." In *Transforming Gender and Food Security in the Global South*, edited by J. Njuki, J. R. Parkins, A. Kaler, 21–46. Abingdon: Routledge.

Morgan, M. 2014. "Measuring Gender Transformative Change." Program Brief AAS-2014-41, AAS, Penang.

Mosedale, S. 2005. "Assessing Women's Empowerment: Towards a Conceptual Framework." *Journal of International Development* 17 (2): 243–257.

Najjar, D., A. Frija, A. El Garhi. 2018. "A Typology Analysis of Agricultural Empowerment Profiles in Rural Egypt with a Particular Focus on Women." Working Paper 1198, Economic Research Forum, Giza.

Narayan, D. 2005. "Conceptual Frameworks and Methodological Challenges." In *Measuring Empowerment: Cross-Disciplinary Perspectives*, edited by D. Narayan, 3–38. Washington, DC: World Bank.

Nazneen, S., A. Darkwah, M. Sultan. 2014. "Researching Women's Empowerment: Reflections on Methodology by Southern Feminists." *Women's Studies International Forum* 45: 55–62.

Newton, J., A. van Eerdewijk, F. Wong. 2019. "What Do Participatory Approaches Have to Offer the Measurement of Empowerment of Women and Girls?" Working Paper, KIT, Amsterdam.

O'Hara, C., and F. Clement. 2018. "Power as Agency: A Critical Reflection on the Measurement of Women's Empowerment in the Development Sector." *World Development* 106: 111–123.

Petesch, P., and R. Bullock. 2018. "Ladder of Power and Freedom: Qualitative Data Collection Tool to Understand Local Perceptions of Agency and Decision-Making." GENNOVATE Resources for Scientists and Research Teams, CIMMYT, Mexico City.

Petesch, P., L. Badstue, G. Prain. 2018. "Gender Norms, Agency, and Innovation in Agriculture and Natural Resource Management: The GENNOVATE Methodology." Technical Report, CIMMYT, Mexico City.

Razavi, S. 2016. "The 2030 Agenda: Challenges of Implementation to Attain Gender Equality and Women's Rights." *Gender & Development* 24 (1): 25–41.

Richardson, R. A. 2018. "Measuring Women's Empowerment: A Critical Review of Current Practices and Recommendations for Researchers." *Social Indicators Research* 137: 539–557.

Rowlands, J. 1997. *Questioning Empowerment: Working with Women in Honduras*. Oxford: Oxfam.

Santoso, M. V., R. B. Kerr, J. Hoddinott, P. Garigipati, S. Olmos, S. Young. 2019. "Role of Women's Empowerment in Child Nutrition Outcomes: A Systematic Review." *Advances in Nutrition* 10 (6): 1138–1151.

Sen, A. 1985. "Well-Being, Agency and Freedom: The Dewey Lectures 1984." *The Journal of Philosophy* 82 (4): 169–221.

Sen, A. 1999. *Development as Freedom*. Oxford: Oxford University Press.

Sodhi, N. S., P. Davidar, M. Rao. 2010. "Empowering Women Facilitates Conservation." *Biological Conservation* 143 (5): 1035–1036.

Sraboni, E., H. J. Malapit, A. R. Quisumbing, A. U. Ahmed. 2014. "Women's Empowerment in Agriculture: What Role for Food Security in Bangladesh?" *World Development* 61: 11–52.

Tsikata, D., and A. K. Darkwah. 2014. "Researching Empowerment: On Methodological Innovations, Pitfalls and Challenges." *Women's Studies International Forum* 45: 81–89.

UN (United Nations). 1995. "Beijing Platform for Action." www.un.org/en/events/pastevents/pdfs/Beijing_Declaration_and_Platform_for_Action.pdf

Van Eerdewijk, A., F. Wong, C. Vaast, J. Newton, M. Tyszler, A. Pennington. 2017. *A Conceptual Model of Women and Girls' Empowerment*. Amsterdam: KIT.

Woodall, J. R., L. Warwick-Booth, R. Cross. 2012. "Has Empowerment Lost Its Power?" *Health Education Research* 27 (4): 742–745.

World Bank. 2012. *World Development Report 2012: Gender Equality and Development*. Washington, DC.

Yount, K. M., A. Peterman, Y. F. Cheong. 2018. "Measuring Women's Empowerment: A Need for Context and Caution." *The Lancet Global Health* 6 (1): e29.

Yount, K., C. Cheong, L. Maxwell, J. Heckert, E. Martinez, G. Seymour. 2019. "Measurement Properties of the Project-Level Women's Empowerment in Agriculture Index." *World Development* 124 (104639).

Chapter 10

TOWARD STRUCTURAL CHANGE: GENDER TRANSFORMATIVE APPROACHES

Cynthia McDougall, Lone Badstue, Annet Mulema, Gundula Fischer, Dina Najjar, Rhiannon Pyburn, Marlène Elias, Deepa Joshi, and Andrea Vos

Almost a quarter of a century after the Beijing Declaration, and with 10 years left to meet the Sustainable Development Goals, *The Guardian* announced the SDG Gender Index's finding that, "Not one single country is set to achieve gender equality by 2030" (Equal Measures 2030 2019, Ford 2019). This aligns with the most recent Global Gender Gap Index, which signals that, on the current trajectory, it will take approximately 170 years to achieve gender equality (WEF 2016)—a wait of seven generations, or two and a half lifetimes for the average woman.[1]

While there has been progress through legislative reforms and targeted interventions in education, health, and social protection, gender inequalities remain particularly pervasive in agriculture-dependent and low-income countries. In the SDG Gender Index, for example, no country in sub-Saharan Africa, Middle East and North Africa, or Latin America and the Caribbean has achieved a *good* category score.[2] Moreover, progress toward gender equality "... is hugely variable, hard to advance at pace and, in places, in retreat... discriminatory gendered norms prevail and resistance to progress is common" (Pantuliano et al. 2019, 2).

Girls and women continue to have insufficient control over economic, social, and political resources, and "stark disparities between women remain, influenced by intersecting social identities such as gender, age, disability, ethnicity and class" (Pantuliano et al. 2019, 2). These inequalities are embedded in complex and dynamic socioeconomic–environmental contexts,

1 The estimated current lifespan for women is 72 years globally (CIA n.d.).

2 On a five-point scale: very poor-poor-fair-good-excellent (Equal Measures 2030 2019). Fragility of context is also a factor: the 10 lowest-scoring countries (Sierra Leone, Liberia, Nigeria, Mali, Mauritania, Niger, Yemen, Congo, Democratic Republic of Congo, Chad) are all on the 2018 Organisation for Economic Co-operation and Development fragile states list.

characterized by climate uncertainties, globalization, and a neoliberal ethos that has embraced yet simultaneously watered down women's empowerment with remarkable momentum (Cornwall 2018).

This limited and uneven progress calls for a critical evaluation of why gender approaches in development—and specifically in agriculture and natural resource management in low-income contexts—have not delivered as intended. This requires an interrogation of how agriculture and natural resource management frame and engage with gender, so that these sectors can more substantively and sustainably address gender inequalities.

This chapter responds to the pressing imperative for these sectors to rethink current manifestations of Gender and Development (GAD). It does so by investigating gender transformative approaches as an emerging response and potential post-GAD way forward. Gender transformative approaches complement and go beyond current "business-as-usual" approaches. The latter *work around* gender constraints and often focus on building women's individual or collective agency or assets. By contrast, gender transformative approaches seek to constructively, and in a context-driven way, *transform* structural barriers, in particular constraining norms, that underpin gender equality. In this way, they go deeper than common gender integration and mainstreaming and tackle the root causes of gender inequalities instead of addressing its symptoms (AAS 2012a). As such, emergent gender transformative strategies embody the ambitious goal of addressing the very foundations of gender equality, seeking to reshape unequal power relations and structures toward more gender equal ones (Morgan et al. 2015, Wong et al. 2019).

As a starting point, this chapter offers a rapid critical review of Women in Development (WID) and GAD approaches as they have been applied in agriculture and natural resource management over the past decades. We then look into gender norms as a leverage point for transformative change (Badstue et al. 2018a). Next, we turn to gender transformative approaches, examining evidence and examples of these in terms of their potential to accelerate progress toward gender equality. The chapter concludes by presenting priority questions for a future research agenda.

The WID to GAD trajectory and critiques

Agriculture and natural resource management, and associated agricultural research for development, have implemented strategies to engage, benefit, and at times empower women for more than half a century. In line with the broader development sphere, these strategies began in a WID approach from the 1960s–1970s. As criticisms of WID became widespread, the sectors transitioned to a GAD approach, which forms the basis for gender mainstreaming today (Razavi and Miller 1995, Okali 2011). Specifically, GAD sought to redress WID's emphasis on women and on getting women into formal development processes.

The approach (at least in theory) turned the focus to gender (rather than women) and to shaping development processes and outcomes (rather than "shaping women" by getting them to take on more responsibilities). It did so first through *roles*-focused framings such as the Harvard Analytical Framework (ILO 1998) then moved—albeit to differing degrees—toward a gender *relations* lens, such as Kabeer's (1994) Social Relations Framework[3] (see March et al. 1999, Okali 2012, Kawarazuka et al. 2017). In practice, efforts to address gender inequalities are lagging behind these theoretical advancements (Cornwall 2000, Baruah 2005, Chant 2016), despite policy and programmatic investments in the agriculture and natural resource management sectors toward mainstreaming and becoming more "gender aware" (see Box 10.1, Milward et al. 2015, Drucza and Abebe 2017). Decades of implementation have now created the opportunity for reflection on the degree to which and why these are—or are not—effective and on track. We present such critical reflection by pointing to two significant areas of critique of common current manifestations of gender approaches: *outcomes* and *how change occurs* (causality).

The first dimension of critique relates to the failures of current gender approaches to deliver gender ***outcomes*** as intended. As Wong et al. (2019, 14) note, "gender integration efforts in development initiatives generally, and in agricultural development in particular, are not as effective as they could be." We identify four key shortcomings in relation to outcomes.

First, current approaches may translate into benefits only for the women directly involved in a given project, and are unlikely to have empowering effects for women *beyond* its reach. This links to a project focus on reducing visible gaps (such as assets or training) for select women ("beneficiaries") rather than addressing broader social constraints that affect all women in the context (beyond project participants).

3 The Social Relations Framework is positioned in a paradigmatic shift in GAD studies in the 1990s, away from the unitary model of the household to the investigation of bargaining processes within households. Kabeer's framework also recognizes how intrahousehold relations are affected by extrahousehold institutions, and hence takes into account four institutional sites: household, community, market, and government (Kabeer 1994).

BOX 10.1 A gender continuum—exploitative, accommodating, transformative

The Gender Integration Continuum positions policies or programs along a continuum ranging from "gender blind" (ignoring gender considerations) to "gender aware" (examining and addressing a range of gender issues, relations, and dynamics). Within the "gender aware" area, the spectrum moves from:

- Gender exploitative: reinforcing or using unequal gender dynamics to achieve project goals—this should be avoided;
- Gender accommodating: recognizing but working around the gender barriers and inequalities, for example engaging women within the homestead; and,
- Gender transformative: fostering examination of gender dynamics and norms and intentionally strengthening, creating, or shifting structures, practices, relations, and dynamics toward equality.

Source: IGWG (2017).

Second, effects on empowerment or gender equality may dissipate or reverse after programs or projects finish. For example, land is a foundational factor in gender equality and a primary lever for women's access to programs, technologies, water, and markets (Agarwal 2003, Gunchinmaa et al. 2011, Meinzen-Dick 2014, Namubiru-Mwaura 2014). Yet the targeting of land resources to women during a project does not necessarily ensure lasting (intergenerational) equitable outcomes. Studies have found that, even after women acquire land—through a project, purchase, redistribution, or other means—a plethora of complex structural factors may subsequently erode their control, including patrilocal residence norms and practices (Gray and Kevane 1999, Hilliard et al. 2016, Doss et al. 2018, Najjar et al. 2020).[4]

Third, some strategies may (unintentionally) reinforce gender stereotypes or barriers (Leder et al. 2017). For example, nutrition-oriented programs that focus on working with women's groups to deliver nutrition and cooking

4 This includes that land may be reallocated from women to male children once the project ends (Jackson 2003, Baruah 2010, Najjar et al. 2019, 2020). In India as well as Egypt, for example, it has been observed that women who acquire and hold land are unwilling or hesitant to pass land on to their daughters (Baruah 2010, Najjar et al. 2019, 2020).

messaging may unintentionally be reinforcing the norm that cooking and unpaid domestic work are "women's work."[5]

Fourth, common current approaches—including and perhaps especially those that target women—may have perverse outcomes. For example, a program in the Gambia targeted women with the task of planting trees because it perceived that they were "environmentally enthusiastic." The work was not only without pay but also increased women's already significant workloads (Schroeder 1993). Perverse outcomes may also include a backlash, tensions, or violence against women (see Kabeer 2005, Slegh et al. 2013).

Next, the second dimension of critique of common current gender approaches relates to ***how social change occurs***. One primary concern is that common approaches address visible gaps (such as access to technology, assets, or knowledge) but fail to engage with underlying structural gender barriers, in particular gender norms (Farnworth et al. 2013, IGWG 2017).[6] Gender accommodative approaches, as the name implies, acknowledge—and work around—gender constraints. For example, an accommodative aquaculture or agriculture project may focus on engaging women within the boundaries of the homestead and in relation to foods for home consumption, as these spaces and the food focus are family and domestic related, and thus already socially acceptable for women. The limitation is that this stays within the boundaries of gender constraints and thus is unlikely to address the underlying formal (policy) or informal (gender norms, attitudes) factors that perpetuate and reproduce these constraints (Kantor 2013, McDougall et al. 2015, IGWG 2017, Wong et al. 2019). In other words, although accommodative strategies may close visible gaps in project activities, the underlying factors that originally limited women and men from engaging and benefiting equally—such as policies, gender norms, or attitudes—are likely still in place.

A related concern is that current approaches risk reverting toward WID's weaknesses in terms of engaging *only* women. Research and interventions have tended to focus on women as atomized units, rather than engaging with complexities and in negotiations of the underlying power relations that serve to reinforce gendered inequalities (Okali 2011, 2012, Alsos et al. 2013). This fails to recognize that agriculture and natural resource management initiatives are complex social change processes in which both men and women are actors and are interconnected (McDougall 2017). Similarly, many interventions

---

5 This may be considered "gender exploitative" because it is taking advantage of existing gender norms in a way that benefits the project (see IGWG 2017).

6 See Peterson (2005) for related epistemological critique.

target women for training and technologies, aiming to strengthen their individual agency within existing social and economic structures, rather than challenging structural factors, such as land tenure or structural adjustment policies and trade agreements, which shape the potential for changes in gender dynamics (Cornwall and Edwards 2016, Galiè and Kantor 2016). Together, this focus on individuals as the unit of analysis in interventions represents a weak mechanism for leveraging change. Specifically, it is overly reliant on the ability of those individuals to translate their improved knowledge and capacities into meaningful choice and strengthened livelihoods. This then risks the outcomes of interventions being limited or short-lived. It may even generate perverse outcomes: men may perceive women-targeting as threatening, which can lead to backlash (Promundo and AAS 2016).

Finally, an associated critique in terms of *how change occurs* relates to scale: it is common for approaches in agriculture and natural resource management to operate at a single scale, often that of the household. On the one hand, this disregards the significance of intrahousehold dynamics (see Okali 2011, Ambler et al. 2018). On the other hand, it misses the fact that gender barriers—and opportunities—are embedded within multiple scales and thus enacting effective and lasting change requires engagement across these (Cole et al. 2014a).[7] For example, Agarwal (1997) showed how individual women's efforts to receive their share of land required "interlinked contestations," such as the establishment of social legitimacy for women's independent land rights in the community *and* equal inheritance laws at the government level (see also Kevane and Gray 1999, Lambrecht 2016). Similarly, Morgan et al. (2015) highlight that gender dynamics influence women's capacity to use a new aquaculture technology at five nested scales, from individual through to macro.

Current gender mainstreaming efforts have lost touch with earlier potent thinking on empowerment and the feminist foundations of gender in development. The seminal framing of empowerment by Kabeer (1999, 2001) and the emphasis on "the relational nature of empowerment... has fallen out of the frame" (Cornwall 2016, 364), for instance in frameworks that emphasize assets and opportunity structures (Alsop et al. 2004). While investments in women's agency may be valuable, an overly narrow approach is likely insufficient for widespread or sustained change.

---

7 Applying a systems thinking lens to this critique (see Meadows 2010), business-as-usual approaches are undercutting their own potential in two ways. First, by neglecting to engage with the feedback loops that work across scales to reinforce (stabilize, perpetuate) or disrupt unequal social systems. Second, by missing the opportunity to engage with critical leverage points for change, in particular norms (McDougall 2017, Manlosa et al. 2019).

Similarly, current gender approaches tend to reinforce an individual and instrumental approach to empowerment. They often focus on economic empowerment in the neoliberal political economy context (Cornwall 2018) for the purpose of leveraging other development goals, such as nutrition or growth. The risk here is of losing the intrinsic value of gender equality, and, along with it, the inherently political mandate of empowerment to address social and gender inequalities (Cornwall 2016). In missing the opportunities for leveraging equality through challenging structural factors, the burden of the work involved in social change is transferred to women (Chant 2016), rather than shared by women and men (and all genders) as members of society and invested in by wider political structures.

These critiques, combined with the increasing establishment of gender on donor and development agendas, have led to a push for more effective engagement with gender in the agriculture and natural resource management sectors over the past decade. Pockets of innovation have emerged, drawing on sectors that were already using gender transformative approaches, in particular reproductive health. Gender transformative strategies were conceptualized, designed, and piloted in various agriculture and natural resource management spheres, including small-scale fisheries and aquaculture (AAS 2012b). A common focus of these emerging strategies is explicit engagement with structural gender barriers, in particular *norms*. The next section sets the stage for understanding gender transformative approaches by elucidating social—and in particular gender—norms and their significance in shaping gender inequality in agriculture and natural resource management.

Gender norms as leverage points for transformation

Social norms are the unwritten rules of behavior regarding what is considered acceptable and appropriate in a given group or society. They "govern social relations and establish expectations as to how we are to act in our everyday affairs... and they determine in significant ways the distribution of the benefits of social life" (Knight and Ensminger 1998, 105). Social norms include perceptions about others that are shared and reproduced within social groups and serve as critical drivers, either enabling or constraining particular social practices.

Social norms regarding gender play a central role in creating and perpetuating gender equalities and inequalities. Gender norms represent socially constituted rules that differentiate women and men's expected roles and conduct (Pearse and Connell 2016). They differ across contexts and interact with other aspects of identity (such as wealth, ethnicity, or religion) and other expectations and practices. Gender norms and the associated power relations operate at multiple levels, from household, social group, and community to agroecological landscapes, market systems, and the overall policy and legislative environment. They are tied to deeply rooted, albeit context-related and dynamic, value systems that inform day-to-day practice in varied and sometimes seemingly contradictory ways. Gender norms are often subconscious and largely maintained by everyday social interactions, and psychological processes that come to define power relations, including women and men's subjectivity (Ridgeway 2009). In many settings and across scales, deep-rooted beliefs about men's intrinsic authority and competence relative to women are continuously "re-inscribed into new organizational procedures and rules that actors develop through their social relations" (ibid., 152), reproducing what is known as hegemonic masculinity (Connell and Messerschmidt 2005).[8]

Normative expectations are reinforced by social sanctions, such as the ridicule of men who show their emotions, or the condemnation and harsh criticism of women who interact in public with men who are not their relatives. The perceived threat of social sanctions against women or men who challenge existing gender norms have been found to compel youth of both genders across a range of agricultural and natural resource management contexts to conform to normative expectations (Elias et al. 2018).

In agricultural and natural resource management contexts, gender norms shape what are considered appropriate pursuits and assets for women and men, the value and recognition placed on each gender group for performing them, and the distribution of benefits derived from these. In Ethiopia, for example, local beliefs framed technologies appropriate for women as those that fit within the homestead and aligned with gendered norms positioning women as responsible for household food and nutritional security (Mulema et al. 2019). Similarly, studies in Africa and Asia found that women's ability to pursue new technologies and engage as agricultural innovators were shaped by norms related to mobility constraints, gendered workloads, and perceptions of men

8 Hegemonic masculinity refers to the socially legitimized practice of men's dominance in society. Specifically, it embodies the dominance of men who represent ways of "being a man" associated with what are considered traditional powerful masculine identities in any given society. Conversely, the term refers to the socially legitimized subordination of women, and other genders, including nonconforming ways of "being a man", perceived as feminine (see Jewkes et al. 2015).

as "farmers" and decision-makers versus women as "helpers" and subordinates (Aregu et al. 2018, Mulema et al. 2019).[9]

In agricultural contexts, these norms intertwine with hegemonic masculinity, connected with the dramatic qualities and visual allure of technology, machinery, and infrastructure (Oldenziel 1999, Brandth and Haugen 2005, Zwarteveen 2008).[10] The "masculine rural" (Campbell and Bell 2000) in Southern African contexts, for example, is associated with the value placed on hard physical labor, toughness, and the need to control nature and equipment. As Cole et al. (2015, 158) describe, "'big man' in a rural, southern African setting... might describe a person who is powerful, chief-like, demands respect, is married (perhaps to multiple women) and head of a household, accumulates wealth through people (e.g., children, spouse), and owns or controls assets such as land, cattle, and farming equipment." This cultural linking of technology, leadership, and masculinity underpins influential gender norms shaping behaviors, opportunities, and constraints for men and women.

Patriarchal norms also have significant influence beyond household and local scales; they manifest in, and shape, whom development programs and policy recognized and enable. Women often remain largely invisible to institutions at program and policy scales, as they are not perceived as "real" farmers, fishers, and agricultural or natural resource management leaders in many contexts (see Zhao et al. 2013, Feldman 2018). As Twyman et al. (2015, 12) note, despite widespread farming of rice by women, "this is a norm held by many researchers, enumerators, community leaders, and male and female farmers, all of whom claim unequivocally that, 'women are not rice producers.'" Gender norms and associated biases that ascribe authority and economic roles to men often mean that women's farming or natural resource management initiatives remain hidden or framed as part of their domestic work. Women's contributions are then underrepresented in data, leading to omissions or weakness in agriculture and natural resource management policy and practice (Kleiber 2015).

Following on from the above, gender norms embedded in development programs and institutions influence the extent to which women and men are able to benefit from new knowledge and technologies delivered through extension systems. Agricultural training and extension systems have been

9 See also the GENNOVATE collection of studies on norms and innovation from across 26 countries: https://gennovate.org/

10 In Australia and Norway, agricultural leadership is also seen as masculine, drawing credibility from masculine notions of on-farm technical expertise, mechanical competence, and physical strength (Brandt and Haugen 2005, Pini 2005).

found to favor men as knowledge recipients and as "knowers," reflecting institutionally held gender norms that position men as "heads of household" and "primary farmers" and imbue them with greater resources and decision-making power (Gilbert et al. 2002, Katungi et al. 2008, Peterman et al. 2010, Davis et al. 2012, Aregu et al. 2017). An analysis of data from 84 GENNOVATE community case studies in 19 countries shows that more than twice the amount of men as women reported receiving encouragement from extension services, and there was an overall difference in the type and quality of men and women's interactions with external partners (Badstue et al. 2018b). Women's unequal access to agricultural information is further reinforced by various locally reinforced norms that exclude women from public spaces and hinder their opportunities to gain knowledge, skills, recognition, and benefits from their agricultural and natural resource management pursuits (Elias 2018).

Patriarchal ownership and inheritance of land in rural societies also evidence the interactions between gender norms across scales. As Doss et al. (2018, 71) highlight:

> "Both the legal systems and patriarchal gender norms may prohibit or make it difficult for women to acquire and retain land. In addition, almost all inheritance systems disadvantage women in terms of inheritance, and when women legally inherit, they often face strong social pressure to relinquish their inheritance."

As an illustration of the tenacity of this challenge, in Ethiopia, multiple iterations of land reform have been enacted to even out land ownership among different groups of people, including adding married women to land certificates (Mulema and Damtew 2016). Yet, although these reform processes have led to modest changes, men still dominate decision-making over land, as formal tenure interacts with local informal dynamics and norms (Tefera 2013; see also Doss et al. 2013). This constrains the productive ability of women in general, and female-headed households are affected more than others in terms of knock-on constraints in accessing inputs and services (Mulema and Damtew 2016; see also Agarwal 1994, 2003). This interaction of structural factors across scales perpetuating land inequalities is similarly illustrated in the Pacific. In the Solomon Islands, for example, "the recursive constitution of property and authority through the state tends to consolidate control over land in the hands of a small number of men, while reproducing state norms and institutions as a masculine domain" (Monson 2017, 385). A critical point is that gender norms are not fixed or immutable; rather, they are negotiated and (re)constructed, sometimes in complex and strategic ways (Locke et al. 2017, Stern et al. 2017). In the heavily male-dominated shrimp production sector in Indonesia, for example, a minority of women engage in a norm-transgressing livelihood activity as shrimp farm operators, even at the cost of condemnation from community members (Sari et al. 2017). In Zambia, some men have taken on caregiving roles traditionally associated with women, while giving up drinking and extramarital behaviors traditionally associated with masculinity, because they perceived the benefits to their family outweighed the risks of social retribution (Bevitt 2017).

In a nuanced example, some women in Egypt who took on irrigation roles, and who were given land titles and training on irrigation technologies, transgressed gender norms related to leadership and technological prowess. Interestingly, these women simultaneously accented their compliance with other norms, such as obedience and propriety, which enabled them to better negotiate their participation in irrigation management and the adaption of the associated technologies (Najjar et al. 2019). The dynamism of norms is of significance to the proposition of gender transformative change: it means that norms may be endogenously questioned in ways that can provide space for negotiation, contestation, and change (Stern et al. 2017). This, in turn, may spur a process in which new normative expectations—and thus gender dynamics—take hold and spread across key reference groups (Bicchieri 2005). In sum, the fact that gender norms are underlying drivers of gendered practices but also dynamic and changeable makes them critical leverage points for enhancing gender equality (McDougall 2017).

Gender transformative approaches: experiences and evidence

In response to the above critiques of current gender approaches, and given expanding awareness regarding the significance of gender norms, a growing number of research-for-development institutes and development agencies have developed and applied a range of gender transformative strategies over the past decade.[11] This section takes a closer look at the approaches and seeks to better understand what outcomes they generate and the mechanisms through which change happens. Table 10.1 draws on existing reviews and the broader

11 These include, but are not limited to, Oxfam Novib's Gender Action Learning System; Send-a-Cow's Transformative Household Methodology; Oxfam's Rapid Care Analysis; World Food Program Community Conversations; Wise Asset-Based Community Development; the Self-Help Africa Family Life Model; CARE's Social Analysis and Action; the Helen Keller Institute's Nurturing Connections; and Promundo and CARE's Journeys of Transformation.

literature to present examples of gender transformative approaches from across health, nutrition, agriculture, and natural resource management sectors.[12] We discuss emerging insights regarding reported *outcomes* and *how change happens* (mechanisms) of the gender transformative approaches. We further unpack the mechanisms of change by presenting and discussing two in-depth examples.

The six selected examples of gender transformative approaches in Table 10.1 offer three emerging insights in terms of outcomes. First, these approaches foster a range of significantly important and interconnected **gender outcomes**:

- Shifts in barriers underlying gender inequalities (gender attitudes including about violence, behaviors associated with harmful masculinities such as drinking alcohol);
- Multiple kinds of improvements in women's empowerment and changes in gender relations (in particular decision-making, division of labor and care work, control over assets, ability to apply knowledge); and,
- Contributions to other development outcomes or intermediate outcomes (production practices; nutrition, HIV and health).

These emerging findings align with the hypothesis that gender transformative approaches may be able to redress limited and superficial gender outcomes. In particular, they represent a significant breadth and *depth* of gender-related outcomes, including changes across the cases in underlying attitudes. As attitudes are measured as a proxy for norms, this suggests the approaches are at least starting to contribute to shifts in some underlying structural gender barriers.

Second, however, the selected cases reveal a **gap in empirical evidence** regarding outcomes from gender transformative approaches compared with those from gender accommodative approaches. Without this evidence base, it is difficult to empirically assess the specific and relative contributions of transformative dimensions across different gender transformative approaches.

The study by Cole et al. (2018; see also Cole et al. 2020) is one of the few that offers a direct empirical, quantitative comparison of an accommodative with a transformative approach. It does so in relation to a technical

12 This selection builds on existing reviews of gender transformative approaches to identify cases with evidenced outcomes (Rottach et al. 2009, Drucza and Abebe 2017, and Wong et al. 2019) as well as specific peer-reviewed publications and gray literature. Some information, such as the duration of interventions, was not available for all cases.

TABLE 10.1 Gender transformative approaches in health, nutrition, agriculture, and natural resource management—mechanisms and outcomes

Sector	Mechanisms and strategies	Key empirical outcomes	Type and size of evaluation or study	Organization and source
Behavior change communication Barotse Floodplain, Zambia				
Small-scale fisheries/ agriculture	• GTA combined with a technical innovation process (participatory action research testing of technologies to reduce postharvest loss) • The GTA (communication for social change) comprised bespoke community theatre (drama skits) raising local gender dynamics, iterating with discussions in participatory action research groups • Time period: 3 months+	**Changes in GTA sites compared with gender accommodative sites:** Increased gender attitude scores (28.6% vs. 11.7% overall), especially for the men who participated (35.7% vs. 13.3%) **Changes in women's empowerment after implementation of GTA:** • Increased women's participation in fishing (5% to 75%) • Increased women's contributions to intrahousehold decisions about the income generated from processing fish (49%) • Fishing gear ownership status changed from sole ownership by men to joint ownership with spouses (44% to 76% reported joint ownership)	**Design:** Quasi-experimental comparing GTA and accommodative (practical gender needs) approaches **Method:** Quantitative survey **N=**6 villages (80 participants)	WorldFish and FISH, AAS with Zambian Department of Fisheries (Source: Cole et al. 2018)
Rapid Care Analysis Ethiopia, Colombia, Malawi, Philippines, Uganda and Zimbabwe				
Health and agriculture	Participatory exercises and focus group discussions focusing on: (1) increasing the recognition of care work; (2) reducing its drudgery; (3) redistributing responsibility for care more equitably; and, (4) ensuring the representation of care workers in interventions. The approach is implemented as part of other community interventions.	**Reported changes after implementation of GTA:** • Increased recognition in households and communities of the burden of unpaid care work on women's lives • Increased recognition among men of the value of care work and the need to redistribute tasks among household members • Changes in intervention and project design within Oxfam to include the burden of unpaid care work on women's lives	Not specified	Oxfam (Sources: Kidder and Pionetti 2013, Oxfam 201 in Druzca and Abebe 2017)

continued

Sector	Mechanisms and strategies	Key empirical outcomes	Type and size of evaluation or study	Organization and source
Gender Action Learning System (GALS) Lilongwe, Malawi				
Agriculture	Six- to twelve-month community-led methodology that can be implemented on its own or integrated into existing development projects. It starts with a Change Catalyst Workshop with participatory visioning and the creating of action plans. This is then scaled up through community action learning and later reviewed by the community.	**Changes in GALS sites compared with non-GALS sites:** • Significant shift toward sharing of on-farm tasks and household tasks between household members • Increased joint realization of benefits from agricultural produce reported between husbands and wives • De facto female-headed households reported increase in social standing and participation in community life	**Design:** External systematic mixed-method evaluation **Method:** Focus group discussions and key informant interviews, GALS and non-GALS participants **N=**260 respondents and 13 focus groups, 2 GALS sites 1 control site	Oxfam and Linda Mayoux (Source: Farnworth 2012 in Wong et al. 2019)
Nurturing Connections Côte d'Ivoire				
Nutrition	Four-month curriculum with four blocks: (1) Let's Communicate; (2) Understanding Perceptions and Gender; (3) Negotiating Power; and, (4) Acting for Change. Each weekly session follows an action learning cycle that begins with a game, or story, followed by a reflection on this activity after which learnings are formulated. The cycle ends with a planning phase where insights are translated into proposed actions. Nurturing Connections is implemented within an agriculture-for-nutrition program (CHANGE).	**Changes in Nurturing Connections (NC) villages compared with non-intervention sites:** • Comparative increase in women's roles in decision-making, including joint decision-making, in four domains: (1) children, spending and youth; (2) nutrition and domestic work; (3) agriculture; and, (4) livestock • Increased spousal communication, especially for women in NC villages • Slightly more equitable viewpoints within households	**Design:** Longitudinal intention-to-treat evaluation **Method:** Survey, analysis: difference in difference. **N=**423 respondents in 8 intervention and 6 non-intervention villages	Helen Keller Institute (Sources: Nordhagen et al. 2017, Hillenbrand 2015 in Wong et al. 2019)

Sector	Mechanisms and strategies	Key empirical outcomes	Type and size of evaluation or study	Organization and source
Agriculture, Nutrition and Gender Linkages (ANGeL) Bangladesh				
Nutrition	ANGeL combines facilitated agricultural production and nutrition behavioral change trainings as well as Helen Keller Institute's Nurturing Connections approach (also described above).	**Changes after implementation of GTA:** • Increased crop diversity in homestead farms • Adoption of improved production practices • Overall: women more likely to apply knowledge gained from trainings • Improved nutrition outcomes • Increased asset ownership for women • Improved gender-equitable attitudes among men and women	**Design:** Internal impact evaluation: randomized control trial **Method:** Baseline and endline household surveys **N=**3,125 households in 16 districts	IFPRI and Helen Keller Institute (Source: IFPRI 2018)
Program H Brazil				
Health (HIV/AIDS prevention)	Questioning traditional norms and masculinities among men using group education activities as well as a behavior change communication campaign. Included role-plays, brainstorming exercises, discussion sessions, and individual reflection on five themes: sexuality and reproductive health; fatherhood and caregiving; from violence to peaceful coexistence; reasons and emotions; and preventing and living with HIV/AIDS. The campaign promoted these themes at community level.	**Changes in GTA sites compared with control/delay sites:** • Young men four to eight times less likely to report sexually transmitted infection symptoms over time in intervention sites • Significant decrease in agreement with inequitable gender norms (10–13 out of 17 items vs. 1 out of 17 in the control area). These positive changes were sustained at the one-year follow-up period	**Design:** Quasi-experimental control design **Method:** Gender-Equitable Men Scale **N=**780 in 2 intervention and 1 control/delayed site	Promundo (Source: Pullerwitz et al. 2006 and Verma 2006 in Rottach et al. 2009)

NOTE: GTA = gender transformative approaches.

innovation (postharvest loss reduction technologies) in three villages in the Barotse Floodplain, Zambia. Using a quasi-experimental design, the study applied an accommodative gender approach in three villages. This entailed timing and location of participatory action research sessions to accommodate women, strategies to include and give space to women in participatory action research sessions, and so forth. In a separate treatment branch of three similar villages, the study applied these practices plus a gender transformative strategy. This comprised drama skits focusing on context-specific gender issues combined with reflexive sessions within participatory action research groups. Comparing findings across study, the gender transformative approach was found to catalyze more significant change in gender attitudes, as well as in the measured indicators of women's empowerment as compared with the gender accommodative approach.

Third, the above gap notwithstanding, the cases illustrate a shared strength in terms of **how to measure complex outcomes.** While the specific measures of gender equality and women's empowerment are not consistent across the cases in Table 10.1, the cases' approaches to measurement have in common that they value multiple research methods and methodologies and employ different ways of knowing. Qualitative and quantitative approaches are used to unveil different experiences of change. As such, the cases reflect an emergent critical questioning in the field regarding the dominance of quantitative methodologies and data as (the only) "real evidence" and increasing recognition of qualitative methods and measures in the field of assessing gender transformative change (Morgan 2014, Hillenbrand et al. 2015b).

In terms of **how change happens**, the examples in Table 10.1 illustrate that gender transformative approaches seek to engage with underlying barriers and focus not only on women. The mechanisms used across the cases are rooted in a combination of reflexive, participatory methods and tools designed to enable participants to be agents in a social change process. These focus in particular on locally driven critical reflection of gender norms and dynamics. Cases 10.1 and 10.2 present two in-depth cases that elucidate strategies and processes in more depth. We then further unpack how gender transformative approaches work iteratively in and across three levels and spheres: *individual capacities; social relations; and, social structures* (see Sarapura and Puskur 2014; also Wong et al. 2019).

At the *individual level*, in both Cases 10.1 and 10.2, the approaches use reflexive processes to develop capabilities and agency in order to critically examine and shift constraining gender norms and practices. When effective, this type of learning is transformative: it shifts mental models, values, and

CASE 10.1 Testing a gender transformative approach combined with a polyculture harvesting technology in Bangladesh

Frequent consumption of nutrient-rich mola, a small indigenous fish, can play a significant role in combating stunting and undernutrition (Belton et al. 2011), both of which are common in Bangladesh. Yet in the Barisal region of southwest Bangladesh, similar to other areas, the harvesting of fish from backyard ponds is a role socially assigned to men. Women—and their spouses—face criticism or ridicule from family and neighbors if women take on this role. Moreover, since harvesting typically requires getting into the pond, women express reluctance to engage in this role—because it would mean their clothing would stay wet all day.

To address these challenges, WorldFish developed and piloted an integrated social and technical strategy as a part of the United States Agency for International Development-funded Aquaculture for Income and Nutrition project. A technical innovation in the form of a gillnet that could be used from the bank was developed to address the practical challenge that women faced. Addressing the normative barriers, however, required a gender transformative strategy at both intrahousehold and community level.

At the household level, this comprised gender consciousness-raising exercises, adapted from Helen Keller International's Nurturing Connections manual (Hillenbrand et al. 2015a). These were facilitated by WorldFish with women and men from the same households, and more powerful household members (often in-laws), over approximately one year, integrated within technical aquaculture and nutrition trainings. Facilitators sought to create a socially and emotionally safe environment for participants to engage candidly and without fear of repercussions. Tools included Hopes & Fears, Power Hierarchies, Who Decides, trust-building exercises, and discussions of gendered behaviors, access to nutrition, and obstacles to change (adapted from Hillenbrand et al. 2015a, Promundo-USA and WorldFish 2016). Several of these were emotionally powerful experiences for participants: some tools surfaced recognition of negative emotional and practical (income or nutrition) effects of gender norms on women and other household members. This sparked dialogue between more and less powerful household members about possibilities for changing gender dynamics.

To reduce normative barriers at the community level, the project piloted similar exercises with community members, including neighbors and village leaders. The tools used included Hopes & Fears, Looking at Our Attitudes, Acting Like a Man/Woman, How Will We Empower Each Other?, Gender Equality Solutions, The Man Box, and a historical timeline of changes in gender relations.

Source: Kruijssen et al. (2016), Choudhury (2019), Österblom et al. (2020), Choudhury and Castellanos (2020).

CASE 10.2 Using *Community Conversations* to transform gender relations in Ethiopia

Research under the CGIAR Research Program on Livestock and Fish looked at gender inequalities and animal health disease constraints in small ruminants and how these affected men and women in smallholder livestock production systems in Ethiopia. The Program found that gender norms and division of labor expose women and men to different levels of risk of zoonotic diseases, with women often more affected. Gender norms constrain women from owning and controlling livestock, which limits their ability to make livestock-related decisions, join local associations such as community-based sheep-breeding cooperatives, and adopt integrated livestock health management practices that improve rural livelihoods and empower women. The Program aimed to address these through gender-related interventions tackling the unequal division of labor, access to and control of livestock resources, and exposure to zoonotic diseases by different household members.

The gender teams from the International Livestock Research Institute and the International Centre for Agricultural Research in the Dry Areas in Ethiopia piloted a community-based transformative approach called Community Conversations. Between 2018 and 2019, a series of modules were facilitated in four villages in three districts in the Southern Nations, Nationalities and Peoples region, and Amhara region. Fifty to sixty men and women small ruminant farmers, researchers, and local development partners participated in a series of four rounds of these conversations. Each session used a combination of interactive learning techniques to aid understanding, learning, and reflection, including pictures and posters, story-telling, and probing questions. Facilitators sought to create a safe space for women and men to freely articulate their views and agreed on indicators of change together with participants. Both qualitative and quantitative methods were used to track change in knowledge, attitudes, and practices.

Source: Lemma et al. (2018), Kinati et al. (2019).

beliefs because it goes beyond knowledge acquisition (Cole et al. 2014b, Wong et al. 2019). Dialogues facilitated in gender transformative processes aim to engage participants at an emotional level to trigger an appreciation of the need for change. For example, Case 10.1 sparked interest in change by surfacing awareness of the effects of constraining gender norms and relations on individual well-being as well as family goals and well-being. This reflects that transformation works through a process of seeing-feeling-changing. This flow of see-feel-change is more powerful than that of analyze-think-change when it comes to catalyzing change (see Kotter and Cohen 2002).

In terms of *social relations*, two points merit attention. First, both in-depth cases—and all examples in Table 10.1—engage men and women *together* in the gender transformative learning processes. Although transformative learning has an individual dimension, it also takes place *among* individuals. The nature of transformative learning and how it occurs is in fact highly social, relational, and interactive (Vernooy and McDougall 2003). The women and men involved "*together* build a more integrated or inclusive perspective of the world. Through the learning process, they jointly transform some part of their worldview, for example their understanding of social relations" (ibid., 116; emphasis added).

Building on this, a second point is that both in-depth cases illustrate participatory interactions about relations and power dynamics. Tools such as 'Power hierarchies' and trust games within facilitated dialogues make these relations and power dynamics—and their outcomes—explicit. The effectiveness of the tools relies on the facilitators being able to create a socially safe and enabling environment that allows reflexive dialogue of this nature (McDougall et al. 2015).

Third, in terms of *social structures*, both cases illustrate the fundamental difference between accommodative and gender transformative approaches: the focus on engaging with underlying structural gender barriers, in particular dynamics around gender norms. While locally focused, both cases additionally illustrate engagement with structures across *multiple scales* and with multiple actors. Case 10.1, for example, explicitly and sequentially engaged at intrahousehold and community scales; Case 10.2 engaged both farmers and other development actors. The latter, in particular, highlights that gender transformative approaches aspire to challenge development actors and agencies, including development and research-for-development organizations, to become critically self-aware of their own gender positions, beliefs, and biases (Sarapura and Puskur 2014, Wong et al. 2019).

Looking ahead: a research agenda to enable more transformative change

A critical unpacking of the quiet, emerging evolution in gender approaches outlined in this chapter suggests that agriculture and natural resource management—and, more broadly, development—are progressing along a spectrum, moving beyond mainstream GAD and accommodative approaches. In other words, the gender, agriculture, and natural resource management trajectory is transitioning into a new era: that of seeking transformative change. Proof of concept and pilot studies are setting the stage for an engagement with gender that is unprecedented in terms of going below the surface to tackle the deeper normative and structural barriers that underpin and perpetuate gender inequalities. These highlight that it is possible to use natural resource management and agriculture interventions as entry points to address structural gender inequalities by engaging community actors in a process of normative change, which will affect women and men's lives beyond agriculture and natural resource management.

While there are solid theoretical arguments and evidence that GAD accommodative approaches are insufficient on their own for enabling agriculture and natural resource management programs to contribute fully to women's empowerment and gender equality, there are still substantive gaps in knowledge and progress is needed in gender transformative approaches. As a contribution to a future research agenda that promotes robust movement toward women's empowerment and gender equality, here we present priority research issues and questions in three critical areas: (1) transformational change with diverse actors and in different contexts; (2) scaling out change at the local level; and, (3) scaling up change beyond the local.

Achieving transformational change with diverse actors in different contexts

The importance of intersectionality is increasingly recognized in gender approaches in agriculture and natural resource management (for example Colfer et al. 2018, Perkins 2019). This needs further progress in gender transformative approaches. As Ndinda and Ndhlovu (2018, 2) note:

> "... we must focus not only on what divides and unites us, but also the complex and interdependent processes that highlight the reasons why women are subordinated. Thus, given diversity of populations, levels of oppression depend on gender, class, race and ethnicity. Such conceptualisations also guard against tunnel vision approaches for investigating SDGs and their implications for redistributive policies and the nature of ownership and control."

In order for a transformative agenda to be inclusive, further investigations are needed regarding how to engage effectively with the multiplicity of gender and identity (Marlow and Martinez Dy 2018, Ndinda and Ndhlovu 2018). This will have important implications in terms of "leaving no one behind"—especially women facing multiple forms of marginalization—within the larger neoliberal trends shaping agriculture and natural resource management.

As well as generating intersectional insights, and refining the ability to measure transformative change (see Chapter 9, this volume), the agriculture and natural resource management sectors will need to generate evidence to guide effective transformative strategies. This includes resolving tensions regarding entry points. Literature suggests that change at the household level is more difficult than change at the community level, and that women simultaneously engage in "scalar politics" to disperse gender struggles at the household and community levels (Howitt 1998, Bassett 2002). Women may thus need to bypass the scale of the household in order to secure productive resources for agricultural innovations. Yet much momentum around gender transformative approaches is focused on household methodologies.

Bringing these together, there are three key gaps/questions:

1. Identifying which gender transformative strategies, at which scales (household, group, district, national) reliably catalyze which outcomes in agriculture and natural resource management. How do these outcomes compare with those from common GAD approaches in the short and the longer term (that is, sustainability), for different actors? How to limit perverse outcomes in agriculture and natural resource management (such as time burdens and backlash)?

2. Clarifying local entry points. Under what conditions do tensions and opportunities between household and community scales exist, and what are the implications for effective and efficient gender transformative strategies?

3. Advancing research outcomes and change needed to unpack variability between and within contexts. How do outcomes of gender transformative approaches vary by context? For different groups within the same contexts (by age, socioeconomic status, other), including more vulnerable women and men? Which factors enhance or limit effectiveness of gender and social change mechanisms for different groups in varying contexts?

This calls for further investment in well-designed pilots across contexts, and with a range of women and men (and other genders, as appropriate). To optimize the utility of the findings, balancing innovation with breadth, these ideally strike a balance between emerging strategies and measurement and the adoption of harmonized methodologies that can enable later meta-analysis.

Scaling-out change at the local level

While a growing number of projects apply gender transformative approaches, to date these efforts have relied on relatively intensive, facilitated, face-to-face interactions with household members. Given the demands of transformative strategies, it seems unlikely that sufficient widespread change will occur through the route of *more* projects in discrete locales. Moreover, as gender transformative approaches gain in popularity among development organizations, there is a significant risk that these complex strategies may be scaled in a reductionist way. In other words, they may be applied as tools without substance, or via organizations lacking the prerequisite capacities, and thus without potential for effective influence (Wong et al. 2019).[13]

These signal a research agenda that address the following three questions:

1. What are the *essential* elements of gender transformative strategies—and are these viable for scaling? To what extent and how can strategies be trimmed and kept affordable, as well as moved across contexts, without becoming token and losing transformative effects? As gender transformative approaches are trimmed down, adapted, and scaled, how can we mitigate the risk of reductionist use or co-optation?

2. Beyond development projects, what is the role and capacity of public and private sector actors in catalyzing and scaling out transformative change, such as through extension systems? Conversely, to what extent can gender transformative strategies be scaled through peer-to-peer (community-to-community, South-to-South) learning models? How does the nature and quality differ from scaling directly through development agencies? For all of the above, how would the required capabilities best be developed?

3. Are there ways to engage women and men effectively in gender transformative processes that do not rely on extensive face-to-face engagements, for example through the use of digital platforms? What risks would this involve and how can these be mitigated?

---

13 Gender transformative approaches could potentially be co-opted by facilitating agencies (and their alliances, including the private sector) in the way that farmer field schools are reported to have been in some contexts, shifting from people- to technology-centric or from empowerment to profit (Sherwood et al. 2012).

Scaling-up: enacting change beyond the local

While household-scale change is essential, achieving lasting, substantive transformation toward gender equality will require change at multiple levels. Specifically, it will require the scaling-up of transformation to beyond-household scales and institutions, including to groups, markets, and policy and legal arenas (Stern et al. 2017, Wong et al. 2019)—and, importantly, to research-for-development organizations themselves. Yet, while some beyond-household approaches are emerging, to date there are relatively few evidenced strategies and pathways to continue "upward" with transformative change. This is an area in critical need of further examination.

In terms of organizational change, the above strategies and pathways rely on the engagement of actors, such as development agencies, government, and civil society organizations (Kantor 2013, Sarapura and Puskur 2014). In order for these organizations to be effective as change agents, they themselves need to manifest gender equality (Goetz 1997, Rao et al. 1999, Cole et al. 2014a, Sarapura and Puskur 2014). Yet development organizations—despite decades of expressed commitment to gender mainstreaming—continue to manifest "tenacious forms of resistance" when it comes to internalizing gender (Verma 2014, 193). Thus, knowledge gaps regarding how to catalyze and sustain effective organizational change are also a priority for further research, so that organizations facilitate first by "walking the talk" of gender equality.

Finally, scaling-up further requires critical engagement with contemporary neoliberal investments and policies that pervade global discourse, including orientation toward a free market, deregulation, and privatization of resources and services. In terms of gender, dual trends and risks are evident. There are risks of decision-making and opportunities excluding women (and marginalized groups), such as within the rapidly expanding, but so far gender-blind, Blue Economy (Cohen et al. 2019, Njuki and Leone 2019, Österblom et al. 2020). Moreover, the very visible global embracing of women's economic empowerment risks co-optation of feminist goals in a way that does little to challenge deeply rooted and persistent power imbalances. Unless explicitly challenged, gender and social inequalities may be further entrenched or even intensified in such investments and trends (Bezner Kerr 2012, Cornwall 2018, Njuki and Leone 2019).

To enable progress in these areas, three key questions emerge:

1. What are the most promising strategies to catalyze gender transformative change beyond the household level—that is, at the community group or network level, and in markets and policy and legal arenas?

What are the risks and how can we mitigate them? What are the effects of different strategies at different scales? How, why, and in what direction does transformation at one scale lead to transformation at another, and what does this indicate in terms of entry points? What factors shape the extent to which transformation occurs, is amplified, or is sustained?

2. How can organizations seeking to catalyze gender transformative change in the development sector transition to and maintain *internal* gender equality cultures and systems as a foundation? What would incentivize such transitions? What lessons are transferable across public, private, civil, community, and development organizations, including agricultural research-for-development organizations?
3. Within the larger neoliberal trends, how do gender transformative strategies affect the ways that different women and marginalized people engage with broader socio-political and economic processes, opportunities, or risks? Are there risks associated with private sector engagement in transformative approaches? What does all the above imply for actors, entry points, and sequencing of strategies to support gender equality in the current neoliberal climate?

Final thoughts

The need to strengthen the contribution of agriculture and natural resource management to gender equality is serious and pressing. Agriculture and natural resource management shape the livelihoods, nutrition, and well-being of the majority of the world's men and women in low-income contexts—yet manifest in deeply gendered outcomes. These dynamics and inequalities have persisted despite wide mainstreaming of GAD. Emerging transformative approaches advance gender in these sectors by pushing back on the limits of how gender is addressed, including against the instrumental trend of essentializing women as "special agents of development." Specifically, transformative approaches represent a shift toward engaging with the underlying constraining social structures and intersectional power dynamics that perpetuate gender inequalities across scales. In doing so, they add value to the sectors by helping unmask and address the systemic faultlines of complex inequalities and institutionalized power and politics, exclusion, and inequality.

More broadly, transforming a persistently unequal world will require analyzing—and consciously strengthening—how gender is interpreted and played out by agriculture and natural resource management institutions as well as in the wider political economy of development. Gender transformative approaches offer a potent opportunity to shift the trajectory and transition from pervasively slow or regressive trends toward substantive and lasting progress. The 2030 transformational agenda provides the mandate and momentum for this transition, calling for "bold and transformative steps which are urgently needed to shift the world onto a sustainable and resilient path" (UN 2015).

Acknowledgments

The authors would like to give their sincere thanks to the CGIAR Gender Platform's internal reviewers Dr. Netsayi Mudege and Dr. Steven Cole, as well as the two anonymous external reviewers, and the book editors. Their input was extremely valuable. We also recognize Wole Kinati and Mamusha Woldegiorgis Lemma for their input in developing the Community Conversations case study, and Afrina Choudhury, Julie Newton, and Dr. Froukje Kruijssen for input to the gillnet case. We sincerely appreciate Surendran Rajaratnam's backstopping throughout. We are extremely grateful to the European Commission and the CGIAR GTA Review Team for sharing their review literature database with us. Last but certainly not least, we are grateful to the CGIAR Collaborative Platform for Gender Research for bringing us together and funding and creating the opportunity for this chapter.

Geography of study sites for publications cited in Chapter 10

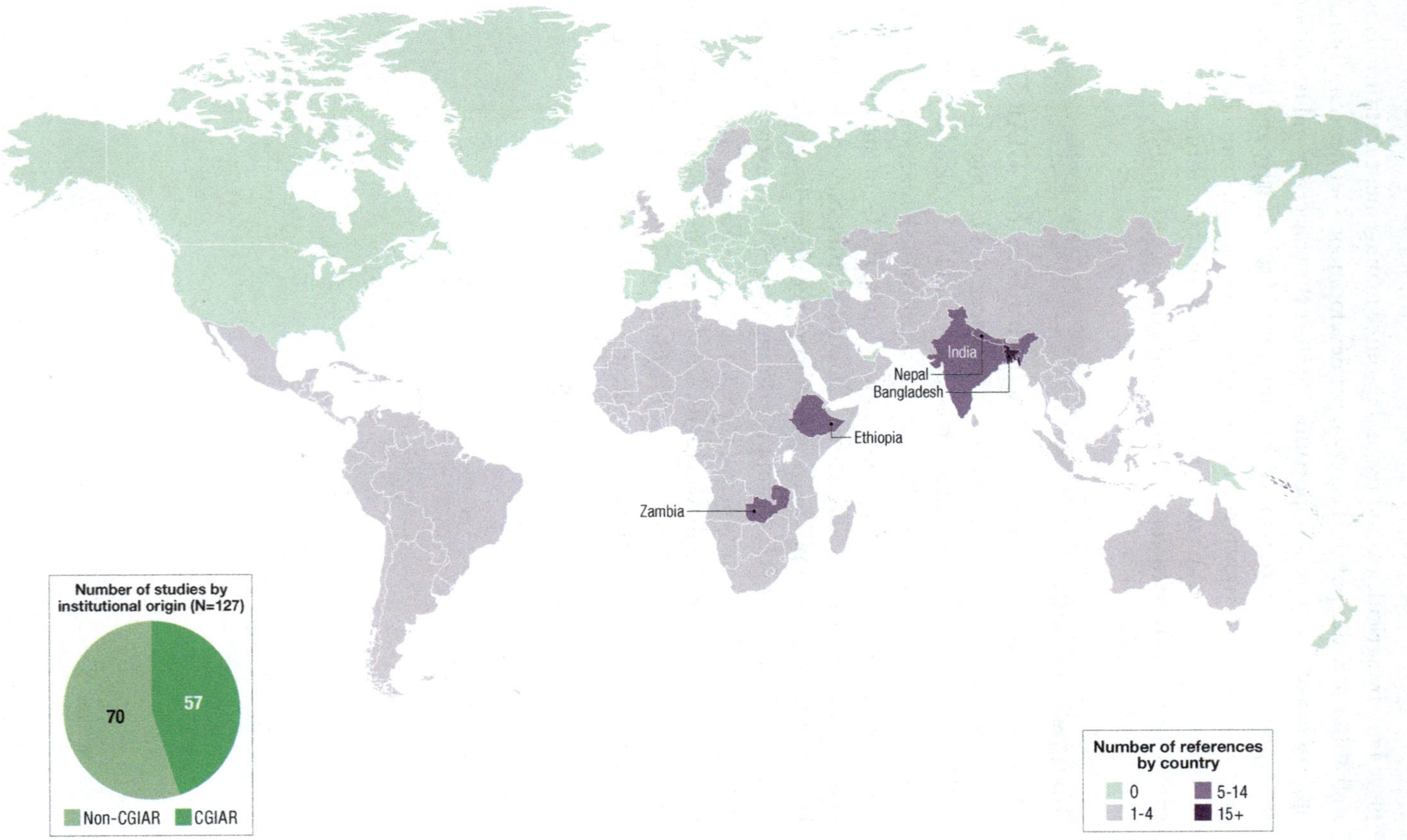

Number of cited studies by research methodology (N=127)

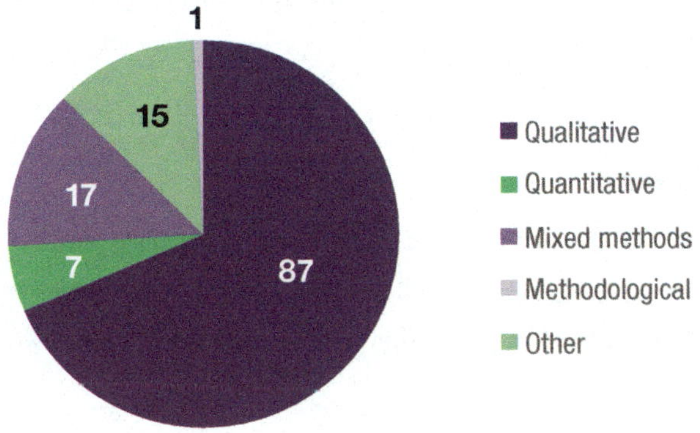

Timeline for references cited

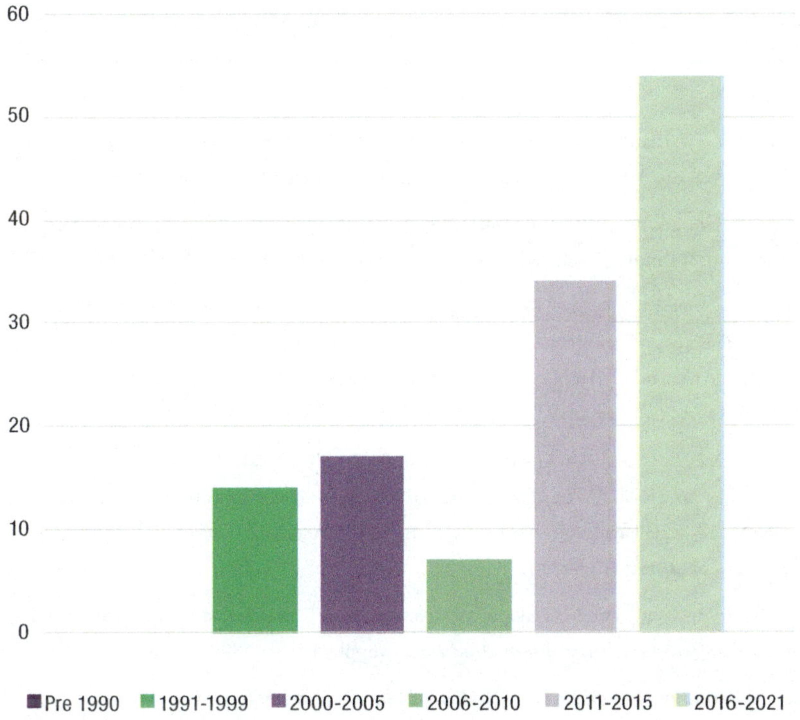

References

AAS (CGIAR Research Program on Aquatic Agricultural Systems). 2012a. *Building Coalitions, Creating Change: An Agenda for Gender Transformative Research in Development.* Penang: AAS.

AAS. 2012b. *CRP AAS Gender Strategy: A Gender Transformative Research Agenda for Aquatic Agricultural Systems*. Penang: AAS.

Agarwal, B. 1994. *A Field of One's Own: Gender and Land Rights in South Asia.* Cambridge: Cambridge University Press.

Agarwal, B. 1997. "'Bargaining' and Gender Relations within and beyond the Houshold." *Feminist Economics* 3 (1): 1–51.

Agarwal, B. 2003. "Gender and Land Rights Revisited: Exploring New Prospects via the State, Family and Market." *Journal of Agrarian Change* 3 (1–2): 184–224.

Alsop, R., N. Heinsohn, A. Somma. 2004. "Measuring Empowerment: An Analytic Framework." In *Power, Rights, and Poverty: Concepts and Connections,* edited by R. Alsop, 120–146. Washington, DC: World Bank.

Alsos, G. A., E. Ljunggren, U. Hytti. 2013. "Gender and Innovation: State of the Art and a Research Agenda." *International Journal of Gender and Entrepreneurship* 5 (3): 236–256.

Ambler, K., C. Doss, C. Kieran, S. Passarelli. 2018. "He Says, She Says: Spousal Disagreement in Survey Measures of Bargaining Power." *Economic Development and Cultural Change.* 68 (2): 765–788.

Aregu, L., S. Rajaratnam, C. McDougall, G. Johnstone, Z. Wah, K. Myat Nwe, M. Akester et al. 2017. *Gender in Myanmar's Small-Scale Aquaculture Sector.* Penang: WorldFish.

Aregu, L., A. Choudhury, S. Rajaratnam, C. Locke, M. McDougall. 2018. "Gender Norms and Agricultural Innovation: Insights from Six Villages in Bangladesh." *Journal of Sustainable Development* 11 (4): 240–287.

Badstue, L., P. Petesch, S. Feldman, G. Prain, M. Elias, P. Kantor. 2018a. "Qualitative, Comparative, and Collaborative Research at Large Scale: An Introduction to GENNOVATE." *Journal of Gender, Agriculture and Food Security* 3 (1): 1–27.

Badstue, L., D. E. Lopez., A. Umantseva, G. Williams, M. Elias, C. R. Farnworth, A. Rietveld et al. 2018b. "What Drives Capacity to Innovate? Insights from Women and Men Small-Scale Farmers in Africa, Asia, and Latin America." *Journal of Gender, Agriculture and Food Security* 3 (1): 54–81.

Baruah, B. 2005. "Gender and Development in South Asia: Can Practice Keep up with Theory?" *Canadian Journal of Development Studies* 26 (Suppl. 1): 677–688.

Baruah, B. 2010. *Women and Property in Urban India.* Vancouver: UBC Press.

Bassett, T. J. 2002. "Women's Cotton and the Spaces of Gender Politics in Northern Côte d'Ivoire." *Gender, Place and Culture: A Journal of Feminist Geography* 9 (4): 351–370.

Belton, B., M. Karim, S. Thilsted, K. Murshed-E-Jahan, W. Collis, M. Phillips. 2011. *Review of Aquaculture and Fish Consumption in Bangladesh.* Dhaka: WorldFish.

Bevitt, K. 2017. "Video Stories to Help Families Overcome Gender-Based Challenges in Zambia." The Fish Tank Blog, March 6. WorldFish.

Bezner Kerr, R. B. 2012. "Lessons from the old Green Revolution for the New: Social, Environmental and Nutritional Assues for Agricultural Change in Africa." *Progress in Development Studies* 12 (2–3): 213–229.

Bicchieri, C. 2005. *The Grammar of Society: The Nature and Dynamics of Social Norms.* Cambridge: Cambridge University Press.

Brandth, B., and M. S. Haugen. 2005. "Doing Rural Masculinity: From Logging to Outfield Tourism." *Journal of Gender Studies* 14 (1): 13–22.

Campbell, H., and M. Mayerfeld Bell. 2000. "The Question of Rural Masculinities." *Rural Sociology* 65 (4): 532–546.

Chant, S. 2016. "Women, Girls and World Poverty: Empowerment, Equality or Essentialism?" *International Development Planning Review* 38 (1): 1–24.

Choudhury, A. 2019. "Merging the Social with the Technical: Using a Gender Transformative Approach in Smallholder Aquaculture Development in Bangladesh." Seeds of Change Conference, Canberra, April 3.

Choudhury, A., and P. Castellanos. 2020. "Empowering Women through Farmer Field Schools." In *Routledge Handbook of Gender and Agriculture*, 251–262. London: Routledge.

CIA (Central Intelligence Agency). n.d. "Life Expectancy at Birth." The World Factbook, accessed September 4, 2019.

Cohen, P. J., E. Allison, N. Andrew, J. Cinner, L. Evans, M. Fabinyi, L. Garces et al. 2019. "Securing a Just Space for Small-Scale Fisheries in the Blue Economy." *Frontiers in Marine Science* 6: 171.

Cole, S. M., R. Puskur, S. Rajaratnam, F. Zulu. 2015. "Exploring the Intricate Relationship between Poverty, Gender Inequality and Rural Masculinity: A Case Study from an Aquatic Agricultural System in Zambia." *Culture, Society and Masculinities* 7 (2): 154–170.

Cole, S. M., B. van Koppen., R. Puskur, N. Estrada, F. DeClerck, J. Baidu-Forson, R. Remans et al. 2014a. "Collaborative Effort to Operationalize the Gender Transformative Approach in the Barotse Floodplain." Working Paper AAS-2014-38. Penang: AAS.

Cole, S. M., P. Kantor, S. Sarapura, S. Rajaratnam. 2014b. "Gender transformative Approaches to Address Inequalities in Food, Nutrition and Economic Outcomes in Aquatic Agricultural Systems." Working Paper AAS-2014-42. Penang: AAS.

Cole, S. M., C. McDougall, A. Kaminski, A. Kefi, A. Chilala, G. Chisule. 2018. "Postharvest Fish Losses and Unequal Gender Relations: Drivers of the Socialecological Trap in the Barotse Floodplain Fishery, Zambia." *Ecology and Society* 2 (23): 18.

Cole, S. M., A. M. Kaminski, C. McDougall, A. S. Kefi, P. A. Marinda, J. Mtonga. 2020. "Gender Accommodative Versus Transformative Approaches: A Comparative Assessment within a Post-Harvest Fish Loss Reduction Intervention." *Gender, Technology and Development* 24 (10): 48-65.

Colfer, C., B. Sijapati Basnett, M. Ihalainen. 2018. "Making Sense of Intersectionality: A Manual for Lovers of People and Forests." Occasional Paper 184. Bogor: CIFOR.

Connell, R. W., and J. W. Messerschmidt. 2005. "Hegemonic Masculinity Rethinking the Concept." *Gender and Society* 19 (6): 829–859.

Cornwall, A. 2000. "Missing Men? Reflections on Men, Masculinities and Gender in GAD." *IDS Bulletin* 31 (2): 18–27.

Cornwall, A. 2016. "Women's Empowerment: What Works?" *Journal of International Development* 28 (3): 342–359.

Cornwall, A. 2018. "Beyond 'Empowerment Lite': Women's Empowerment, Neoliberal Development and Global Justice." *cadernos pagu* 52: e185202.

Cornwall, A., and J. Edwards. 2016. "Introduction: Negotiating Empowerment." *IDS Bulletin* 47 (1A): 1–9.

Davis, K., E. Nkonya, E. Kato, D. Mekonnen, M. Odendo, R. Miiro, J. Nkuba. 2012. "Impact of Farmer Field Schools on Agricultural Productivity and Poverty in East Africa." *World Development* 40 (2): 402–413.

Doss, C., C. Kovarik, A. Peterman, A. Quisumbing, M. van den Bold. 2013. "Gender Inequalities in Ownership and Control of Land in Africa: Myths Versus Reality." Discussion Paper 01308, IFPRI, Washington, DC.

Doss, C., R. Meinzen-Dick, A. Quisumbing, S. Theis. 2018. "Women in Agriculture: Four Myths." *Global Food Security* 16: 69–74.

Drucza, K., and W. Abebe. 2017. "Gender Transformative Methodologies in Ethiopia's Agricultural Sector." Technical Report: Annexes, CIMMYT, Addis Ababa.

Elias, M. 2018. "Gendered Aspirations and Occupations among Rural Youth, in Agriculture and beyond: A Cross-Regional Prescriptive." *Journal of Gender, Agriculture and Food Security* 3 (1): 82–107.

Equal Measures 2030. 2019 "SDG Gender Index." Accessed September 3, 2019. https://data.em2030.org/em2030-sdg-gender-index/

Farnworth, C., M. Fones Sundell, A. Nzioki, V. Shivutse, M. Davis. 2013. *Transforming Gender Relations in Agriculture in Sub-Saharan Africa*. Stockholm: SIANI.

Feldman, S. 2018. "Feminist Science and Epistemologies: Key Issues Central to GENNOVATE's Research Program." GENNOVATE Resources for Scientists and Research Teams, CDMX, CIMMYT, Mexico City.

Ford, L. 2019. "Not One Single Country Set to Achieve Gender Equality by 2030." *The Guardian*, June 3.

Galiè, A., and P. Kantor. 2016. "From Gender Analysis to Transforming Gender Norms: Using Empowerment Pathways to Enhance Gender Equity and Food Security in Tanzania." In *Transforming Gender and Food Security in the Global South*, edited by J. Njuki, J. Parkins, A. Kaler. London: Routledge.

Gilbert, R., W. Sakala, T. Benson. 2002. "Gender Analysis of a Nationwide Cropping System Trial Survey in Malawi." *African Studies Quarterly* 6 (1&2): 223–243.

Goetz, A.-M. 1997. "Introduction: Getting Institutions Right for Women in Development." In *Getting Institutions Right for Women in Development*, edited by A.-M. Goetz, 1–28. New York: Zed Books.

Gray, L., and M. Kevane. 1999. "Diminished Access, Diverted Exclusion: Women and Land Tenure in Sub-Saharan Africa." *African Studies Review* 42 (2): 15–39.

Gunchinmaa, T., D. Hamdamova, B. van Koppen. 2011. "Gender in Irrigated Farming: A Case Study in the Zerafshan River Basin, Uzbekistan." *Gender, Technology and Development* 15 (2): 201–222.

Hillenbrand, E., D. Lindsey, R. Ridolfi, A. von Kotze, K. Nahar, T. Zaman, S. Das et al. 2015a. *Nurturing Connections: Adapted for Homestead Food Production and Nutrition*. Dhaka: HKI Bangladesh.

Hillenbrand, E., N. Karim, D. Wu. 2015b. *Measuring Gender Transformative Change: A Review of Literature and Promising Practices*. Penang: WorldFish and AAS.

Hilliard, S., E. Bukusi, S. Grabe, T. Lu, A. Hatcher, Z. Kwena et al. 2016. "Perceived Impact of a Land and Property Rights Program on Violence Against Women in Rural Kenya: A Qualitative Investigation." *Violence against Women* 22 (14): 1682–1703.

Howitt, R. 1998. "Scale as Relation: Musical Metaphors of Geographical Scale." *Area* 30 (1): 49–58.

IFPRI (International Food Policy Research Institute). 2018. "Agriculture, Nutrition, and Gender Linkages (ANGeL)." Outcome Brochure. IFPRI, Washington, DC.

IGWG (Interagency Gender Working Group). 2017. *The Gender Integration Continuum*. Accessed August 20, 2019. https://www.igwg.org/wp-content/uploads/2017/05/Gender-Continuum-PowerPoint_final.pdf

Jackson, C. 2003. "Gender Analysis of Land: Beyond Land Rights for Women?" *Journal of Agrarian Change* 3 (4): 453–480.

Jewkes, R. 2015. "Hegemonic Masculinity: Combining Theory and Practice in Gender Interventions." *Culture, Health & Sexuality* 1028 (Suppl. 2): 112–127.

Kabeer, N. 1994. *Reversed Realities: Gender Hierarchies in Development Thought*. London and New York: Verso.

Kabeer, N. 1999. "Resources, Agency, Achievements: Reflections on the Measurement of Women's Empowerment." *Development and Change* 30 (3): 435–464.

Kabeer, N. 2001. "Reflections on the Measurement of Women's Empowerment." In *Discussing Women's Empowerment: Theory and Practice*, edited by A. Sisask, 17–57. Stockholm: Sida.

Kabeer, N. 2005. "Is Microfinance a 'Magic Bullet' for Women's Empowerment? Analysis of Findings from South Asia." *Economic and Political Weekly* 40 (44): 4709–4718.

Kantor, P. 2013. "Transforming Gender Relations: A Key to Lasting Positive Agricultural Development Outcomes." Working Paper AAS-2013-12. Penang: AAS.

Katungi, E., S. Edmeades, M. Smale. 2008. "Gender, Social Capital and Information Exchange in Rural Uganda." *Journal of International Development* 20 (1): 25–52.

Kawarazuka, N., C. Locke, C. McDougall, P. Kantor, M. Morgan. 2017. "Bringing Analysis of Gender and Social–Ecological Resilience together in Small-Scale Fisheries Research: Challenges and Opportunities." *Ambio* 46 (2): 201–213.

Kevane, M., and L. Gray. 1999. "A Woman's Field Is Made at Night: Gendered Land Rights and Norms in Burkina Faso." *Feminist Economics* 5 (3): 1–26.

Kidder, T. and C. Pionetti. 2013. *Participatory Methodology: Rapid Care Analysis: Guidance for Managers and Facilitators*. Oxford: Oxfam GB.

Kinati, W., M. Lemma, A. Mulema, B. Wieland. 2019. *The Role of Community Conversations in Transforming Gender Relations and Reducing Zoonotic Risks in the Highlands of Ethiopia.* Nairobi: ILRI.

Kleiber, D., L. M. Harris, A. C. J. Vincent. 2015. "Gender and Small-Scale Fisheries: A Case for Counting Women and Beyond." *Fish and Fisheries* 16 (4): 547–562.

Knight, J., and J. Ensminger. 1998. "Conflict over Changing Social Norms: Bargaining, Ideology, and Enforcement." In *The New Institutionalism in Sociology*, edited by M. C. Brighton, and V. Nee, 105–106. Stanford, CA: Stanford University Press.

Kotter, J., and D. Cohen. 2002. *The Heart of Change.* Cambridge, MA: Harvard Business School Press.

Kruijssen, F., G. Audet-Belanger, A. Choudhury, C. Crissman, J. Dalsgaard, C. Dawson, M. Dickson et al. 2016. "Value Chain Transformation: Taking Stock of WorldFish Research on Value Chains and Markets." Working Paper AAS-2016-03, AAS, Penang.

Lambrecht, I. B. 2016. "'As a Husband I Will Love, Lead, and Provide.' Gendered Access to Land in Ghana." *World Development* 88: 188–200.

Leder, S., F. Clement, E. Karki. 2017. "Reframing Women's Empowerment in Water Security Programmes in Western Nepal." *Gender & Development* 25 (2): 235–251.

Lemma, M., W. Kinati, A. A. Mulema, Z. Bassa, A. Tigabe, H. Desta, M. Mekonnen et al. 2018. "Report of Community Conversations about Gender Roles in Livestock." Nairobi and Addis Ababa: ILRI.

Locke, C., P. Muljono, C. McDougall, M. Morgan. 2017. "Innovation and Gendered Negotiations: Insights from Six Small-Scale Fishing Communities." *Fish and Fisheries* 18 (5): 943–957.

Manlosa, A. O., J. Schultner, I. Dorresteijn, J. Fischer. 2019. "Leverage Points for Improving Gender Equality and Human Well-Being in a Smallholder Farming Context." *Sustainability Science* 14 (2): 529–541.

March, C., I. Smith, M. Mukhopadhyay. 1999. *A Guide to Gender-Analysis Frameworks.* Oxford: Oxfam.

Marlow, S., and A. Martinez Dy. 2018. "Annual Review Article: Is It Time to Rethink the Gender Agenda in Entrepreneurship Research?" *International Small Business Journal: Researching Entrepreneurship* 36 (1): 3–22.

McDougall, C. 2017. "Gender and Systems Research: Leveraging Change." In *Sustainable Intensification in Smallholder Agriculture: An Integrated Systems Research Approach*, edited by I. Oborn, B. Vanlauwe, M. Phillips, R. Thomas, W. Brooijmans, K. Atta-Krah, 275–288. London: Earthscan and Routledge.

McDougall, C., S. Cole, S. Rajaratnam, J. Brown, A. Choudhury, J. Kato-Wallace, A. Manlosa, K. Meng, C. Muyaule, A. Schwartz, H. Teioli. 2015. "Implementing a Gender Transformative Research Approach: Early Lessons." In *Research in Development: Learning from the CGIAR Research Program on Aquatic Agricultural Systems*, edited by B. Douthwaite, J. Apgar, A. Schwartz, C. McDougall, S. Attwood, S. Senaratna Sellamuttu, T. Clayton, 41–56. Penang: AAS.

Meadows, D. 2010. "Leverage Points: Places to Intervene in a System." *The Solutions Journal* 1 (1): 41–49.

Meinzen-Dick, R. 2014. "Property Rights and Sustainable Irrigation: A Developing Country Perspective." *Agricultural Water Management* 145: 23–31.

Milward, K., M. Mukhopadhyay, F. F. Wong. 2015. "Gender Mainstreaming Critiques: Signposts or Dead Ends?" *IDS Bulletin* 46 (4): 75–81.

Monson, R. 2017. "The Politics of Property: Gender, Land and Political Authority in Solomon Islands." in *Kastom, Property and Ideology: Land Transformations in Melanesia*, edited by S. McDonnell, M. Allen, C. Filer, 383–404. Canberra: ANU Press.

Morgan, M. 2014. *Measuring Gender Transformative Change.* Penang: WorldFish.

Morgan, M., A. Choudhury, M. Braun, D. Beare, J. Benedict, P. Kantor. 2015. "Understanding the Gender Dimensions of Adopting Climate-Smart Smallholder Aquaculture Innovations." Working Paper AAS-2015-08, AAS, Penang.

Mulema, A. A. and E. Damtew. 2016. *Gender-Based Constraints and Opportunities to Agricultural Intensification in Ethiopia: A Systematic Review.* Nairobi: ILRI.

Mulema, A. A., W. Jogo, E. Damtwe, K. Mekonnen, P. Thorne. 2019. "Women Farmers' Participation in the Agricultural Research Process: Implications for Agricultural Sustainability in Ethiopia." *International Journal of Agricultural Sustainability* 17 (2): 127–145.

Najjar, D., B. Baruah, A. El Garhi. 2019. "Women, Irrigation and Social Norms in Egypt: 'The More Things Change, the More They Stay the Same?'" *Water Policy* 21 (2): 291–309.

Najjar, D., B. Baruah, A. El Garhi. 2020. "Gender and Asset Ownership in the Old and New Lands of Egypt." *Feminist Economics* 26 (3): 119–143.

Namubiru-Mwaura, E. 2014. *Land Tenure and Gender: Approaches and Challenges for Strengthening Rural Women's Land Rights.* Washington, DC: World Bank.

Ndinda, C., and T. P. Ndhlovu. 2018. "Gender, Poverty and Inequality: Exploration from a Transformative Perspective." *Journal of International Women's Studies* 19 (5): 1–12.

Neusteurer, D. 2016. "Green economy und ihre rolle im neoliberalen kapitalismus." *Socijalna Ekologija* 25 (3): 311–324.

Njuki, J., and M. Leone. 2019. *Positioning Women Smallholder Fisher Folk to Benefit from the Blue Economy.* Ottawa: IDRC.

Nordhagen, S., C. Bastardes Tort, A. Kes, L. Winograd. 2017. "Nurturing Connections? Evaluating the Impact of a Women's Empowerment Curriculum in Côte d'Ivoire." Working Paper, ICRW and HKI, Washington, DC.

Okali, C. 2011. "Searching for New Pathways towards Achieving Gender Equity: Beyond Boserup and 'Women's Role in Economic Development'." ESA Working Paper 11-09, Agricultural Development Economics Division, FAO, Rome.

Okali, C. 2012. "Gender Analysis: Engaging with Rural Development and Agricultural Policy Processes." Working Paper 26, Future Agricultures Consortium.

Oldenziel, R. 1999. *Making Technology Masculine: Men, Women, and Modern Machines in America, 1870-1945.* Amsterdam: Amsterdam University Press.

Österblom, H., C. Colette, D. Wabnitz, E. Allison, S. Haond, J. Bebbington, N. Bennett et al. 2020. *Towards Ocean Equity.* Washington, DC: WRI.

Pantuliano, S., C. Harper, N. Jones, P. Domingo, R. Marcus, F. Samuels et al. 2019. *Gender Equality and Women and Girls' Empowerment: Our Approach and Priorities.* London: ODI.

Pearse, R., and R. Connell. 2016. "Gender Norms and the Economy: Insights from Social Research." *Feminist Economics* 22 (1): 30–53.

Perkins, P. 2019. "Climate Justice, Gender and Intersectionality." In *Routledge Handbook of Climate Justice*, edited by T. Jafry, M. Mikulewicz, K. Helwig, 349–358. Abingdon and New York: Routledge.

Peterman, A., J. Behrman, A. Quisumbing. 2010. *A Review of Empirical Evidence on Gender Differences in Non-Land Agricultural Inputs, Technology, and Services in Developing Countries.* Rome: FAO.

Peterson, V. S. 2005. "How (the Meaning of) Gender Matters in Political Economy." *Key Debates in New Political Economy* 10 (4): 499–521.

Pini, B. 2005. "The Third Sex: Women Leaders in Australian Agriculture." *Gender, Work & Organization* 12 (1): 73–88.

Promundo-US and the CGIAR Research Program on Aquatic Agricultural Systems. 2016. *Promoting Gender-Transformative Change with Men and Boys: A Manual to Spark Critical Reflection on Harmful Gender Norms with Men and Boys in Aquatic Agricultural Systems.* Washington, DC, and Penang.

Rao, A., R. Stuart, K. Kelleher. 1999. "Introduction." In *Gender at Work: Organizational Change for Equality*, edited by A. Rao, R. Stuart, K. Kelleher, 1–24. West Hartford. CT: Kumarian Press.

Razavi, S., and C. Miller. 1995. "From WID to GAD: Conceptual Shifts in the Women and Development Discourse." Occasional Paper 1. Geneva: UNRISD.

Ridgeway, C. L. 2009. "Framed before We Know It: How Gender Shapes Social Relations." *Gender and Society* 23 (2): 145–160.

Rottach, E., S. R. Schuler, K. Hardee. 2009. *Gender Perspectives Improve Reproductive Health Outcomes: New Evidence.* IGWG.

Sarapura, S., and R. Puskur. 2014. "Gender Capacity Development and Organizational Culture Change in the CGIAR Research Program on Aquatic Agricultural Systems: A Conceptual Framework." Working Paper AAS-2014, AAS, Penang.

Sari, I., C. McDougall, S. Rajaratnam. 2017. *Women's Empowerment in Aquaculture: Two Case Studies from Indonesia*. Penang and Rome: FAO and WorldFish.

Schroeder, R. A. 1993. "Shady Practice: Gender and the Political Ecology of Resource Stabilization in Gambian Garden/Orchards." *Economic Geography* 69 (4): 349–365.

Sherwood, S., M. Schut, C. Leeuwis. 2012. "Learning in the Social Wild: Encounters between Farmer Field Schools and Agricultural Science and Development in Ecuador." In *Adaptive Collaborative Approaches in Natural Resoruces Governance: Rethinking Participation, Learning and Innovation*, edited by H. R. Ojha, A. Hall, R. Sulajman, 102–137. Abingdon: Routledge.

Slegh, H., G. Barker, A. Kimonyo. 2013. "'I Can Do Women's Work': Reflections on Engaging Men as Allies in Women's Economic Empowerment in Rwanda." *Gender & Development* 21 (1): 15–30.

Stern, E., L. Heise, L. McLean. 2017. "The Doing and Undoing of Male Household Decision-Making and Economic Authority in Rwanda and Its Implications for Gender Transformative Programming." *Culture, Health & Sexuality* 20 (9): 976–991.

Tefera, T. 2013. "Land Ownership: The Path towards Rural Women Empowerment: A Case from Southern Ethiopia." *International Journal of Sociology and Anthropology* 5 (8): 330–338.

Twyman, J., J. Muriel, M. A. García. 2015. "Identifying Women Farmers: Informal Gender Norms as Institutional Barriers to Recognizing Women's Contributions to Agriculture." *Journal of Gender, Agriculture and Food Security* 1 (2): 1–17.

UN (United Nations). 2015. *Transforming Our World: The 2030 Agenda for Sustainable Development.* United Nations Sustainable Knowledge Platform A/RES/70/1.

Verma, R. 2014. "Business as Unusual: The Potential for Gender Transformative Change in Development and Mountain Contexts." *Mountain Research and Development* 34 (3): 188–196.

Vernooy, R., and C. McDougall. 2003. *Principles for Good Practice in Participatory Research: Reflecting on Lessons from the Field.* In *Managing Natural Resources for Sustainable Livelihoods: Uniting Science and Participation*, edited by B. Pound, S. Snapp, C. McDougall, A. Braun, 113–141. London and Ottawa: Earthscan and IDRC.

WEF (World Economic Forum). 2016 *The Global Gender Gap Report 2016.* Geneva: WEF.

Wong, F., A. Vos, R. Pyburn, J. Newton. 2019. "Implementing Gender Transformative Approaches in Agriculture." Discussion Paper, CGIAR Collaborative Platform for Gender Research , Amsterdam.

Zhao, M., M. Tyzack, R. Anderson, E. Onoakpovike. 2013. "Women as Visible and Invisible Workers in Fisheries: A Case Study of Northern England." *Marine Policy* 37 (1): 69–76.

Zwarteveen, M. 2008. "Men, Masculinities and Water Powers in Irrigation." *Water Alternatives* 1 (1): 111–130.

Contributors

Rhiannon Pyburn is a Senior Expert supported by the NL-CGIAR Partnership working with the CGIAR Research Program on Policies, Institutions, and Markets, and a Senior Gender and Agriculture Advisor at KIT Royal Tropical Institute in Amsterdam, Netherlands.

Anouka van Eerdewijk is a Senior Gender Advisor at KIT Royal Tropical Institute in Amsterdam, Netherlands.

Mariola Acosta is a PhD Candidate at Wageningen University, Netherlands, and Visiting Researcher at the Alliance of Bioversity International and the International Center for Tropical Agriculture (CIAT), based in Gainesville, Florida, USA.

Sounkoura Adetonah is a Senior Researcher at the International Institute of Tropical Agriculture (IITA), based in Cotonou, Benin.

Lone Badstue is a Rural Development Sociologist and Research Theme Leader for Gender and Social Inclusion at the International Maize and Wheat Improvement Center (CIMMYT), based in in Texcoco, Mexico.

Nitya Chanana is a Research Consultant in the CGIAR Research Program on Climate Change, Agriculture and Food Security (CCAFS), Borlaug Institute for South Asia (BISA), International Maize and Wheat Improvement Center (CIMMYT) in New Delhi, India.

Steven Cole is a Senior Scientist and a Gender Research Coordinator at the International Institute of Tropical Agriculture (IIAT), based in Dar es Salaam, Tanzania.

Cheryl Doss is an Associate Professor and Senior Departmental Lecturer in Development Economics in the Department of International Development at the University of Oxford, UK.

Marlène Elias is a Senior Scientist and a Gender Lead for the Alliance of Bioversity International and the International Center for Tropical Agriculture (CIAT), based in Rome, Italy.

Gundula Fischer is a Social Scientist and Gender Expert at the International Institute of Tropical Agriculture (CIAT), based in Arusha, Tanzania.

Alessandra Galiè is a Senior Scientist Gender at the Policies, Institutions and Livelihoods Program at the International Livestock Research Institute (ILRI) in Nairobi, Kenya.

Bryce Gallant is a Gender, Youth and Inclusion Intern in the CGIAR Research Program on Water, Land and Ecosystems (WLE) in Colombo, Sri Lanka.

Tatiana Gumucio is a Postdoctoral Research Scientist at the International Research Institute for Climate and Society (IRI) at Columbia University, in New York, USA.

Jessica Heckert is a Research Fellow in the Poverty, Health, and Nutrition Division at the International Food Policy Research Institute (IFPRI) in Washington, DC, USA.

Sophia Huyer is a Gender and Social Inclusion Leader in the CGIAR Research Program on Climate Change, Agriculture and Food Security (CCAFS) at the International Livestock Research Institute (ILRI), based in Brighton, Canada.

Markus Ihalainen is a Senior Research Officer at the Center for International Forestry Research (CIFOR), based in Dakar, Senegal.

Deepa Joshi is a Gender, Youth and Inclusion Lead at the International Water Management Institute (IWMI) and in the CGIAR Research Program on Water, Land and Ecosystems (WLE) in Colombo, Sri Lanka.

Nozomi Kawarazuka is a Scientist at the International Potato Center (CIP), based in Hanoi, Viet Nam.

Arun Khatri-Chhetri is a Science Officer for Low Emissions Development in the CGIAR Research Program on Climate Change, Agriculture and Food Security (CCAFS) at the University of Vermont, Burlington, USA.

Katrina Kosec is Senior Research Fellow at the International Food Policy Research Institute (IFPRI) and Adjunct Professor at Johns Hopkins University, based in Washington, DC, USA.

Hazel Malapit is a Senior Research Coordinator in the Poverty, Health, and Nutrition Division at the International Food Policy Research Institute (IFPRI) in Washington, DC, USA.

Pricilla Marimo is an Associate Scientist at Bioversity International, based in Kampala, Uganda.

Sarah Mayanja is a Research Associate at the International Potato Center (CIP), based in Kampala, Uganda.

Cynthia McDougall is a Principal Scientist and Gender Research Leader at WorldFish in Penang, Malaysia.

Ruth Meinzen-Dick is a Senior Research Fellow at the International Food Policy Research Institute (IFPRI) and Co-lead of the Flagship on Natural Resource Governance for the CGIAR Program on Policies, Institutions, and Markets in Washington, DC, USA.

Rohini Ram Mohan is a Gender Specialist at the International Rice Research Institute (IRRI), based in New Delhi, India.

Iliana Monterroso is a Scientist and Co-coordinator of Gender and Social Inclusion Research with the Center for International Forestry Research (CIFOR), based in Guatemala City, Guatemala.

Netsayi Noris Mudege was a Gender Research Coordinator at the International Potato Center (CIP), and is now a Senior Scientist at WorldFish, based in Lusaka, Zambia.

Gaudiose Mujawamariya is a Researcher at AfricaRice, based in Antananarivo, Madagascar.

Annet Abenakyo Mulema is a Science Officer–Gender and Social Inclusion in the Climate Change, Agriculture and Food Security Research Program (CCAFS) at the International Livestock Research Institute (ILRI) in Nairobi, Kenya.

Catherine Mungai is a Partnership and Policy Engagement Specialist in the CGIAR Research Program on Climate Change, Agriculture and Food Security (CCAFS) in the International Livestock Research Institute (ILRI) in Nairobi, Kenya.

Dina Najjar is a Gender Research Scientist at the Sustainable Intensification and Resilient Production Systems Program (SIRPSP) at the International Center for Agricultural Research in the Dry Areas (ICARDA), based in Rabat, Morocco.

Eileen Bogweh Nchanji is a Gender Specialist in the African Regional Office at the Alliance of Bioversity International and the International Center for Tropical Agriculture (CIAT) based in Nairobi, Kenya.

Esther Njuguna-Mungai is a Senior Scientist—Gender Research, at the International Crops Research Institute for the Semi-Arid Tropics (ICRISAT), East African Regional Office, in Nairobi, Kenya.

Gloria Otieno is a Genetic Resources and Food Security Policy Specialist at the Alliance of Bioversity International and the International Center for Tropical Agriculture (CIAT), based in Kampala, Uganda.

Mathieu Ouedraogo is a Senior Scientist—Participatory Action Research, for the CGIAR Research Program on Climate Change, Agriculture and Food Security (CCAFS), West Africa, at ICRISAT, based in Bamako, Mali.

Ravula Padmaja is a Senior Scientist for Gender Research with Markets, Institutions, Nutrition, and Diversity in the Research Program on Innovative Systems for the Drylands at the International Crops Research Institute for the Semi-Arid Tropics (ICRISAT) in Hyderabad, India.

Ana Maria Paez Valencia is a Social Scientist and Gender Specialist at World Agroforestry in Nairobi, Kenya.

Kokila P. Perera is a former Gender and Inclusion Consultant for the CGIAR Research Program on Water, Land and Ecosystems (WLE) at the International Water Management Institute (IWMI) in Colombo, Sri Lanka.

Valentina Peveri is a Food and Environmental Anthropologist, currently serving as an Adjunct Professor at The American University of Rome (AUR) and as a consultant for Includovate, based in Bologna, Italy.

Vivian Polar is a Senior Gender and Innovation Specialist in the Research Program on Roots, Tubers and Bananas (RTB) at the International Potato Center (CIP) in Lima, Peru.

Ranjitha Puskur is a Senior Scientist leading the Livelihoods, Gender and Nutrition Research Program at the International Rice Research Institute (IRRI), based in Nairobi, Kenya.

Ayesha Qaisrani is a Senior Research Associate at the Sustainable Development Policy Institute (SDPI) in Islamabad, Pakistan.

Agnes Quisumbing is a Senior Research Fellow in the Poverty, Health, and Nutrition Division of the International Food Policy Research Institute (IFPRI) and Leader of the Institute's Gender Cross-cutting Theme in Washington, DC, USA.

Maren Radeny is a Science Officer in the CGIAR Research Program on Climate Change, Agriculture and Food Security (CCAFS) at the International Livestock Research Institute (ILRI), based in Kigali, Rwanda.

John Recha is a Climate-Smart Agriculture and Policy Scientist at the CGIAR Research Program on Climate Change, Agriculture and Food Security (CCAFS) at the International Livestock Research Institute (ILRI) in Nairobi, Kenya.

Jessica Scott is a Research Fellow in the Value Chains and Nutrition Division at WorldFish, based in Yangon, Myanmar.

Sumer Shaikh is a Research Affiliate at the Center for International Forestry Research (CIFOR) and Press Secretary at the Sierra Club in Los Angeles, USA.

Elisabeth Simelton is a Climate Change Scientist at World Agroforestry (ICRAF), based in Hanoi, Viet Nam.

Vanya Slavchevska is a Scientist–Gender and Social Inclusion in the Crops for Nutrition and Health Lever of the Alliance of Bioversity International and the International Center for Tropical Agriculture (CIAT), based in Cali, Colombia.

Katie Tavenner is a Senior Program Officer for Gender and Agriculture at the research incubator Includovate, based in Turrialba, Costa Rica.

Béla Teeken is an Associate Scientist at the International Institute of Tropical Agriculture (IITA) in Ibadan, Nigeria.

Jennifer Twyman is a Visiting Researcher at the Alliance of Bioversity and the International Center for Tropical Agriculture (CIAT) and former Gender Research Leader at CIAT, based in Cali, Colombia.

Ronnie Vernooy is a Genetic Resources Policy Specialist at the Alliance of Bioversity International and the International Center for Tropical Agriculture (CIAT), based in Wageningen, Netherlands.

Andrea Vos is a Junior Advisor on Gender and Agriculture at KIT Royal Tropical Institute in Amsterdam, Netherlands.

Jummai Othniel Yila is a Scientist–Gender Research at the International Crops Research Institute for the Semi-Arid Tropics (ICRISAT), based in Bamako, Mali.

